DISCARDED BY
MEMPHIS PUBLIC LIBRARY

Optoelectronics, Fiber Optics, and Laser Cookbook

About the author

Thomas Petruzzellis, a professional electronic specialist and instructor at Binghamton University, has written extensively for industry publications, including *Electronics Now, Modern Electronics, QST, Microcomputer Journal,* and *Nuts & Volts*. He is the author of the *Alarm, Sensor, and Security Circuit Cookbook.*

Other titles of interest

Alarm, Sensor, Security Cookbook by Thomas Petruzzellis
Lasers, Ray Guns, and Light Cannons: Projects from the Wizard's Workbench by Gordon McComb
Gadgeteer's Goldmine by Gordon McComb

Optoelectronics, Fiber Optics, and Laser Cookbook

Thomas Petruzzellis

McGraw-Hill

New York San Francisco Washington, D.C. Auckland Bogotá
Caracas Lisbon London Madrid Mexico City Milan
Montreal New Delhi San Juan Singapore
Sydney Tokyo Toronto

Library of Congress Cataloging-in-Publication Data

Petruzzellis, Thomas
Optoelectronics, fiberoptics, and laser cookbook / Thomas
Petruzzellis.
p. cm.
Includes index.
ISBN 0-07-049839-3 (hc : acid-free paper).—ISBN 0-07-049840-7
(pc : acid-free paper)
1. Electronics—Amateurs' manuals. 2. Optoelectronics—Amateurs'
manuals. 3. Fiber optics—Amateurs' manuals. 4. Lasers—Amateurs'
manuals.
TK9965.P48 1997
621.381'045—dc21 97-423
CIP

McGraw-Hill

A Division of The ***McGraw-Hill*** *Companies*

Copyright © 1997 by The McGraw-Hill Companies, Inc. All rights reserved. Printed in the United States of America. Except as permitted under the United States Copyright Act of 1976, no part of this publication may be reproduced or distributed in any form or by any means, or stored in a data base or retrieval system, without the prior written permission of the publisher.

1 2 3 4 5 6 7 8 9 0 FGR/FGR 9 0 2 1 0 9 8 7

ISBN 0-07-049840-7 (PBK)
ISBN 0-07-049839-3 (HC)

The sponsoring editor for this book was Scott Grillo, the editing supervisor was Sally Glover, and the production supervisor was Claire Stanley. It was set in ITC Century Light by McGraw-Hill's Professional Book Group composition unit, Hightstown, N.J.

Printed and bound by Fairfield Graphics.

McGraw-Hill books are available at special quantity discounts to use as premiums and sales promotions, or for use in corporate training programs. For more information, please write to the Director of Special Sales, McGraw-Hill, 11 West 19th Street, New York, NY 10011. Or contact your local bookstore.

Information contained in this work has been obtained by The McGraw-Hill Companies, Inc. ("McGraw-Hill") from sources believed to be reliable. However, neither McGraw-Hill Hill nor it's authors guarantees the accuracy or completeness of any information published herein and neither McGraw-Hill nor its authors shall be responsible for any errors, omissions, or damages arising out of use of this information. This work is published with the understanding that McGraw-Hill and its authors are supplying information but are not attempting to render engineering or other professional services. If such services are required, the assistance of an appropriate professional should be sought.

.This book is printed on recycled, acid-free paper containing a minimum of 50% recycled, de-inked fiber.

Dedication

This book is dedicated to Betsy:
My wife
My friend
My confidante
She is a beacon of light atop the rocky coast
who guides me through the storms of life
and keeps me on the straight course.

Contents

Introduction

Electromagnetic waves are all around us every hour of every day, but humans are aware of only a small band of frequencies known as *visible light*. The visible light spectrum is tucked between the infrared and ultraviolet portions of the electromagnetic spectrum. Compared to the entire electromagnetic spectrum, visible light is an extremely tiny slice of wavelengths, between ultraviolet at 400 nanometers and near infrared at 800 nanometers.

This book peeks into the combined technologies of optics and electronics, or *optoelectronics*. Electronic devices that emit or detect optical or near-optical radiation are called *optoelectronic components*. Optoelectronic components include LEDs and other light sources such as lasers, gas lamps, and ultraviolet lamps, as well as a wide range of optical detectors such as photocells, phototransistors, solar cells, and photodiodes. Purely optical components used with optoelectronic parts include such things as lenses, mirrors, optical beam splitters, prisms, and optical fibers. All of these electronic and optical components comprise the fascinating field of optoelectronics.

Fiber optics are becoming and will continue to be an ever-increasingly important aspect of optoelectronics. The technology of fiber optics includes light sources and drivers, optocouplers, photo detectors, optointerfaces, optical fibers, lasers, and purely optical components such as lenses. Many optoelectronic and fiber-optic components can be used together, as you will see. Fiber-optic technology can be found in communications, control, alarm, and monitoring applications. In this book I discuss the various types of fibers and their characteristics, as well as how they can be used. I also introduce a number of useful fiber-optic circuits and applications and cover future fiber-optic trends.

This book also presents the wondrous world of laser technology. Lasers too will change our world into the next century. I discuss various types of lasers, as well as classical experiments and a number of laser applications and circuits. Several interesting and diverse laser projects and experiments will assist you in gaining a broad sense of laser applications and their future possibilities. This book presents the combined technologies of optoelectronics, fiber optics, and lasers in an "all-in-one" expansive and encompassing book that I believe to be the first of its kind.

The applications for optoelectronics, fiber optics, and lasers are widespread and include lighting, communications, sensing and control, alarm, and metering circuits. As technology marches toward the twenty-first century, reliance on optoelectronics, fiber optics, and lasers will increase phenomenally. This book delves into the related optoelectronic, fiber-optic, and laser technologies to provide a broad range of optical, infrared, and ultraviolet circuits that can be used for sensing, metering, control, alarm, and communications projects. The book was designed to be an informative resource that you can use well into your electronics future. I present many new and novel circuits, demonstrations, and experiments that I hope you will find interesting and enlightening.

Chapter 1 covers the characteristics of various light sources, including LEDs and lasers in the optical and near-optical spectrum, and infrared and ultraviolet light. Chapter 1 also compares the numerous types of light sensors, such as photoresistors, phototransistors, and solar cells, their response characteristics, and how they can be used.

Chapter 2 delves into all the various light components such as lenses, beam splitters, mirrors, prisms, filters, collimators, and polarizers, as well other light-handling components. This chapter also covers the care and handling of optical components. Chapter 2 also introduces optical fibers and compares stepped and graded index fibers. Integrated source sensors are also introduced.

Chapter 3 begins with do-it-yourself optocouplers and optoisolators. Then many different types of integrated-circuit optocoupler and optoisolators are presented and compared. I cover a number of new optointegrated circuits that are available for experimenters, such as integrated optoisolators, optorelays, SCR and Triac optoisolators, opto line drivers, opto SCR, and relay drivers. Chapter 3 also presents a two-way optotransceiver system, a current loop interface, an optically isolated TTL-to-RS-232 interface, a two-way IR RS-232 serial computer interface, and a high-speed serial interface you can use to transfer data between two computers systems.

Chapter 4 looks into the topic of light-metering circuits. Circuits in this chapter include simple light-meters, light comparators, a linear light-meter, a programmable light-meter, and a bargraph light-meter. You'll see an ultra-sensitive light-meter, and then move on to simple radiometers, sunlight photometers, and an advanced digital UV sunlight photometer. Finally I describe a digital ozone meter and show you how to use it to measure a column of ozone in the atmosphere.

Chapter 5 is the largest chapter in this book. It covers a broad range of optical sensors and many different types of sensing circuits, including an optical position sensor, an optical slot door/window sensor, a film-strip sensor, and optical rotary encoders. Next you move on to a bifurcated light pipe sensor that can be used for direction and counting applications in tight or confined spaces. Included in this group of sensors is one of my favorites, the position-sensitive detector, which is available in both single- and dual-axis versions. The next group of sensors includes an optical tachometer, a light-to-frequency converter, and a light-to-frequency/frequency-to-voltage data transfer system. The next project in Chapter 5 is an optical temperature transmitter/receiver system you can use to send temperature readings through the air. Hunters' Companion is the next project, which hunters can use to find downed game, such as deer, in the woods. The pyroelectric IR hot/cold leak detector can be used to locate hot or cold leaks along windows or doors, or to sense cold leaks around freezer doors. When combined with a IR Fresnel lens, this circuit can be used as a long-range heat/cold leak detector. The last project in this chapter is an experimental lightning monitor that you can use to turn off equipment during an electrical storm.

Chapter 6 discusses photoresistive and phototransistor optoisolator control circuits, including light/dark sensors, a linear bargraph relay driver, a solar-activated SCR switch, and number of optical trigger relay circuits. Many different ac and dc control circuits are also presented, including optically isolated SCR and Triac control circuits, which can be used to control fans, motors, lights, and appliances. Heavy-duty SCR and Triac controllers are also presented. In Chapter 6 you find a fiber-optic ac power control circuit, an IR remote-control system, and a five-channel IR tone-activated remote-control system.

Chapter 7 delves into different types of light alarm detection systems, from a simple line-powered outage alarm to a laser perimeter alarm. Alarm projects in this chapter include a personal alarm/strobe, a drawer/cupboard alarm, a reflective light alarm chip, and a balanced light alarm, which can be incorporated into a night-vision IR TV camera alarm system. The next several projects are centered around IR pyroelectric sensors, which you can use to form your own burglar alarm. The next group of projects includes an IR beam-break alarm system, two portable wireless IR/RF security systems, and a long-range laser perimeter alarm system.

Chapter 8 deals exclusively with free-space communication systems. The chapter begins by comparing simplex and duplex communication systems. Projects include light-wave receivers, a sensitive light-wave listener, a simple light-wave communication system, a short/medium-range signaling system, a 30-kHz IR AM voice communication system, and a pulse-frequency-modulated IR communication system. The next project is a wireless speaker system. The final two projects center around an infrared phase-locked loop laser voice and data communication system and a touch-tone control system that can be used with the communication system to remotely control equipment or appliances.

Chapter 9 covers a broad range of fiber-optic sensors and communications applications. First I discuss fiber-optic connectors and describe how you can build your own fiber-optic connectors. Then the discussion moves to fiber-optic sensors and how you can make them. Three indirect-mode fiber-optic sensors are presented first. Then I present pressure-sensing fiber sensors, unclad fiber sensors, fiber-optic liquid sensors, and finally fiber-optic vibration sensors. The next several projects include a fiber-optic voice communications link, fiber-optic LED drivers, a fiber-optic data communications link, and finally a high-speed fiber-optic data link. Future trends in fiber optics are also discussed at the end of the chapter.

Chapter 10 presents a number of different reflective light projects such as the photophone, invented by Alexander Graham Bell. From the photophone you then move to reflective laser light design makers, reflective light wheel alignment, and bar code systems. Then I present the optical lever, which is the basis of the next few projects, which include an optical electroscope, an optical galvenometer, and an optical magnetometer, which can be used to monitor geomagnetic storms and magnetic anomalies. The final project in this chapter is a remote laser listener that you can use to monitor conversations from a distance through windows.

Chapter 11 covers many different types of laser light projects, from simple to complex, such as a laser earthquake monitor. The chapter begins with helium-neon and semiconductor lasers and power supplies and moves on to light-handling techniques. Then a number of laser applications and experiments, including reflection, refraction, absorption, beam spreading, and filters are presented. Classical experiments such as diffraction through a razor blade and the double-slit experiment are covered in this chapter. The Schlieren Effect is also presented, as well as the Michelson laser interferometer and how you can use it to perform a number of interesting experiments. Then you will see a brief discussion of holograms, as well as a few other practical laser applications. The last project in Chapter 11 is a fiber-optic laser seismometer.

The last chapter, 12, covers distance-measuring experiments and methods of determining distance from point to point using light. First I discuss the optical lever and how it can be used with a mirror and a protractor to measure distances remotely. Experiments include using plane trigonometry and a right triangle to compute vertical heights and horizontal distances, and trigonometry and a time-distance formula to measure both the distance to a distant ship as well as the speed of the ship. Next you will measure the curvature of the earth and illustrate an optical triangulation sensor. Finally I discuss how a laser range finder operates and how you can build one.

I hope you will find this new book both informative and inspirational, that it will spark your imagination and allow you to create new applications and circuits using the topics and ideas discussed. This book provides both neophytes and seasoned experimenters with a resource they can use well into the future.

1
CHAPTER
Light sources and detectors

The electromagnetic spectrum encompasses an enormous range of wavelengths, with gamma rays at 1010 micrometers (µm) to radio waves at the opposite end of the spectrum at 106 centimeters (cm)(see Figure 1-1). The optical and near-optical spectrum is only a very narrow band of frequencies between 400 and 800 nanometers (nm). Infrared wavelengths can be found between 800 nm and 1 µm at one edge of the visible spectrum, while ultraviolet wavelengths lie at the other edge of the optical range, from 400 to 10 nm, as shown in Figure 1-2. Table 1-1 illustrates the conversion units associated with the optical spectrum.

Light is part particle and part wave. Particles are called *photons*, while the wave aspect of light is part of the electromagnetic spectrum. Imagine this dual character of light as photons bubbling up and down as they travel through space. Looking at the path of light sideways shows the photon as a *sine wave*, a wave that moves up and down in a smooth gradual motion. Sine waves are comprised of crests and troughs. The distance between two consecutive crests determines the wavelength and thus the frequency. Figure 1-3 depicts the relationship between wavelength and frequency. If the distance between two crests is small, the wavelength is small but the frequency is high. If, on the other hand, the distance between the crests is increased, then the frequency becomes lower as the wavelength increases. The frequency of an electromagnetic wave is the number of cycles that occur in one second. Frequency is expressed in hertz (Hz) or cycles per second, while wavelength is represented by the Greek lambda character (λ). Light, which is part of the electromagnetic spectrum, travels at 300,000 kilometers per second or 186,000 miles per second (C), as do radio waves. If either the frequency or length of a wave is known, the unknown value can be calculated from the following formulas:

$$\text{Frequency (Hz)} = \text{C} / \text{Wavelength } (\lambda)$$

or

$$\text{Wavelength } (\lambda) = \text{C} / \text{Frequency (Hz)}$$

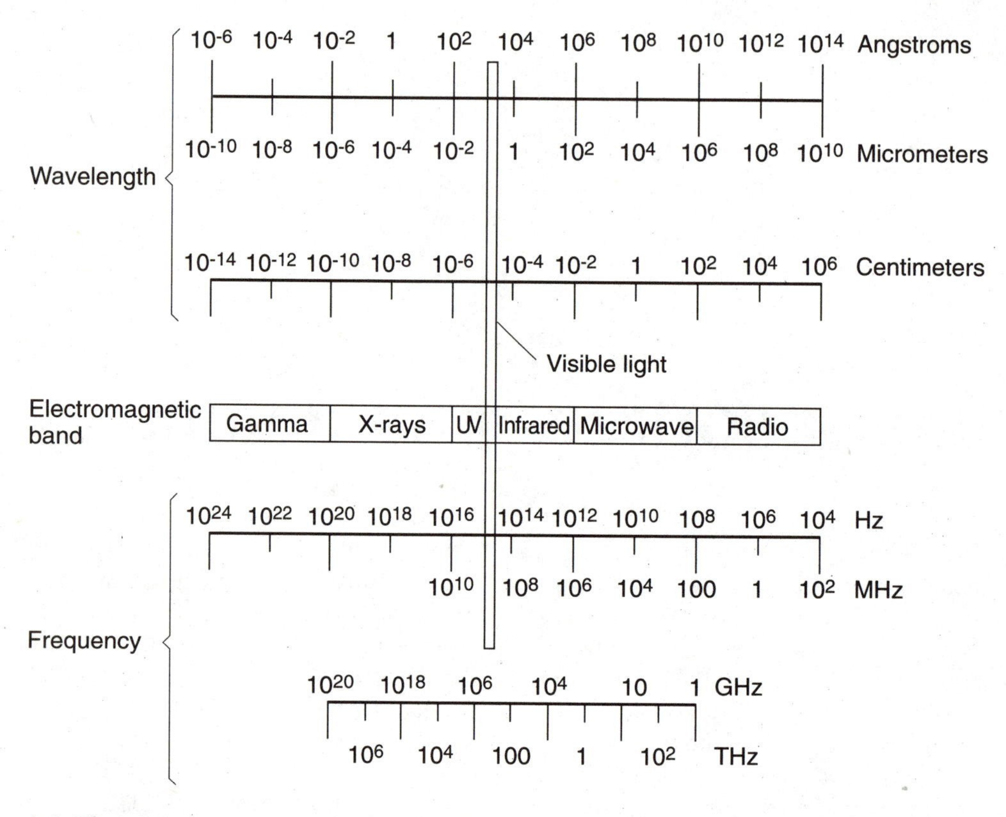

1-1 Electromagnetic spectrum.

In order to experiment with light, you must first take a look at the range of light sources, as well as the different types of detectors. The field that merges electronics and optics is called *optoelectronics*.

Sources of light

The first and most important source of light is sunlight, which encompasses all colors of visible light plus a range of both infrared and ultraviolet light rays. Sunlight is a wide-spectrum light source that supplies both light so we can see the world and light for photosynthesis, which allows plants to grow. Sunlight also provides warmth and vitamin D, which are very beneficial to humans and animals alike. Sunlight can also be harnessed to produce electricity with photovoltaic or solar electric cells or simply hot-air solar collectors or shallow heating ponds. You will make use of sunlight in a number of projects in this book.

The solar spectrum is shown in Figure 1-4. This diagram illustrates the components of the earth's atmosphere and how certain gasses absorb specific wavelengths of light. You can learn much about the atmosphere by monitoring sunlight, as you will see later. You can also construct detectors to measure the amount of sunlight falling

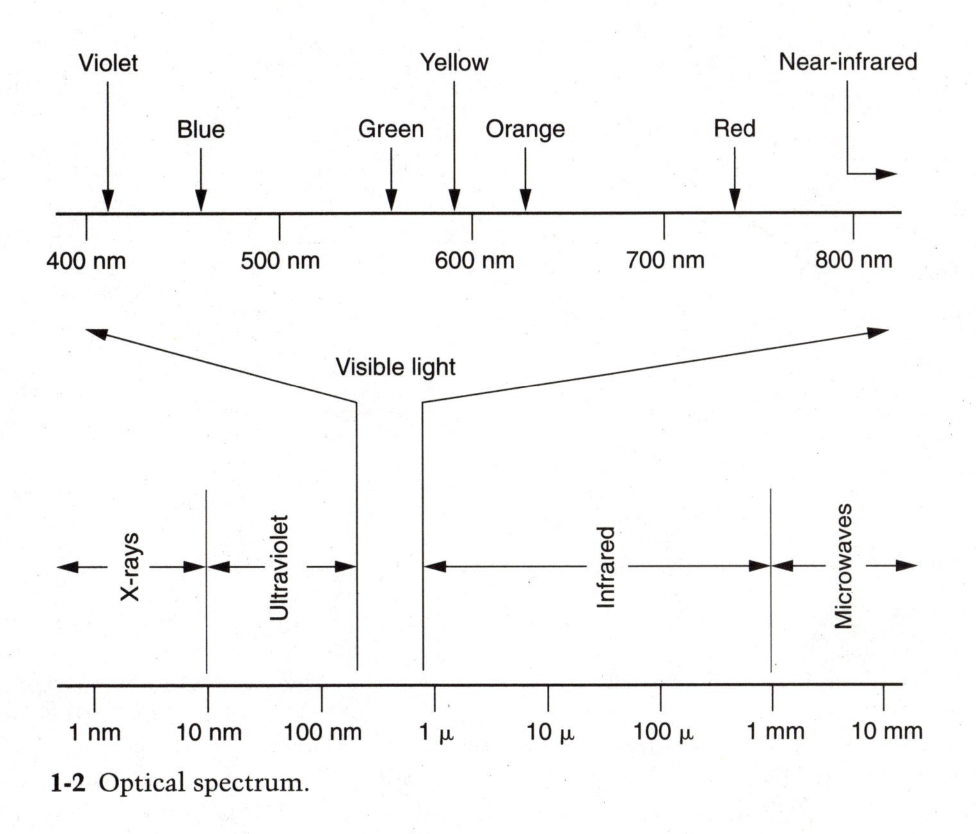

1-2 Optical spectrum.

Table 1-1 Conversion units for use with the optical spectrum.

Prefix	Abbreviation	Power of 10	Value
terra	T	10^{12}	Thousand billion
giga	G	10^{9}	Billion
mega	M	10^{6}	Million
kilo	K	10^{3}	Thousand
deci	d	10^{-1}	Tenths
centi	c	10^{-2}	Hundredths
milli	m	10^{-3}	Thousanths
micro	μ	10^{-6}	Millionths
nano	n	10^{-9}	Billionths
pico	p	10^{-12}	Thousand billionths

1 micrometer = 1 / 1,000,000 meters, 10^{-6} m, 1,000 nm, or 10,000 A units
1 nanometer = 1 / 1,000,000,000 meters, 10^{-9} m, 1 / 1,000 m, or 10 A units
1 angstrom = 1 / 10,000,000,000 meters, 10^{-10} m, 1 / 10,000 m, or 1 / 10 nm

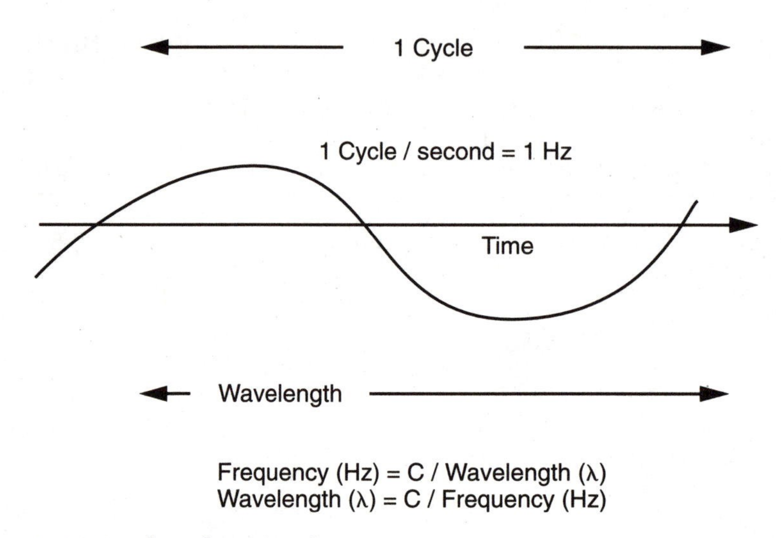

1-3 Wavelength versus frequency.

or the surface of the Earth, as well as the amount of infrared and ultraviolet light, ozone, and other gasses.

A second source of light is the ubiquitous incandescent lamp. An incandescent lamp is composed of a thin tungsten wire or filament encased in an evacuated glass envelope. An electrical current is passed through the filament, causing it to become incandescent or white hot. You can increase the brilliance and life of an incandescent lamp by filling the glass envelope with a gas such as argon, nitrogen, or krypton. The new ultra-bright halogen lamp has a quartz glass envelope filled with a halogen gas

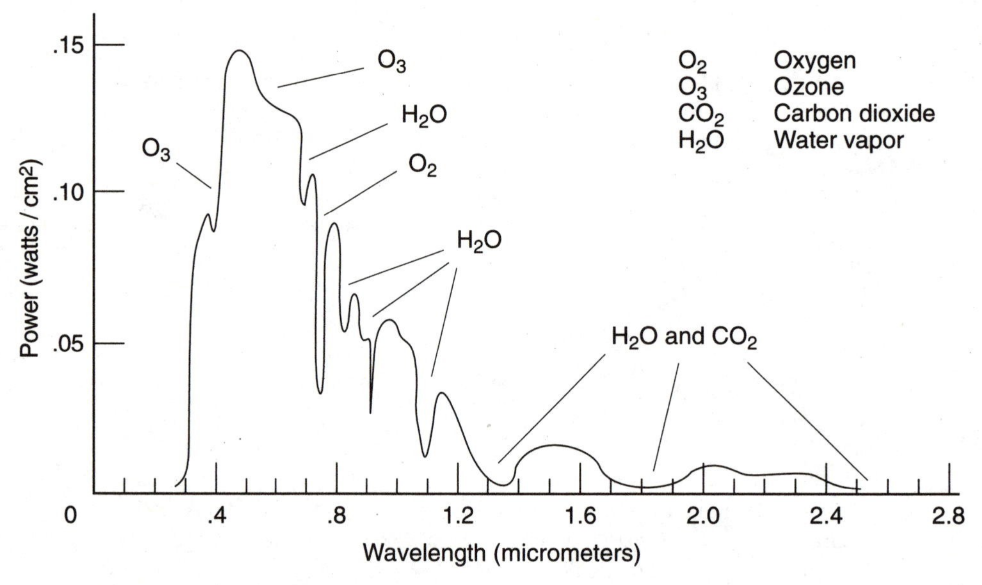

1-4 Solar spectrum.

such as iodide or bromine. Incandescent lamps are available in many sizes and shapes as well as operating voltages.

Another type of lamp is a gas discharge lamp. The most basic gas discharge lamp is the small neon glow lamp found in power indicators and testing circuits. The neon lamp is a glass envelope filled with a neon gas that glows orange when the input voltage across the two electrodes in the envelope exceeds 60 to 70 volts. The ionization or breakdown voltage of neon occurs and an electrical discharge is established between the electrodes. Note that the neon bulb does not have a filament, but two post electrodes only, which gives the neon lamp a very long life. Neon gas discharge lamps make great indicators and can be coupled with light detectors to form a rather unusual type of optocoupler, which can be used to detect various ac voltages. Neon lamps are inexpensive and will last many years if used with a current-limiting resistor. Other types of gas discharge lamps are flash tubes and xenon, mercury vapor, and sodium vapor lamps.

The next light source is the light-emitting diode, or LED. LEDs are used extensively throughout this book since they are inexpensive and have a long life. The LED is a PN junction semiconductor diode that emits visible light when forward biased. Visible LEDs emit relatively narrow bands of light, presently red, green, yellow, and blue. LEDs are inexpensive, very efficient, easy to implement, have a long life, consume far less power than incandescent lamps, and can be turned on and off very rapidly with electronic circuits. They are current dependent and their light output is directly proportional to the forward current through the LED. LEDs must be used in conjunction with a resistor, as shown in Figure 1-5, in order to limit the current through the LED. An LED used without a current-limiting resistor would soon burn out. A forward current of 20 milliamperes (mA) is generally considered safe and is generally provided by using a standard 220-ohm resistor with the LED.

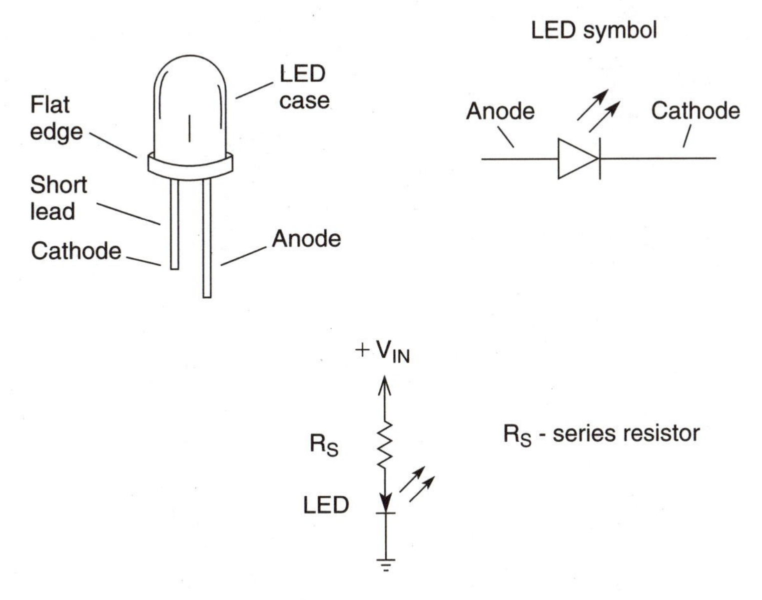

1-5 Light-emitting diode.

The diagram in Figure 1-6 illustrates a tricolor LED. Three colors are constructed by packaging a red and green LED in the same clear epoxy housing. The two semiconductors are usually connected in reverse—parallel, as shown, with two leads protruding from the package. Applying a plus voltage to pin AA will turn the LED red, while a minus voltage to pin AA will turn the green LED on. When an ac signal is applied to the tricolor LED, the package will light up with a yellow color.

Light-emitting diodes are now available in the super-bright variety, which are currently used in burn-out-proof tail-light arrays for automobiles. When placed behind a plastic diffuser, a single super-bright LED can be used as a bicycle tail light. A cluster of super-brights used with a diffuser can be made into a modern-day traffic signal. Linear arrays of super-brights are currently used in copy machines and pseudolaser printers. Sophisticated applications of super-bright LEDs include optical fiber and free-space light-wave communication systems. The new super-brights emit most of their light in the 660-nm range, compared to the 880 nm of conventional visible light LEDs. Super-brights typically operate with 50 to 60 mA of current, considerably higher consumption than that of conventional LEDs.

Infrared LEDs emit light just beyond the visible red wavelengths. Infrared LEDs are commonly used in remote-control circuits, alarm systems, and communication

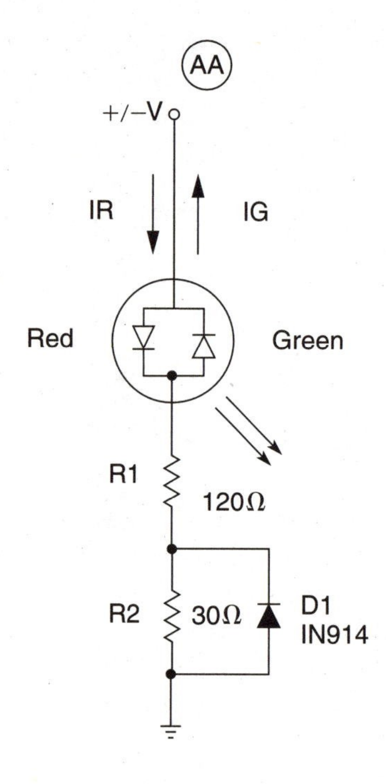

V	Color
+	Red
−	Green
AC	Yellow

1-6 Tricolor LED.

links. Infrared LEDs are often pulsed or modulated in the 30- to 40-kHz range to avoid interference from visible light sources. The diagram in Figure 1-7 shows various light sources and where they are situated in the visible light spectrum. When LEDs are paired with electronic light sensors, it is important to consider the spectral sensitivities of the photodetector in order to match the light source for optimum sensitivity. LEDs have revolutionized modern electronics. LEDs are used in all aspects of electronics from indicators to optocouplers, isolating and linking various types of circuits.

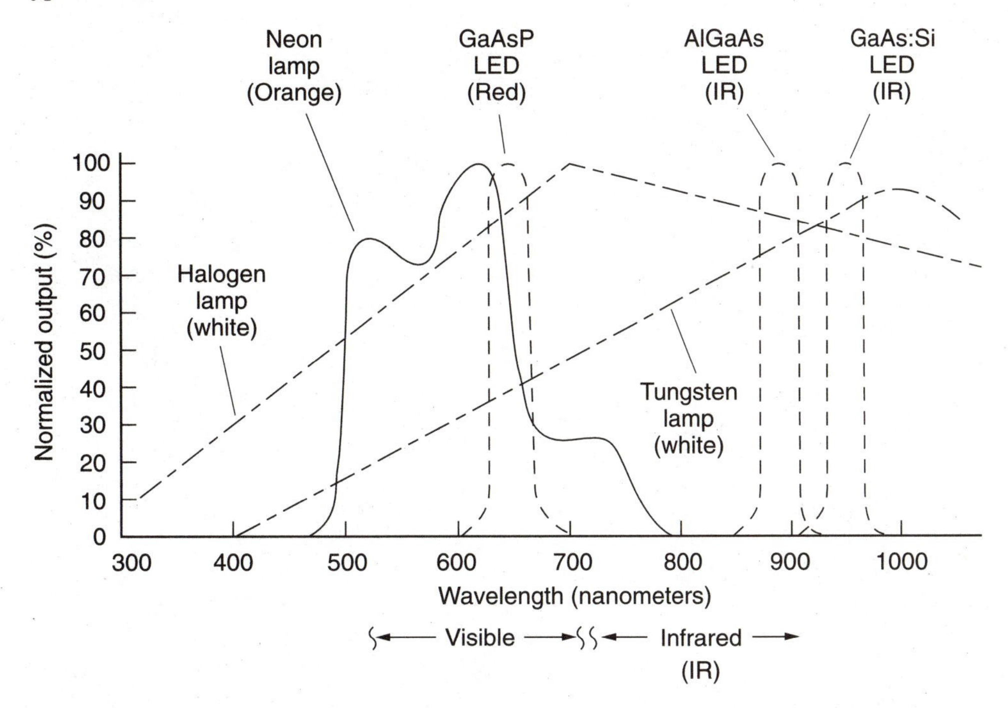

1-7 Light spectrum.

The final light source is a laser. One of the unique properties of lasers is their ability to emit light of a specific color or wavelength, in contrast to sunlight or incandescent lamps, which emit a broad range of wavelengths all at the same time. A laser, on the other hand, usually emits light at one specific frequency or wavelength or a number of specific wavelengths. Light emission at each individual wavelength is a called a *main line*. Most helium-neon lasers emit light at one specific wavelength, while argon generates two distinct main lines at 488 and 514 nm, as shown in Table 1-2. Lasers are *coherent* light sources since they generate light of one color or a few colors, as opposed to sunlight or incandescent lamps, which are *noncoherent* light sources. There are four main types of lasers: solid, liquid, gas, and semiconductor. The earliest laser was classified as a solid laser. A synthetic ruby rod was excited by a flash tube rapidly firing white light into the ruby rod. One end of the ruby rod was fully silvered, while the light output end of the ruby was partially silvered to allow a back-and-forth pumping action to take place within the ruby rod. It is interesting to note that concurrent with the development of the ruby laser was an early version of the solid-state laser, although crude at that point in time.

The laser most commonly used by experimenters and researchers to date has been the helium-neon gas laser. The helium-neon laser, illustrated in Figure 1-8, is a glass vessel filled with 10 parts helium with 1 part neon pressurized to about 1 mm/Hg. Electrodes are placed at the ends of the tube to provide a means to electrify or ionize the gasses in the tube. Mirrors are mounted at either end of the tube, which

Table 1-2 Wavelengths of lasers.

Type	Wavelength
Argon	488 and 514.5 nm
Carbon dioxide	10,600 nm
Dye laser	300 to 1000 nm
Excimer	193 to 315 nm
Cadmium	442 and 325 nm
Helium-neon	632.8 nm
Nitrogen	337 nm
Krypton	647 nm
Nd:YAG	1064 nm
Ruby	694.3 nm
Semiconductor diode	780, 840, and 904 nm

forms an optical resonator. One mirror is totally reflective, while the mirror at the other end of the output side of the laser tube is only partially reflective. The helium-neon laser is commonly used in supermarket code scanners as well as surveying, holography, medicine, welding/cutting, and laboratory experiments. The helium-neon laser is a relatively low-cost laser that is readily available through surplus outlets. Helium-neon lasers emit a characteristic red beam at 632.8 nm. A typical helium-neon laser circuit diagram is shown in Figure 1-9.

Gas lasers include helium, cadmium, argon, krypton, and carbon dioxide, the last commonly used for cutting lasers. Chemical lasers, usually of extremely high power, are used by the military to shoot down aircraft and missiles. A chemical laser often consists of a flammable mixture of hydrogen and fluorine and is ignited into a flame to start the "lasing" action.

Eximer lasers are often used to emit high-energy ultraviolet light. They commonly use argon or krypton, which electrically reacts with a halogen such as bromine or chlorine. Liquid or dye lasers use molecular organic dyes as the lasing material. The dye is directed through a cavity, pumped by an optical source such as a CO_2 laser. The unique characteristic of a dye laser is that you can tune its output wavelength by varying the mixtures of dyes, and therefore change the color of the laser beam from blue to red.

Semiconductor or solid-state lasers are widely used in a host of consumer electronic items, including compact-disc players, bar-code readers, fiber-optic telephony, printers, and copiers. A laser diode consists of a PN junction with specially cleaved and mirrored surfaces, as shown in Figure 1-10. Current is applied to the

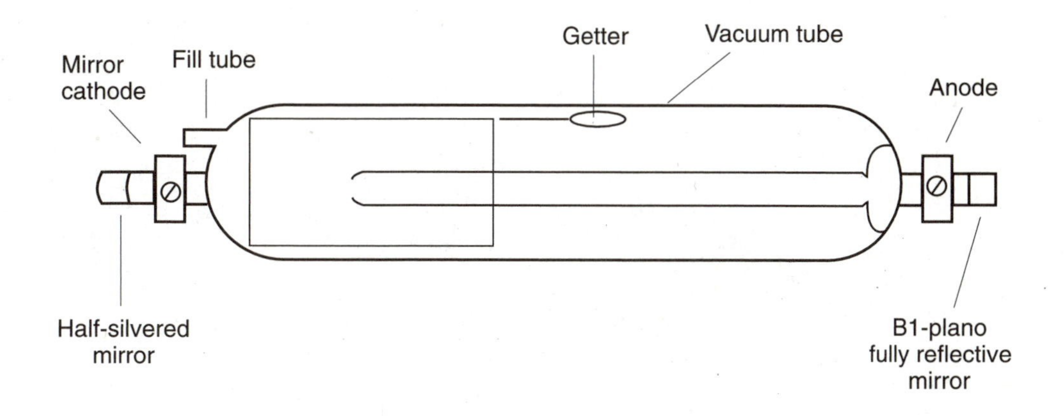

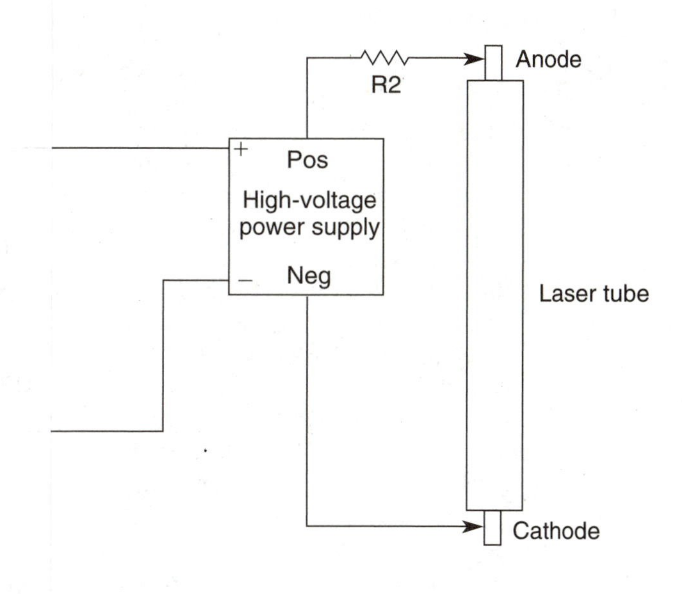

MSCPLIC

ESHELBY Agency
09/08/2003
07:30 PM

Please return the items by the due date(s) listed below, to any library location. Call 751-7360 or online at www.memphislibrary.org for renewals. AV & 7-day Popular - no renewals.

Optoelectronics, fiber optics, and laser
0115245322447
Due: 09/29/2003

Electronic troubleshooting
0115245338930
Due: 09/29/2003

Schaum's outline of theory and problems
0115245339185
Due: 09/29/2003

emiconductor laser diode to light. The light is bounced action between the two mirrored surfaces inside the op- om each facet or mirror within the cavity forms a fan- gence of 15 by 30 degrees. Virtually all the radiation ce of a laser diode can be collected and collimated into low-cost laser diodes are designed to emit a 3- to 4-milliwatt beam in the 780-nm range. Low-cost continuous-wave (CW) laser diodes can be purchased for about $20 and often less from surplus dealers. All the laser projects in this book will be of the helium-neon and solid-state laser diode varieties. Laser diodes are also available in infrared types. These laser diodes are often used in fiber-optic data and telephony circuits, as well as burglar alarm systems.

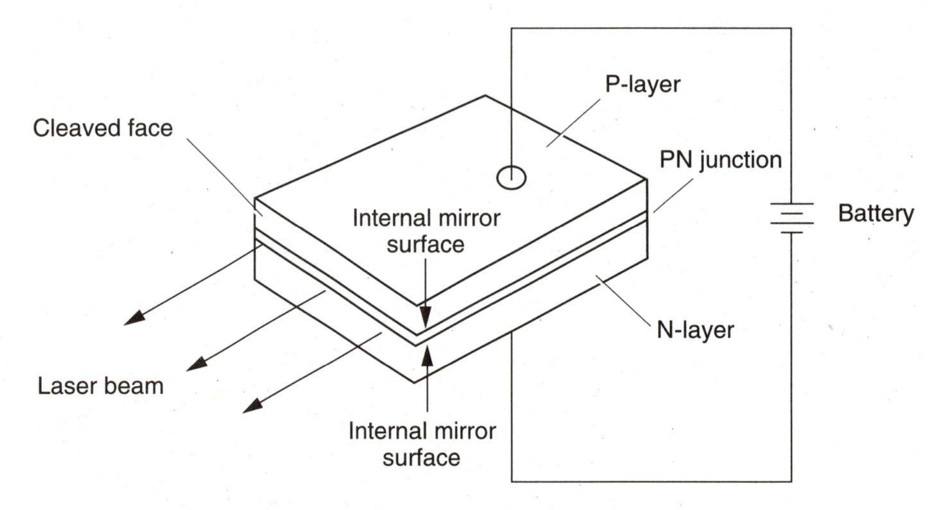

1-10 Semiconductor laser diode.

Types of sensors or detectors

Now that I have discussed the most common types of light sources, the next topic will cover a variety of different light sensors that can be used in optocouplers, control circuit alarm systems, sensing systems, and communication systems. Combining light sources and optical detectors is very important to optoelectronics and is commonly used to interface and isolate one electronic circuit from another.

The first light sensor is a photoresistive or cadmium sulphide (CdS) cell. As the name implies, this device is a resistive detector. As light impinges on the sensor, the resistance varies from 1 megohm when the sensor is dark to a few hundred ohms when light falls on the detector. The photoresistive sensor was once perhaps the most commonly used light sensor. The cadmium sulphide cell is easy to implement, inexpensive, and quite sensitive. The CdS cell is primarily sensitive to green light. Photoresistive cells are considered slow by today's electronics, however, since they require a second or two to return to their normal high resistance.

Photoresistive sensors are therefore well suited to relay and control circuits rather than interfacing data circuits. Photoresistive cells are generally a bit larger than most solid-state sensors, about $\frac{3}{8}$ to $\frac{1}{2}$ inch in diameter. The diagram in Figure 1-11 shows a typical photoresistive sensor symbol and simple voltage divider circuits that are commonly used. The voltage divider on the left turns on with the presence of light, while the divider on the right can control devices with the absence of light.

Phototransistors are another type of light sensor that are commonly used. A phototransistor is shown in Figure 1-12. The phototransistor is considerably faster in response than the photoresistive cells described previously. Typical responses for phototransistors are on the order of up to one microsecond. Most people do not know that all transistors and FETs are light-sensitive, but phototransistors exploit this effect and are more efficient. Most phototransistors are generally NPN devices with only two leads. Phototransistors are generally placed in a clear epoxy case, which allows light to fall on the base region of the device. The base region of a pho-

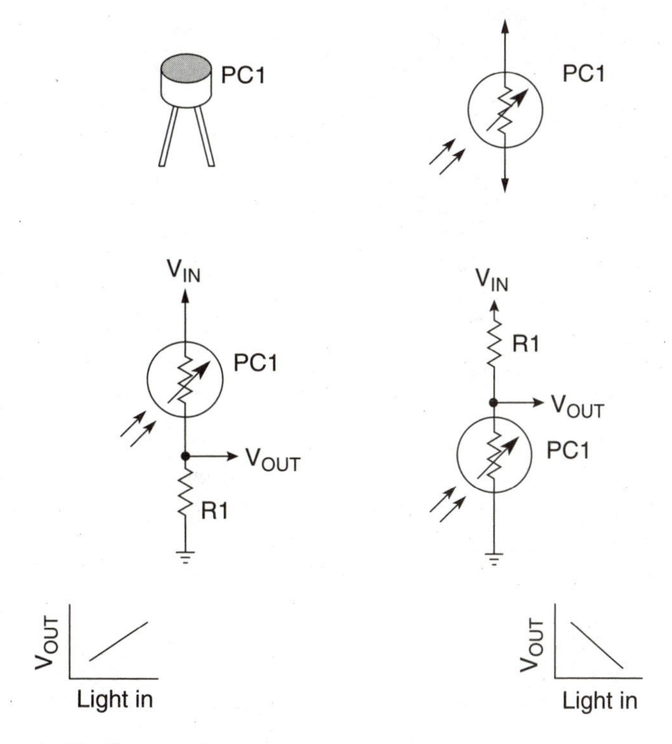

1-11 Photoresistor.

totransistor is much larger than that of a conventional transistor. Phototransistors are considered photoconductive devices, since the detector allows current from an external source to flow in response to light. The easiest way to implement a phototransistor is to connect a series bias resistor, as shown in the figure. The series resistor value should be between 100 kilohms and 1 megohm for high-sensitivity applications. To use a phototransistor for fast signal response, typically a 10-kilohm value would be used for the series resistor. Note the trade-off between sensitivity and speed. A phototransistor is usually mounted in a light shield consisting of a tube that is painted black to keep unwanted or stray light from reaching the detector. Phototransistors are available in visible and infrared types. Figure 1-13 illustrates the differences in spectral response between CdS photoresistive sensors and a typical phototransistor. Phototransistors are also available with a second "on-chip" transistor to amplify the phototransistor's output. This type of phototransistor is called a Darlington phototransistor; it is much more sensitive than a conventional phototransistor and is often used in communication and control circuits.

The fastest photodetector response can be obtained from a PIN diode detector, which is often used in high-speed pulse-modulation and telephony circuits, as well as high-speed data interfaces. PIN diodes are available in germanium, silicon, and gallium arsenide (GaAs) types. Germanium PIN diodes were the earliest photodiodes, and they were very noisy compared to modern-day silicon and GaAs diodes. The silicon PIN diode is spectrally sensitive to light in the 600- to 1000-nm range, while ger-

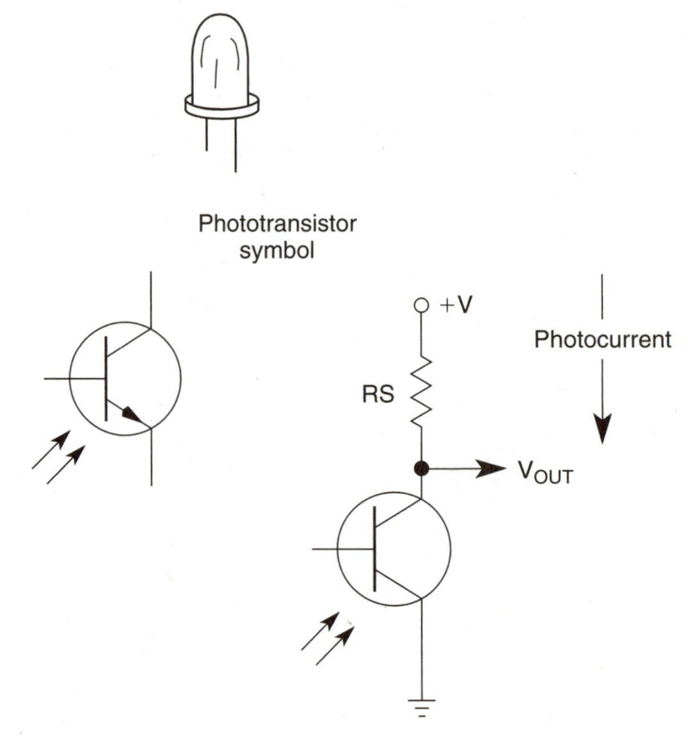

1-12 Phototransistor.

manium and GaAs PIN diodes operate more efficiently between 1600 and 1800 nm. The latest photo PIN diode detectors have a bandwidth of up to 350 MHz. The speed and bandwidth of a photoelectric detection circuit is governed to a large extent by the RC time constants in the detection circuit as well as in the postdetection filtering network after the detector.

If, for example, you need to measure UV sunlight for atmospheric ozone detection, you would need a gallium phosphide (GaP) photodiode since these optical diodes do not respond to infrared light as do most other photodiodes. GaP photodiodes are more sensitive to the UV spectrum, and the sensitivity drops off at the edge of the infrared region. Ultraviolet-sensitive silicon diodes are available for optical power meters, but unfortunately they too are sensitive to the infrared region.

The next light detector is the solar cell or photovoltaic cell, which is currently available in many sizes, shapes, and current outputs. As the name implies, this type of detector produces a voltage output and is most often used for detectors and battery chargers in remote regions where conventional power does not exist. Solar cells are connected in series to increase their voltage or in parallel to increase current output if they are used in power-charging circuits. Solar cells are also widely used as light detectors and are sensitive to visible light as well as near-infrared radiation. A typical solar cell responds to changes in light intensity within 20 microseconds, which permits the detection of voice-modulated light-wave signals, as you will see in the following chapters. The solar cell symbol and pictorial diagrams are shown in Figure 1-14.

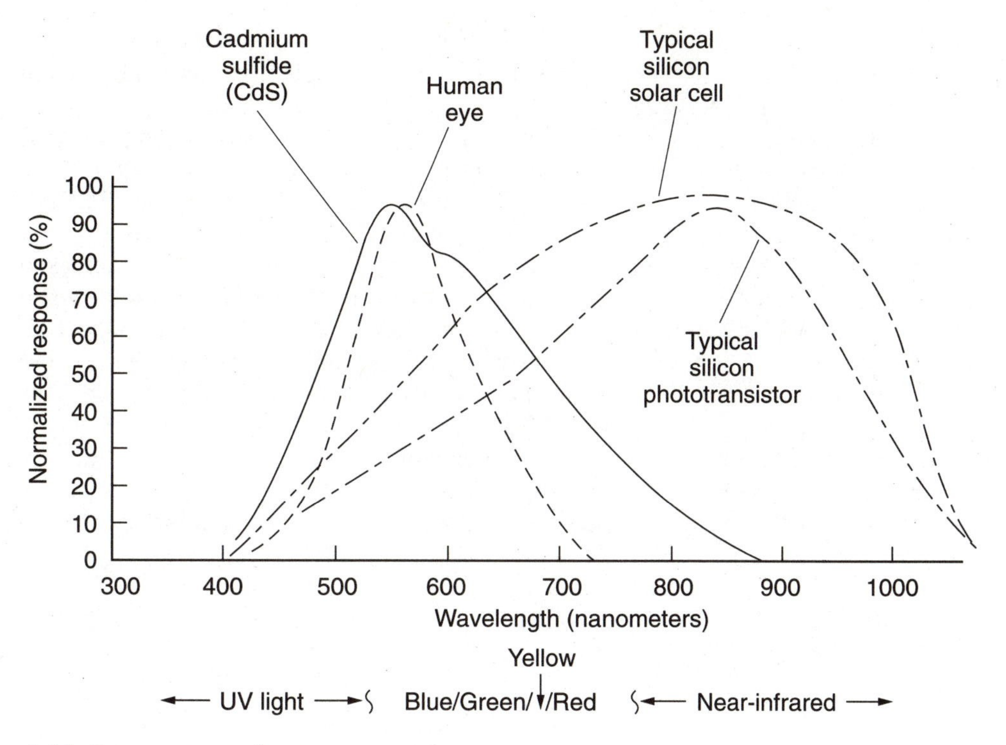

1-13 Sensor spectral response.

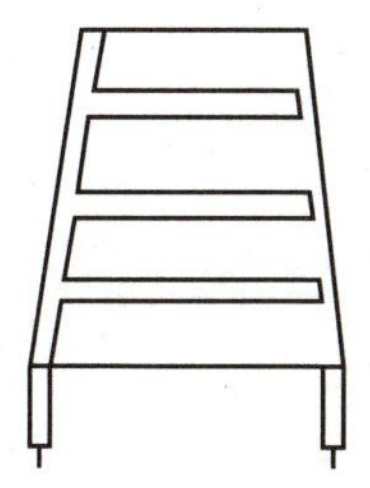

1-14 Solar cell.

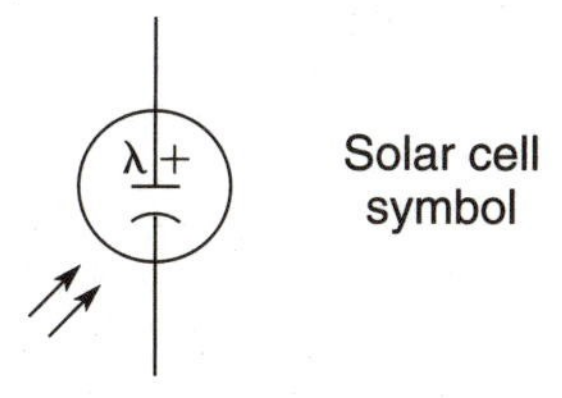

Solar cells are gradually becoming less expensive and might eventually provide a significant production of electricity. Presently, photovoltaic cells are commonly used in watches, calculators, rechargeable flashlights, and battery chargers.

The last photodetector is a photomultiplier tube. Photomultiplier tubes (PMTs) are highly sensitive light detectors, often used in laboratory instruments such as spectrophotometers. They are especially sensitive to infrared wavelengths. PMT detectors have a number of photoanodes, arranged so the electrons emitted by the cathode are attached to the positively charged anodes, from which an increased number of electrons are released and attached to the next successive anode, which is at a higher positive potential than the preceding anode as the electrons make their way through the tube. A greater number of electrons are released as the process continues through the PMT detector until the electrons reach the final anode. This process of multiplication results in a very high potential. This way a very high current flow can be achieved from a tiny initial current. Unfortunately, PMTs require a high-voltage power supply to excite the tube anodes and the price of the tubes has become quite high, making these detectors less accessible to the hobbyist.

2

CHAPTER

Optics

In order to control and use the effects of light rays efficiently, you also need to use some of the various types of optical components. Optical components allow you to conduct, bend, focus, reflect, or in general change the characteristics of light. You can find many of the optical components described in this chapter around the house or shop, or you could easily fabricate or purchase them inexpensively from science supply companies or surplus outlets (see the Appendix).

Refraction

The first method of controlling light rays is through refraction. Refraction is simply the bending of light. Light rays bend more when traveling through high-density media such as glass. The amount of bending is called the *index of refraction*. Light always refracts as it enters or exits a medium of different density. The "normal line" shown in Figure 2-1 shows the effects of refraction in relation to the change of density through a dense media. An incident beam refracts *toward* the normal line when the beam passes from a low- to a higher-density media.

A vacuum has a refractive index of 1, glass has a refractive index of about 1.6, and air has an index of 1.0003. Barometric pressure, temperature, and impurities can affect the index of refraction. Lenses are refractive media designed in such a way to bend light in a particular way. The refractive index is determined by the chemical makeup of the lens. Two common lens types are Crown and Schott glass, with refractive indexes of 1.52 and 1.79, respectively. The shape of a particular lens more than any other factor determines how light will be refracted. Any piece of flat glass is a refractive medium. If you were to make the glass thicker in the middle or thicker at the edges, then you could control the rays of light. There are six major types of lenses, including the plano or flat lens, the convex lens (which curves outward), the concave lens (which curves inward), and the meniscus (which curves in on one side and curves out on the other side). There are also combinations, such as plano convex, double convex, and convex meniscus, which

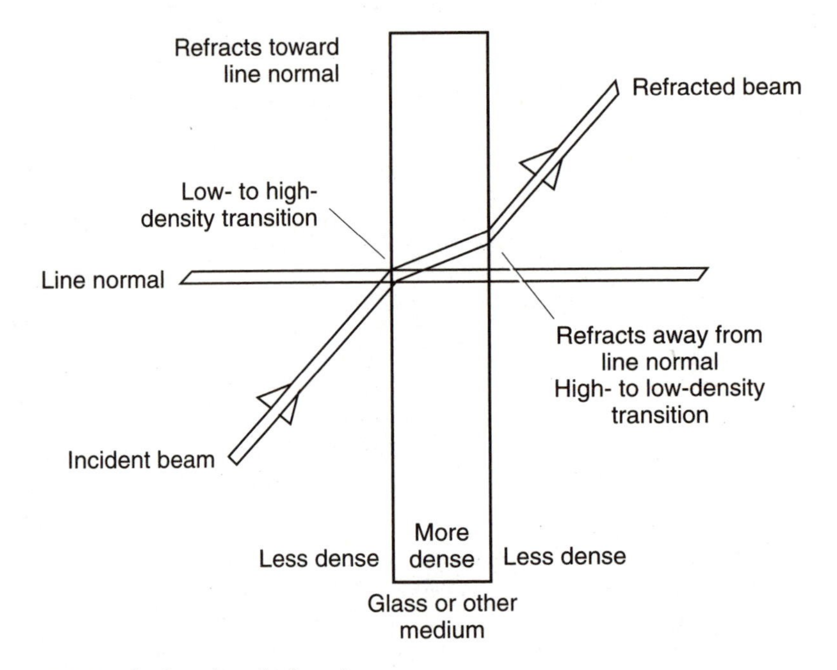

2-1 Light bending/refraction.

all form positive-type lenses. Negative lenses are the concave meniscus, plano concave, and double concave lenses (see Figure 2-2).

Lens diameter and focal length

The lens diameter and focal length are also important characteristics that can affect light rays. The focal length is important since it dictates how much the lens is refracting light. A short focal length means that light rays are brought to a point very quickly and are thus heavily refracted by the lens. A long focal point length means the refraction is mild and the light rays are gradually brought to a point. Light is shown from above and the lens is moved up and down until a focused spot of light is

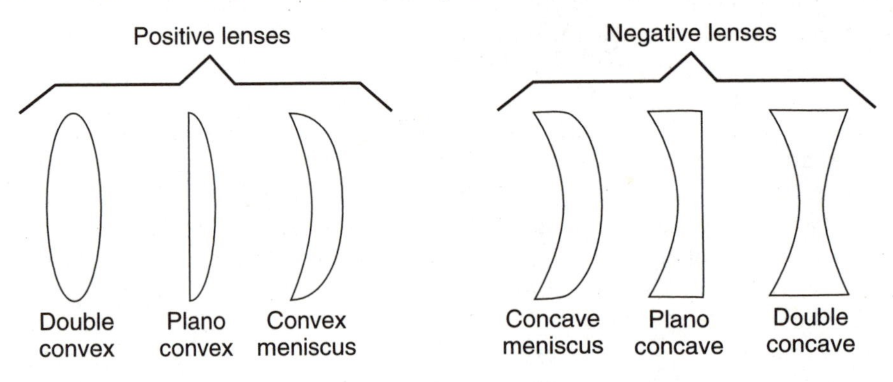

2-2 Major lens types.

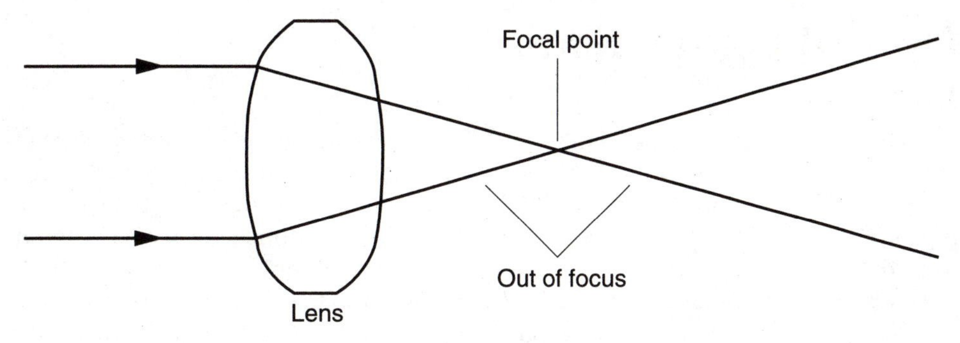

2-3 Focal distance.

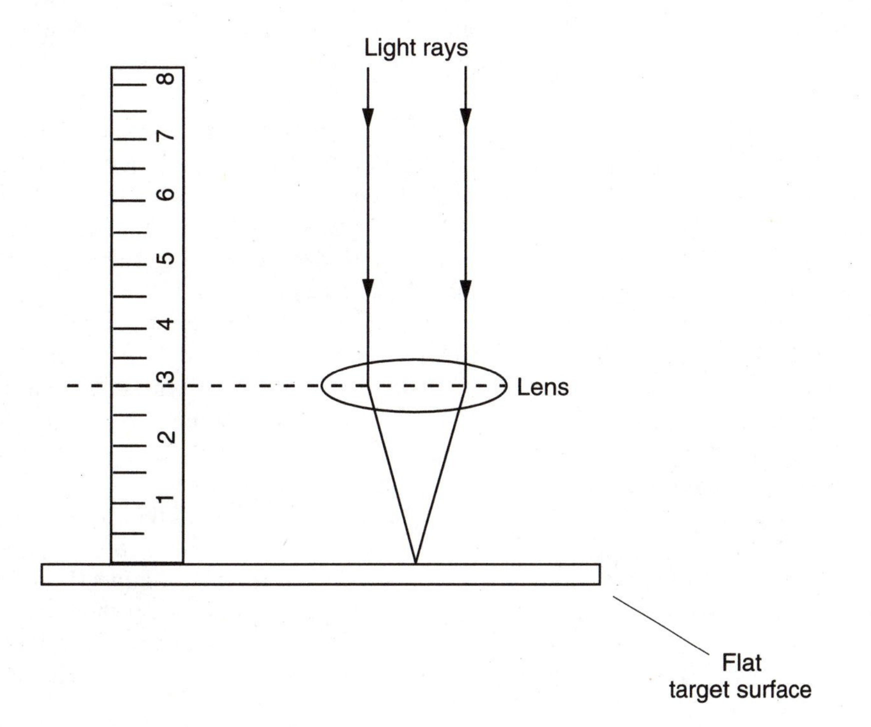

2-4 Measuring focal distance.

projected to the flat target surface. The focal point of a lens is usually quite crucial, as can be seen in Figure 2-3. You can easily determine the focal length of a particular lens by holding a lens parallel to a flat surface, with a ruler or measuring device perpendicular to the target surface, as shown in Figure 2-4. The focal point is a finite or specific point where the light rays converge and the light is out of focus both ahead of the focal point and after the focal length distance.

The diameter of a particular lens determines the light-gathering capability. The larger the lens, the more light it can collect (up to a point). The projects in this book will focus on using smaller, inexpensive lenses available through surplus outlets.

Optical components

Lenses

The least expensive types of lenses are composed of bare glass with no coatings of any type. An optical system that uses a number of uncoated lenses can suffer from an appreciable amount of light loss due to reflection. These stray reflections have to go somewhere and they often strike the inner walls of the optical device. Reflections from uncoated lenses decrease contrast and can cause flaring and ghosting. Lens coatings vary from lens to lens. Expensive lenses can have multicoated layers of a specific thickness in quarter wavelengths. When buying used or surplus lenses, you should check to see if the particular lens has been coated. It is possible to determine if a lens has an antireflective coating by holding the lens at a 45-degree angle and looking at the reflected light. Coated optics usually have a purplish coloring.

Mirrors

Mirrors are used extensively in optical experiments, especially with laser systems, to redirect a light beam or to mix different light beams together. The general principle of reflection states that the angle of incidence is equal to the angle of reflectance. If you bounce a light beam off a mirror at a 45-degree angle to the normal line, then the reflected rays will be present at a 45-degree angle on the opposite side of the normal line, as shown in Figure 2-5. Most household mirrors have a silver coating that is applied to keep the mirror from tarnishing. This type of mirror is called a *rear-surface mirror*, as shown in the left diagram of Figure 2-6. The diagram at the right illustrates a *front-surface mirror*. The most common front-surface mirror is constructed of aluminum with a protective coating over the mirror's surface. Gold-coated front-surface mirrors provide the maximum amount of reflection to all wavelengths, but gold-surface mirrors are not often used due to the cost.

Dielectric coatings are often used on mirrors that are designed for laser systems. A dielectric coating is usually very thin and semitransparent, and is often applied in layers. The coating reflects light since its index of refraction is higher than that of the glass underneath. The amount of reflection varies depending on the angle of incidence, coating type, and thickness of the coating. Dielectric coatings are sensitive to wavelength, so you should pick the mirror coating for the particular type of laser light you are working with.

Beam splitters

The beam splitter is another important optical component, especially for laser experiments. Beam splitters, as the name implies, divide one beam of light into two. Beam splitters are also used as beam combiners when positioned carefully. They can be used to combine light from two sources into one single beam of light. Beam splitters come in two varieties—the glass-plate type and the cube type—as shown in Figure 2-7. Cube beam splitters are fabricated by cementing two right-angle prisms together so their common hypotenuses touch. A reflective or polarizing layer is usually added at their junction. Antireflective coatings are typically applied to incoming as well as outgoing light faces of a beam splitter to reduce light loss. Most cube beam

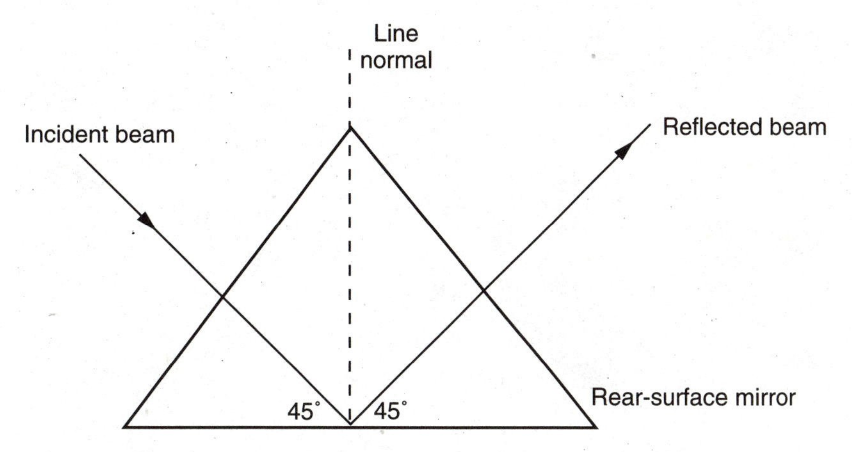

2-5 Rear-surface mirror reflection.

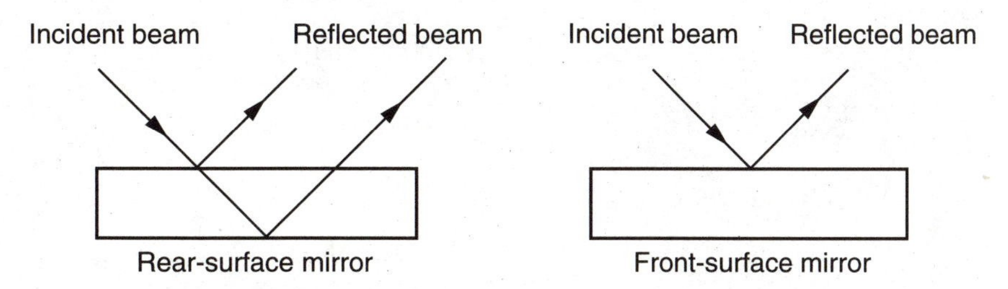

2-6 Front-surface versus rear-surface mirrors.

splitters divide light equally and are called 50/50 splitters. Glass-plate splitters use a flat piece of a coated glass to reflect and pass light. Although it is possible to use uncoated glass as a beam splitter, it is best to use a genuine beam splitter. Glass-plate beam splitters usually suffer from satellite or multiple images, and hence you get two reflected beams instead of one. The most efficient beam splitters are the cube type.

Porro prisms

A porro prism or retro reflector resembles a right-angle prism, as shown in Figure 2-8. The right-angle prism is often used to turn light around 180 degrees in order to send the light back to where it came from. Light enters the hypotenuse of the prism, strikes one side, bounces off the opposite side, and is redirected out the hypotenuse.

Filters

The next optical component is the filter. Filters are very important to light-beam manipulation. Filters accept light at certain wavelengths and block all other light. The color of the filter typically determines the wavelength of light that is accepted or passed. A red filter, for example, passes red light but blocks all other colors. Filters can be constructed either for maximum blockage or for varying amounts of light

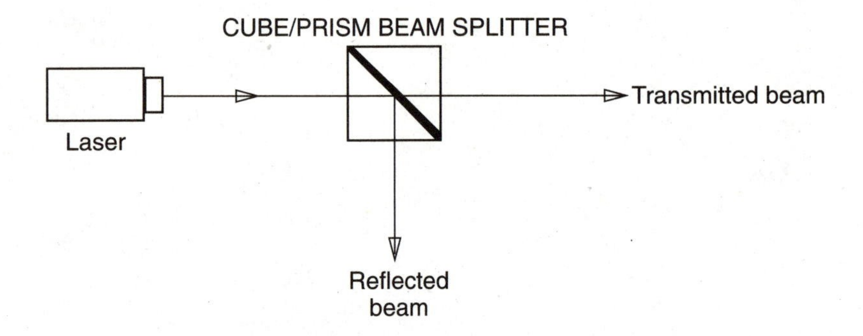

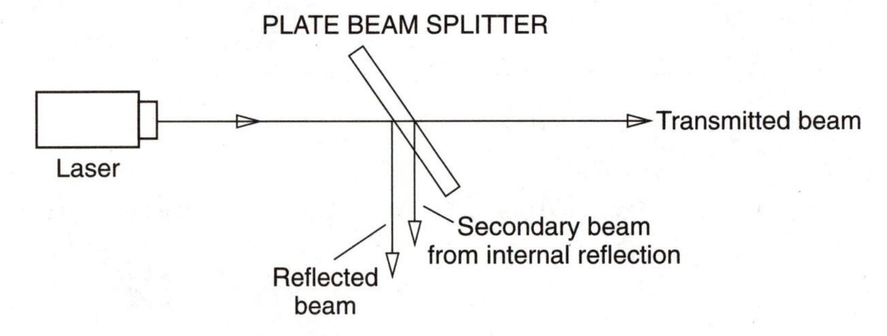

2-7 Beam splitters.

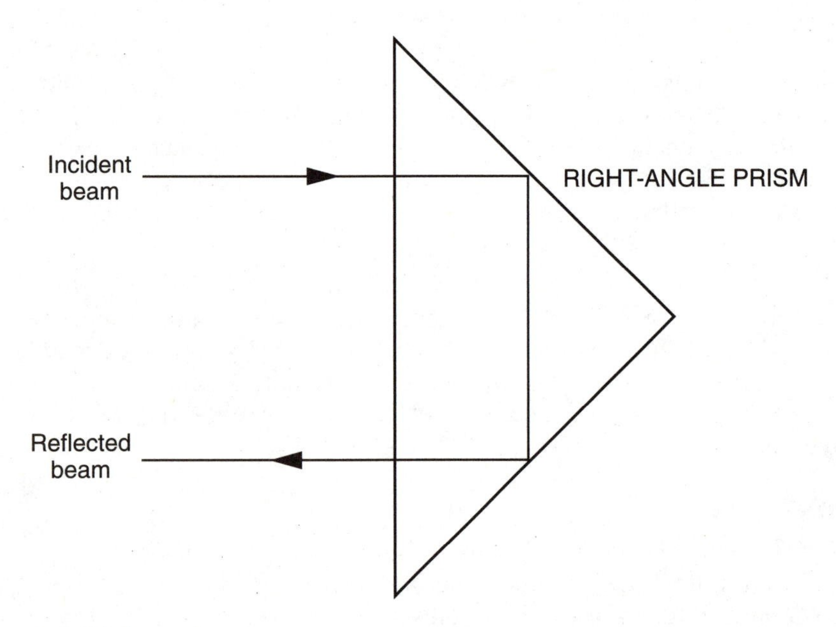

2-8 Right-angle prism.

passage. Many filters are designed to pass infrared light only and to block the visible light spectrum, while other types of filters might pass only ultraviolet light and nothing else. Filters are generally divided into three types: color gel, interference, and dichroic.

Gel filters are constructed of dyes that are mixed into a Mylar or plastic binder. Depending on the dye used, the filter is made to pass only certain wavelengths of light. A good gel filter could conceivably have a bandpass of 40 to 50 nm.

Interference filters consist of several dielectric and sometimes metallic layers that each block a certain range of wavelengths. One layer might block light under 500 nm, while a second layer might block light at 550 nm. The band of light between 500 and 550 nm is passed, therefore, giving the interference filter the bandpass classification. A color gel filter is shown in Figure 2-9, while a two-layer interference filter is depicted in Figure 2-10.

Dichroic filters use chemical dyes to absorb specific wavelengths of light. Dichroic filters are often constructed of cordierite crystals. Dichroism is sometimes used to create polarizing materials.

The last type of filter is the polarizing type, shown in Figure 2-11. You can readily experiment with polarized light using two polarizing filters. Light is composed of two components—a magnetic field component and an electrical field component—hence the name *electromagnetic*.

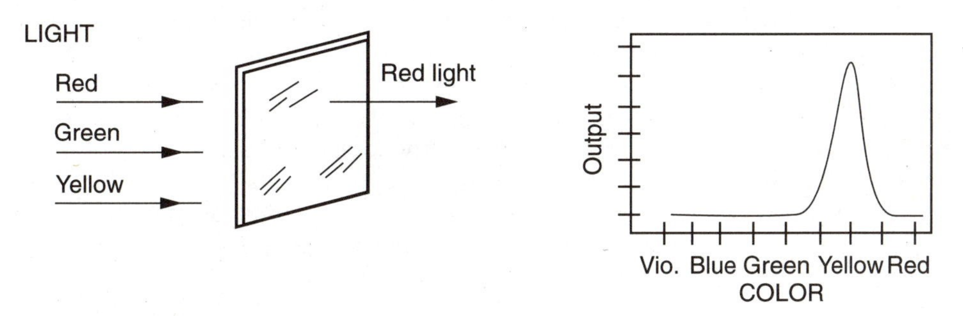

2-9 Color gel bandpass filters.

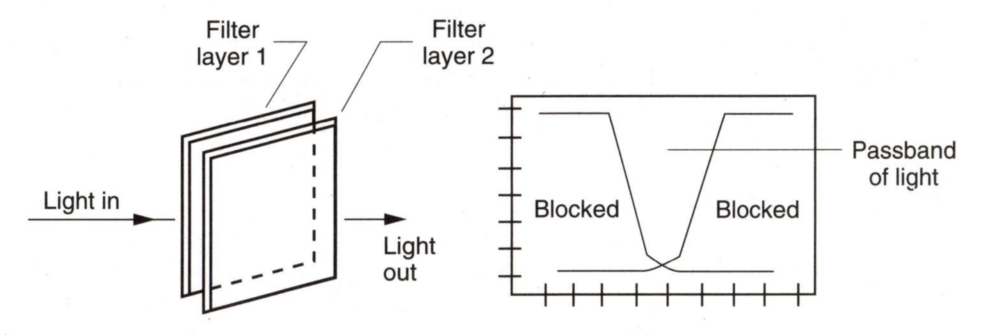

2-10 Interference filters.

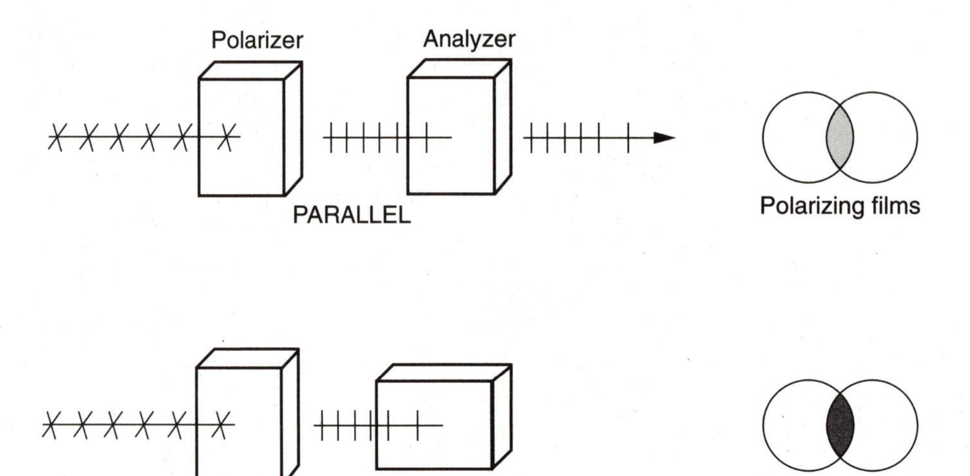

2-11 Polarizers.

These components or vectors of light travel together, but at 90-degree angles. One vector travels up and down, while the other vector travels left and right. In sunlight, the phases of both vectors are constantly changing, but they always remain out of phase by 90 degrees. This type of light is called *unpolarized sunlight*. Flashlights and other incandescent lamps are of the unpolarized type, while laser light is always polarized. Polarized light is created from unpolarized light with the use of a polarizing filter. Polarizing filters are typically made from organic or chemical dyes. Generally the two filters, one named the *polarizer* and the other the *analyzer*, are sandwiched together and one of the filters rotates from 0 to 90 degrees. The intensity of the light passing through the filters will increase and decrease every 90 degrees of rotation.

Prisms

Prisms are another optical component often used in optical systems. Prisms refract and disperse white light into its color components. Dispersion allows a prism with white light shining on it to create a rainbow, as shown in Figure 2-12. Multiline laser beams can also be used to disperse light via a prism. An argon laser, for example, could be passed through a prism to separate green and blue colors. Prisms can also be used to redirect a light beam to another angle or to polarize a beam of laser light and direct it in one or more directions.

Diffraction gratings

A diffraction grating diffracts light in a controlled manner. Diffraction gratings allow holograms to work and give a colorful rainbow look to compact discs. Diffraction gratings come in two main types: reflective and transmissive. Transmission gratings are constructed by etching a piece of clear film with a precision tool or laser. The amount of diffraction and the size of the interference fringes are determined by the number of lines made. Reflective gratings are made with a metallic

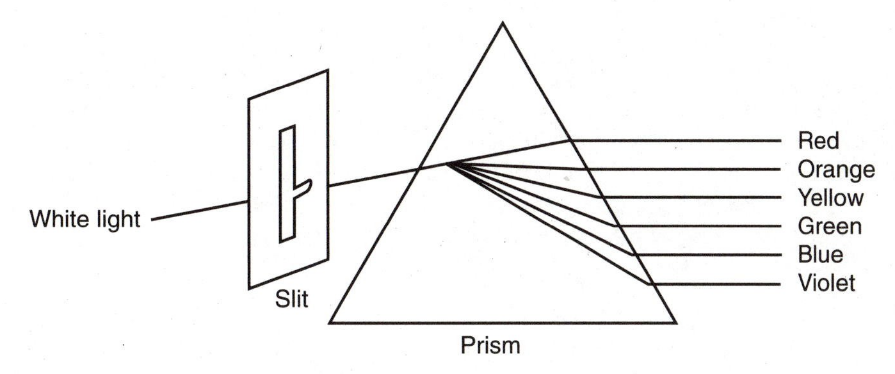

2-12 Rainbow prism.

sheet, and you view the diffraction with reflected light. An example of reflective grating is a CD-ROM. A diffraction grating can also be used to split a laser beam into many smaller beams.

Ronchi Rulings are a coarse type of diffraction grating, but they are made with more precision. A Ronchi Ruling has far fewer rulings, usually 50 to 400 lines per inch rather than the typical 15,000 lines per inch for other diffraction gratings. Ronchi Rulings are scribed or etched in glass and are held to higher tolerances than other diffraction gratings.

Pinholes

Pinholes are often used in optical work to make spatial filters or to diffract light. Spatial filters are used to "clean up" a beam spot by taking just the center portion of a light beam and excluding the perimeter or noise (see Figure 2-13). Spatial filters are often a pinhole coupled to a microscope objective lens.

An optical slit is a pinhole that has been enlarged to make a long narrow rectangle. It is designed to produce diffraction of a laser beam. The width rather than the length of the slit is the most important measurement. Optical slits can be fabricated with precision tooling, and often consist of two razor blades spaced close together to achieve the desired effect. (See Chapter 11 for more on on laser-beam diffraction.)

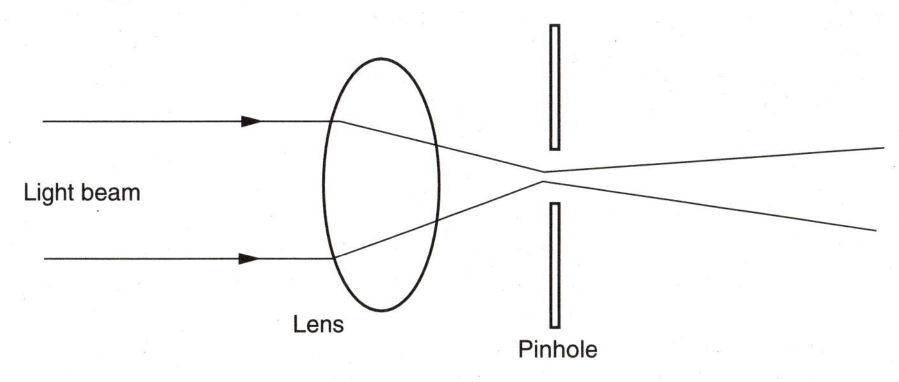

2-13 Spatial filter.

Light shields

An often-used light component is the light shield or collimator. A simple light shield is a metal or cardboard tube painted flat black on the inside to help reduce stray or reflected light from reaching a light detector. A light shield also helps guide light into a photodetector. A simple light shield is shown in front of a phototransistor in Figure 2-14. A small lens is often used with a light detector inside the light shield tube. You can increase the useful distance between source and sensor even further by using a lens assembly.

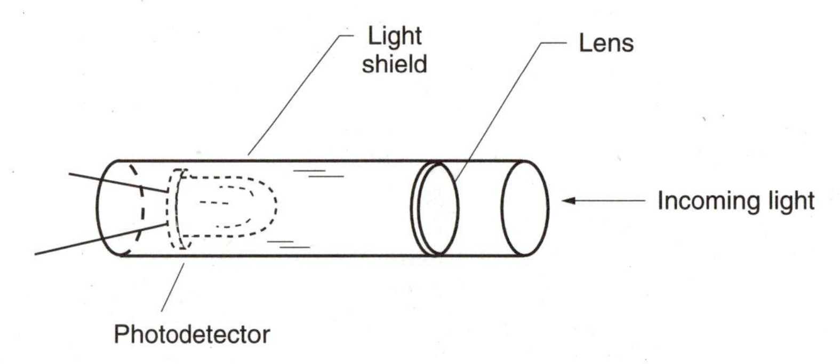

2-14 Light collimator.

Optocouplers/optoisolators

One of the most used optical components is the optocoupler or optoisolator, which can take many forms. Optocouplers are often called source/sensor pairs and they have many important applications in modern electronics. As the name implies, optocouplers are used to isolate one circuit from another, but at the same time they allow coupling from one circuit to another via light. Generally optocouplers can be found with LEDs and sensors integrated in a small 8- or 14-pin integrated-circuit (IC) package, as shown in Figure 2-15. The diagram on the left illustrates an LED source with a phototransistor sensor, while the diagram to the right shows an LED with a triac, which can control ac loads. Light-emitting diodes can be coupled to phototransistors, photodiodes, SCRs, and triacs. Optocouplers can also take the form of separate components such as neon bulbs and photocells or incandescent lamps and photoresistors for special applications, which you will see in the next chapter.

Fiber optics

The single most important optical component that has revolutionized our modern-day life is the optical fiber. Optical fibers are of two main categories, glass or plastic. Glass or silica fibers are generally highly transparent and pass the most amount of light. High-quality optical-grade fibers allow doctors to remotely view internal bodily functions. These high-quality fibers belong to a class of transparent fibers that are quite costly. A second class of highly translucent silica fibers allow a great deal of

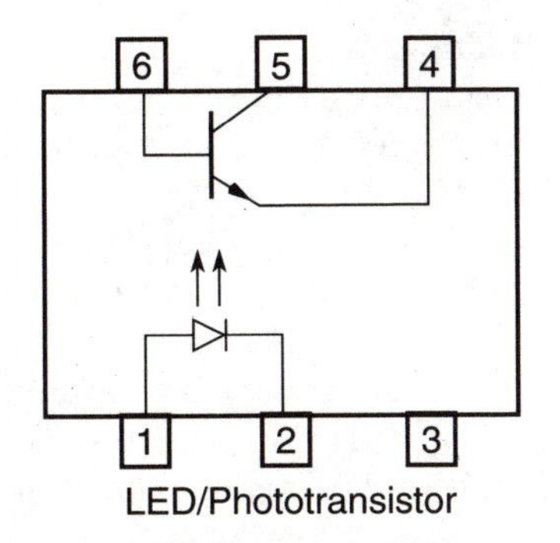

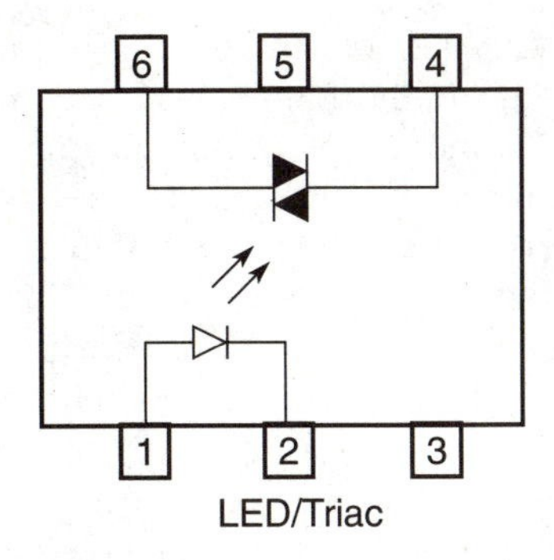

2-15 Integrated source-sensors.

light to pass, but do not allow an actual image to be seen through the optical fiber. These fibers are less costly than optical-grade fibers, but more expensive than plastic fibers. Highly translucent silica fibers are used most often for high-grade communication links. Plastic fibers are of the lowest quality but still pass a measurable amount of light. Plastic fibers can be more easily bent and twisted than silica fibers and they are also easier to interface to connectors.

Optical fibers conduct light either by internal reflection, as shown in Figure 2-16, or by continually refocusing incoming light rays toward the center of the fiber. Optical fibers are referred to as either *stepped index* or *graded index* types.

A stepped index fiber denotes that there is an abrupt change in the refractive index of the core and cladding. *Cladding* is the outermost layer surrounding the internal core of the fiber, as shown in the figure. A graded index fiber means that there is a more gradual change in the core index of refraction. This is accomplished

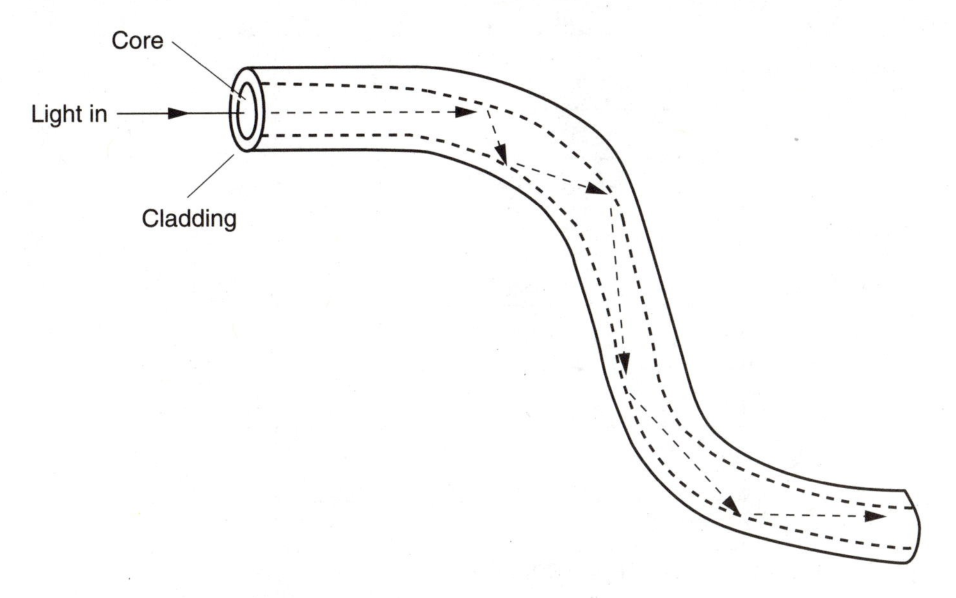

2-16 Optical fiber.

by changing the core material in a graduated manner from the center outward toward the cladding. Figure 2-17 illustrates the differences between stepped index and graded index fibers. As you might imagine, graded index fibers are the more costly of the two types and are used extensively in telephony circuits. The projects and experiments in this book will use the less-expensive stepped index fibers.

Fiber optics is quietly revolutionizing the communications industry around the world. Presently AT&T is busily working to convert the six major east-west transcontinental telephone trunk lines to fiber optics. Fiber optics has helped enormously to reduce noise in telephone circuits as well as to greatly increase the number of simultaneous conversations per cable. Long-distance telephone-line maintenance has become much less costly with the replacement of copper wire by fiber lines. In the past, transcontinental phone cables consisted of 22 coaxial copper cables along with many additional wires for monitoring and alarm circuits, in a 3½-inch shielded lead cable that was often pressurized with nitrogen to keep the lines dry. These massive cables had to be amplified every two miles from the east to the west coast. If one of the cables was accidentally broken, it would take more than a day to repair, at a loss of $3000 a minute. The equivalent fiber-optic cable is about one inch in diameter, needs amplification every 30 miles, and can be repaired in a half an hour or less. Fiber-optic cables provide considerable savings in installation, maintenance, and repair over copper wires. There are also an increased number of circuits and increased bandwidth in optical fibers.

So you can readily see how fiber optics is slowly revolutionizing communications, which is only one of its many applications. It has also revolutionized control

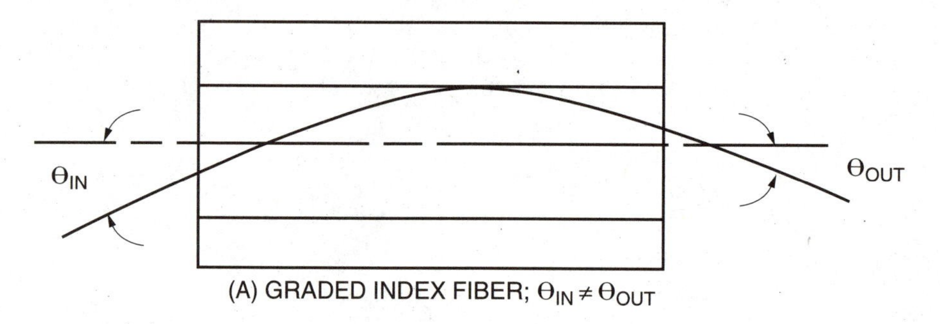

(A) GRADED INDEX FIBER; $\Theta_{IN} \neq \Theta_{OUT}$

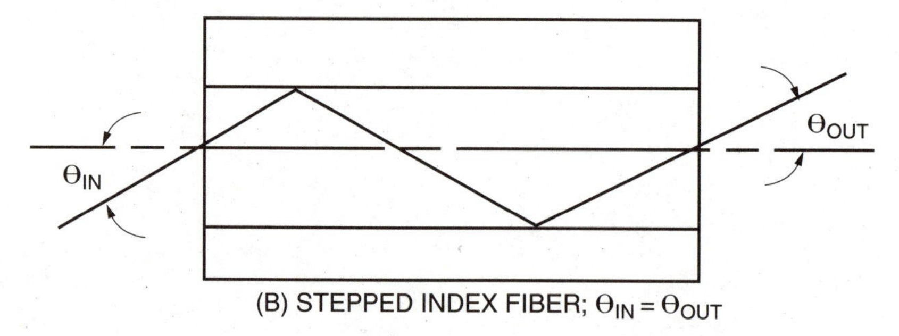

(B) STEPPED INDEX FIBER; $\Theta_{IN} = \Theta_{OUT}$

2-17 Stepped-index versus graded-index optical fibers.

circuits that can remotely control high-voltage circuits or circuits in hazardous environments. Fiber optics has also greatly assisted the medical profession, allowing doctors to examine people without invasive procedures.

Selecting optical equipment

New optical components such as lenses, filters, and beam splitters can be very expensive. Try to obtain good optical components from equipment taken out of service from one of the numerous surplus dealers listed in the Appendix. If you are a serious laser experimenter, however, you will want to locate the best quality optics you can afford. Try to locate prime surplus components if available. Prime surplus components were originally of the highest quality but can often cost far less than the original price because they are no longer needed. Optics that are damaged in some way are referred to as *seconds*, and the amount of damage can range from minor to serious. There are three types of imperfections: scratches, digs or gouges, and chips. Scratches can often be hairline and sometimes cannot be seen with the naked eye. Lasers and microscopes are often needed to detect this type of scratch. If a scratch is at the edge of a lens or is shallow, the lens can often still be used. If the scratch is shallow, it can possibly be "rubbed out." If the scratch is in the center of the lens, it is deemed unusable. Gouges or digs can also be found in lower-grade surplus lenses. If the dig or gouge is in the center of the lens, it is worthless, but if it is near the edge, the lens can often still be used. Many times surplus lenses contain edge chips and are still usable. Reject any lenses where the chip extends 10 percent beyond the edge area towards the center of the lens.

Surplus mirrors often contain scratches or gouges, which will most definitely distort a laser beam. These cannot be used. Again, common sense prevails. If the scratches or gouges are around the edge of the area of the mirror, it can still be functional. If the mirror is coated, look for small pinholes or other defects in the coating. If the mirror contains visible coating defects, select another one.

Laser optics, lens filters, mirrors, and beam splitters should be kept as clean as possible. Dust and grime will seriously affect the performance of any laser light system. Keep optical components wrapped in tissue when not in use, and store them in plastic bags. Avoid touching optical components such as lenses and mirrors with your fingers. Hold lenses, mirrors, and beam splitters either with tissue or by their edges, since oils from the skin act as an acid to etch the optical coatings over time. Use approved lens-cleaning fluid or pure alcohol to clean your lenses and other optical components. Use the cleaner solution sparingly to lightly wet a tissue used for cleaning. You can also use compressed air to remove tissue fibers or dust that accumulates.

Choose your optical component storage area carefully. Select an area free of dust and moisture. Place all small optical components in plastic bags and label them. A small pocket of air left in the bag will act as an added cushion. Then place the plastic bags in a storage box or container for long-term storage.

3 CHAPTER

Optocouplers and optointerfacing

This chapter covers a broad range of optocouplers, from do-it-yourself types to various integrated-circuit optocouplers. Then you will move on to new and more sophisticated signal-processing optocoupler ICs. Also presented are a number of simple optointerfacing circuits and drivers to help you become familiar with using LEDs and photodetectors. Later in the chapter I will present numerous practical interface circuits, including a current loop interface, an optically isolated RS-232-to-TTL interface, and two wireless infrared RS-232 computer links.

Optoisolators and optocouplers can take many forms, as you will see, from separate components such as LEDs and light bulbs to photodetectors such as phototransistors and solar cells. Optoisolators can be easily constructed from individual components and are often combined with heat-shrink tubing. A second popular method of constructing your own optoisolator is to drill out a block of wood or dowel to accept the source/sensor pair, as shown in Figure 3-1.

Optocouplers are divided into three major classifications, as shown. The first type is the enclosed source/sensor pair, which is commonly used for electronic isolation between two circuits, data transfer, and control circuits. The second type is the transmission slot pair. This optocoupler is formed by housing an LED and phototransistor in a wood block or dowel, which is slotted to accept a shutter vane or slotted wheel that can control light between the source and sensor. The slotted transmission pair combination is often used for object detection, limit switches, optical potentiometers, and vibration sensors. The third type of optocoupler is the reflective light-pair combination. An LED or light source and a detector are placed parallel to each other and arranged to act as a reflective pair. Opposite the source/sensor pair is a white or reflective surface. As an object passes between the source/sensor pair and the reflective surface, the light intensity received by the sensor is changed or modulated. The reflective pair combination is often used for object detection, counting, tachometers, limit switches, and reflective monitors. A vari-

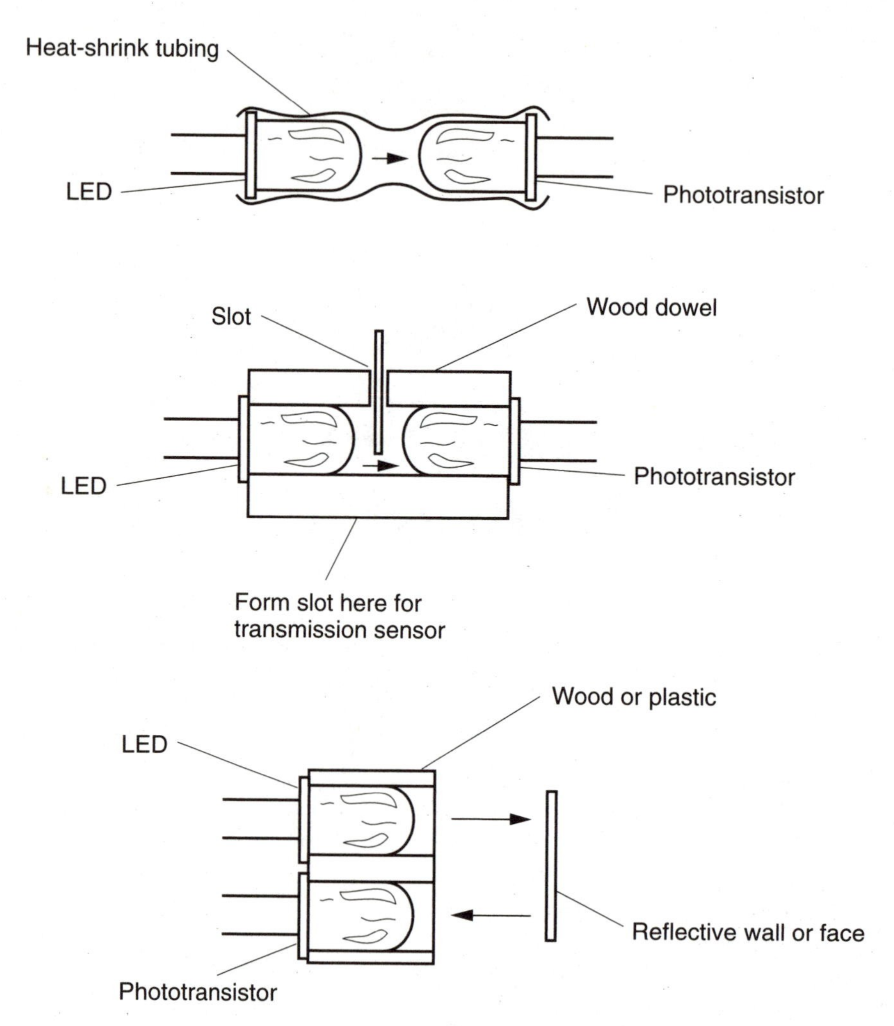

3-1 Do-it-yourself optocouplers.

ation of the reflective source/sensor pair is the bifurcated light pipe. In this reflective source/sensor pair, two fiber-optic light pipes are fused together into one. Each of the separate light pipes contains either an LED or photodetector. The remaining single fused end is polished and can take many forms as a light-probe sensor tip, which can be used to inspect inaccessible locations quite easily. The bifurcated light pipe will be discussed in detail in Chapter 5.

Miniaturization of semiconductors has provided engineers and experimenters with optocoupler integrated circuits that can contain many different source/sensor combinations, in 6-, 8-, 14-, and 16-pin IC packages. The diagrams in Figures 3-2 through 3-5 illustrate some of the more ubiquitous optical integrated circuits. Optocoupler ICs can take many forms, from simple single-channel source/sensor pairs to isolation amplifiers to multiple-channel isolated relays.

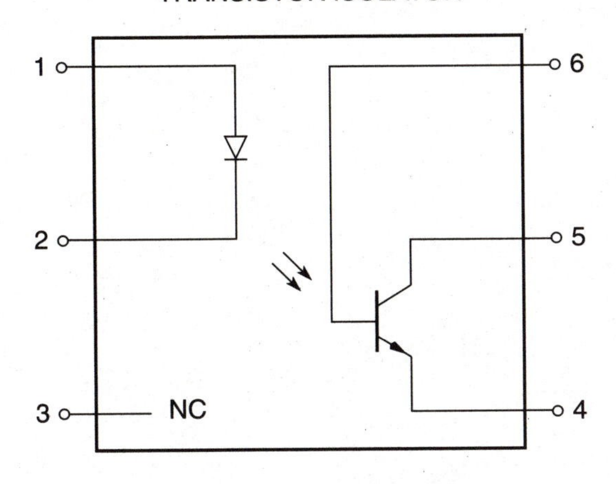

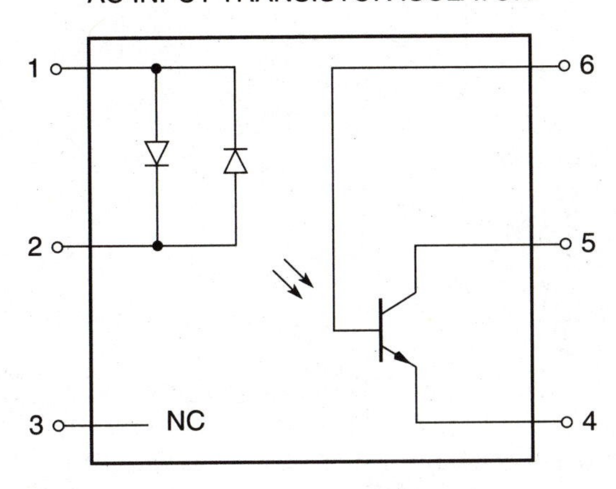

3-2
Transistor optoisolators.

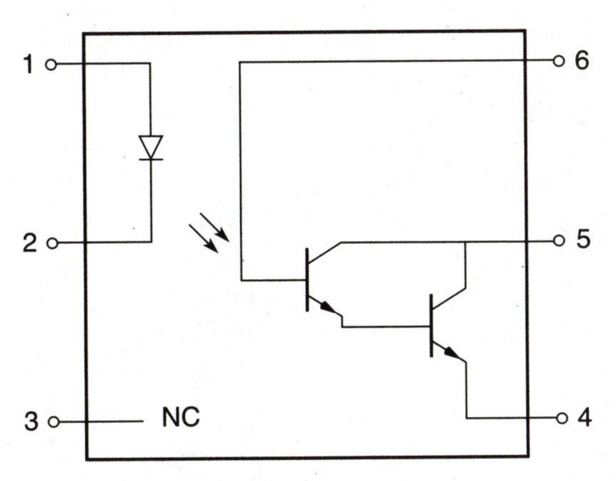

3-3
Darlington optoisolator.

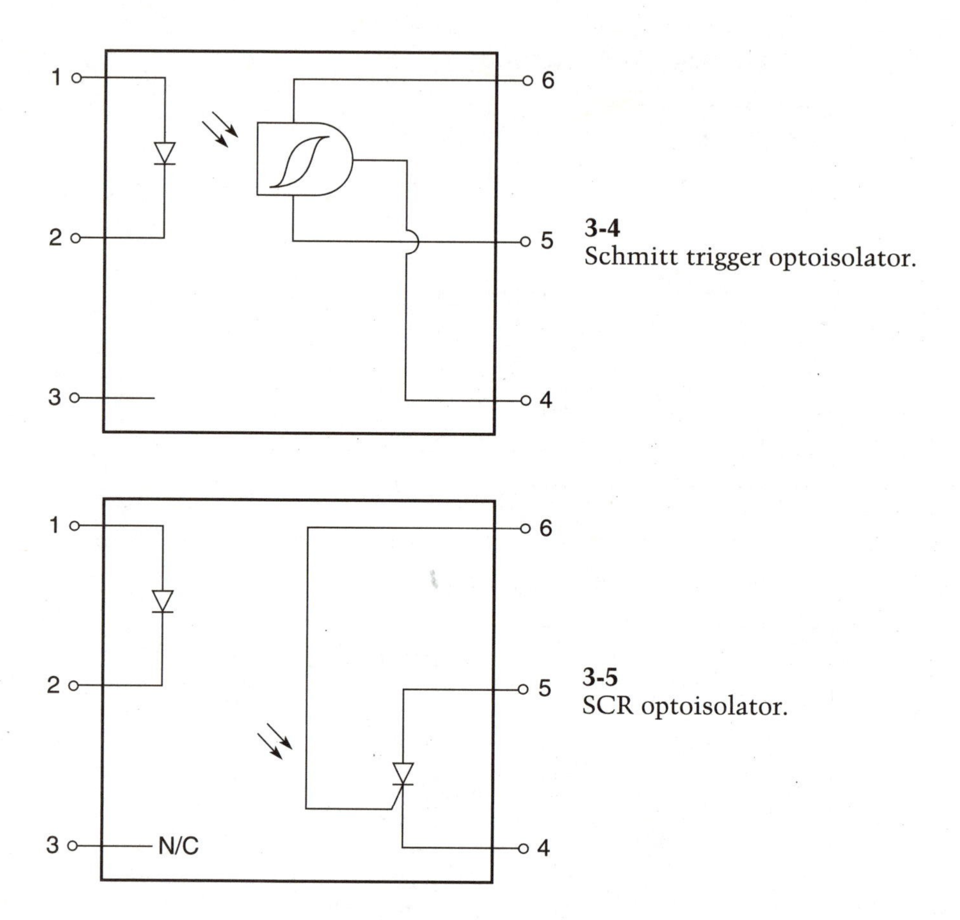

3-4
Schmitt trigger optoisolator.

3-5
SCR optoisolator.

The top diagram in Figure 3-2 illustrates an LED-transistor optoisolator in a six-pin IC package. The transistor optoisolator is available from a variety of manufacturers; you can compare their responses in Table 3-1. In the simple LED-transistor optoisolator, an LED and phototransistor are packaged together. The optical transistor placed in the package is generally an NPN type with only collector and emitter leads protruding. The base of the optotransistor is made quite large to accept the maximum amount of light from the LED.

The ac input transistor optoisolator, as the name implies, can be used with an ac input signal source. Typically both the conventional dc transistor and ac coupled transistor optoisolator require a series resistor between the input and the optoisolator's LED to limit the current through the LED. Without the series resistor, the LED would soon burn out. Table 3-2 lists five examples of ac input optoisolators.

The next optoisolator is the photo-Darlington type, shown in the diagram in Figure 3-3. The photo-Darlington isolator contains an LED and a two-stage transistor amplifier in a small package. The photo-Darlington produces much more light amplification from low-amplitude signals. Table 3-3 shows a number of photo-Darlington optoisolators. Note the range of rise/fall times of these devices. Most optocoupler devices have faster rise times and slower decay times. The H11B series of

Table 3-1 Transistor optoisolator.

Device	BV/CEO (volts)	Rise/fall (microsec.)
4N25	30	3/3
4N26	30	3/3
4N27	30	3/3
4N28	30	3/3
H11A1	30	2/2
H11A2	30	2/2
H11A3	30	2/2
H11A4	30	2/2
MCT2	30	10/10
MCT210	30	5/5
MCT26	30	5/5
MCT270	30	10/10
MCT4	30	2/2
MCT-5200	30	12/12
TIL111	30	10/10

Table 3-2 ac input transistor optoisolator.

Device	BV/CEO (volts)	Rise/fall (microsec.)
H11AA1	30	5/5
H11AA2	30	5/5
H11AA3	30	5/5
H11AA4	30	5/5

optoisolators (see the table) are measurably slower in response times. When selecting an optoisolator, take care to choose a specific application to ensure the proper response times.

The Schmitt trigger optoisolator, shown in Figure 3-4, combines an LED and a Schmitt trigger gate in one package. Schmitt trigger devices accept noisy or marginal signals and clean them up to provide a well-formed output pulse. Table 3-4 illustrates a number of Schmitt trigger optoisolators.

The diagram in Figure 3-5 depicts an SCR optoisolator, which is generally used to latch or "turn on" dc circuits. In normal operation, the LED is biased and the SCR is latched until the current path is broken open. SCR optoisolators are often used to control computer circuits or to drive either relays that can handle high-current loads or additional high-current SCRs. Examples of SCR isolators are shown in Table 3-5.

The triac optoisolator, shown in Figure 3-6, is used to switch ac or line current loads. It can provide switching and isolation with moderate current-handling capabilities. Triac optoisolators are often used to drive heavy-duty triacs or relays, which

Table 3-3 Darlington optoisolator.

Device	BV/CEO (volts)	Rise/fall (watts/sec.)
4N29	30	5/40
4N30	30	5/40
4N31	30	5/40
4N32	30	5/100
H11B1	25	5/100
H11B2	25	125/100
H11B3	55	125/100
H24B1	25	100/60
MCA230	30	10/100
MCA231	30	10/100
MCA255	30	10/100

Table 3-4 Schmitt trigger optoisolator.

Device	VCC (volts)	Rise/fall (microsec.)
H11L1	15	1/2
H11L2	15	1/2
H11L3	15	1/2
H11N1	15	.3/.3
H11N2	15	.3/.3
H11N3	15	.3/.3

Table 3-5 SCR optoisolator.

Device	Blocking voltage	Typical (T) (microsec.)
4N39	200	1
4N40	200	1
H11C1	200	50
H11C2	200	50
H11C3	200	50
H11C4	400	50
H11C5	400	50
H11C6	400	50

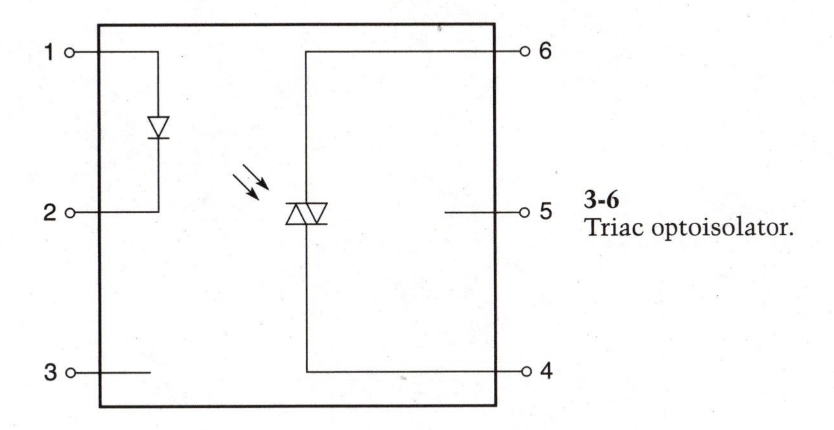

3-6 Triac optoisolator.

Table 3-6 Triac optoisolator.

Device	Blocking voltage	Holding current
MOC 3010	250	100
MOC 3011	250	100
MOC 3012	250	100
MOC 3020	400	100
MOC 3021	400	100
MOC 3022	400	100
MOC 3023	400	100

then control large current ac load devices such as motors, fans, or compressors. Table 3-6 illustrates a range of triac optoisolators.

New low-power, high-isolation optocoupler relays are currently available in single and dual configurations, as shown in Figure 3-7. The new AT&T Microelectronics optocoupler relays can switch up to 200 volts at up to 50 milliamperes (mA) of current. These relays provide high off resistance and low on resistance (100 ohms). The new solid-state optocoupler relays provide clean, bounce-free operation, and are available in both a single-pole version (LH1541AT) and a dual-relay package (LH1544AB). The new optocoupler relays are constructed from a GaAlAs LED coupled to MOSFET switches. Applications for the new optocoupler relays include data acquisition, telephony, test equipment, and sensing applications such as thermocouple switches.

Optoisolator interfaces

The basic optoisolator in Figure 3-8 applies a high input to the LED and provides a low signal at the optoisolator's transistor output. Power applied to VCC1 is a 5-volt signal through a series resistor to the LED, while the secondary stage can be powered from a 5- to 12-volt power supply. An R2 resistor is used to bias the optoisolator's transistor.

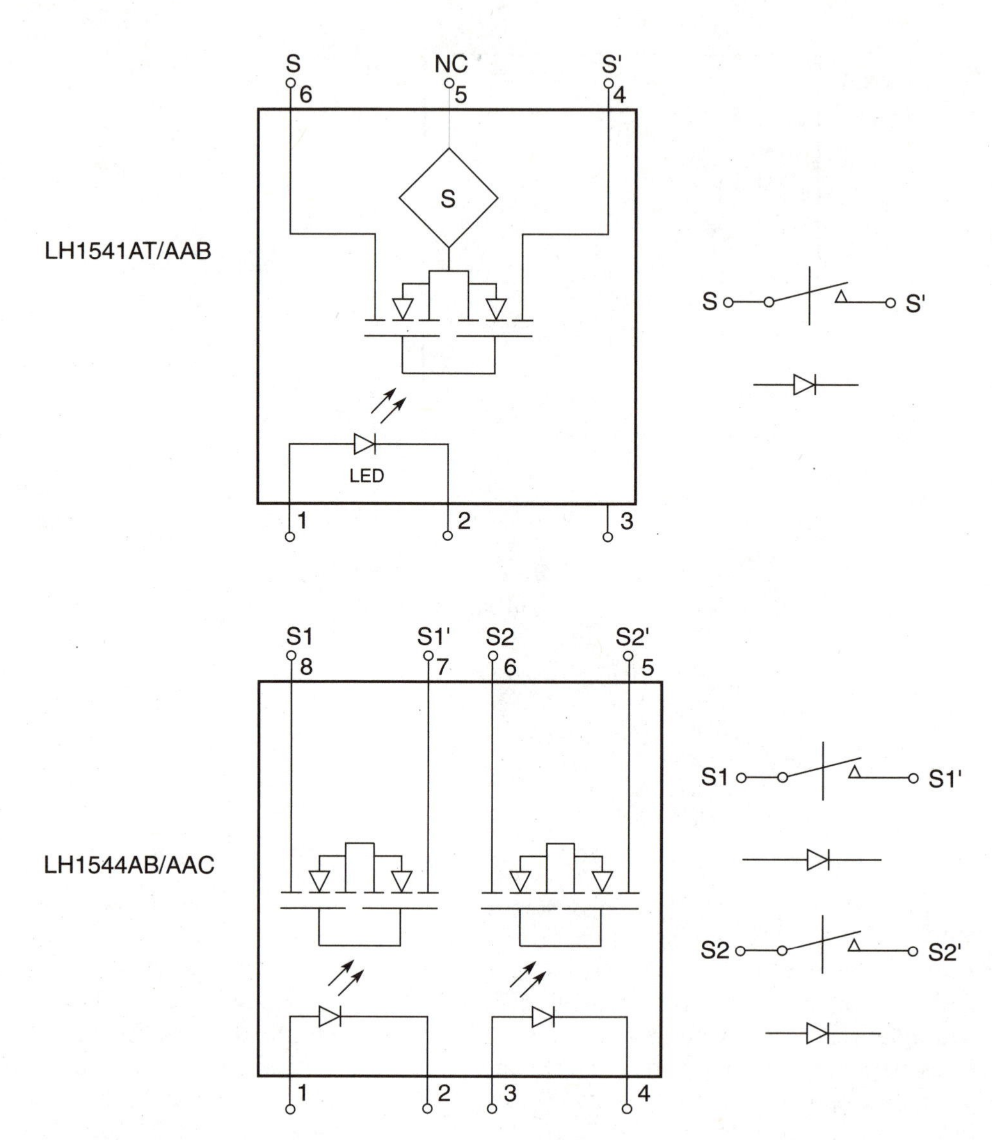

3-7 Solid-state relay isolators.

Inverted output optointerface

The optoisolator depicted in Figure 3-9 is a transistor-boosted inverted output coupler. A series resistor at R1 is used to limit the current through the LED. The optoisolator's transistor is used to drive a second transistor at Q1. A 5-volt signal at the LED produces a 5-volt output at the collector of Q1. Both transistors are biased via two 4.7-kilohm resistors. Power supplied to the LED at V_{CC1} is 5 volts, and power to V_{CC2} is supplied to R2 and R3 and can be from 5 to 12 volts dc. A load or low-current relay can be substituted for the resistor at R3.

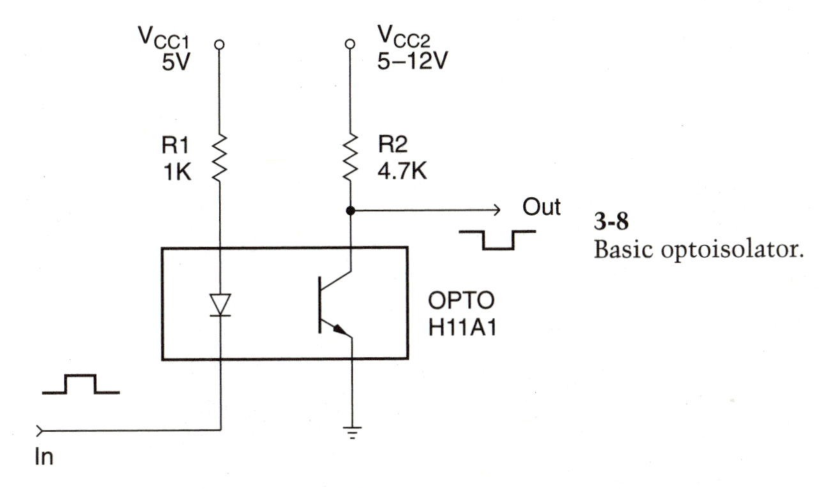

3-8
Basic optoisolator.

Noninverted output optoisolator

A noninverted transistor-boosted output optointerface circuit is shown in Figure 3-10. A high at the input to the LED activates the 2N2222 transistor at Q1. A low output is then produced at the emitter of Q1 at R3. Power to the LED is 5 volts, but you can increase it by changing the value of the series resistor from 220 ohms to 1 kilohm. Power from VCC2 can be supplied from any 5- to 12-volt dc power source. A load or relay can be used in place of resistor R3.

TTL-to-TTL optoisolator

The optoisolator in Figure 3-11 shows a TTL-to-TTL optocoupler interface. Shown is an inverter driving the LED with a high-input signal. The inverter could be

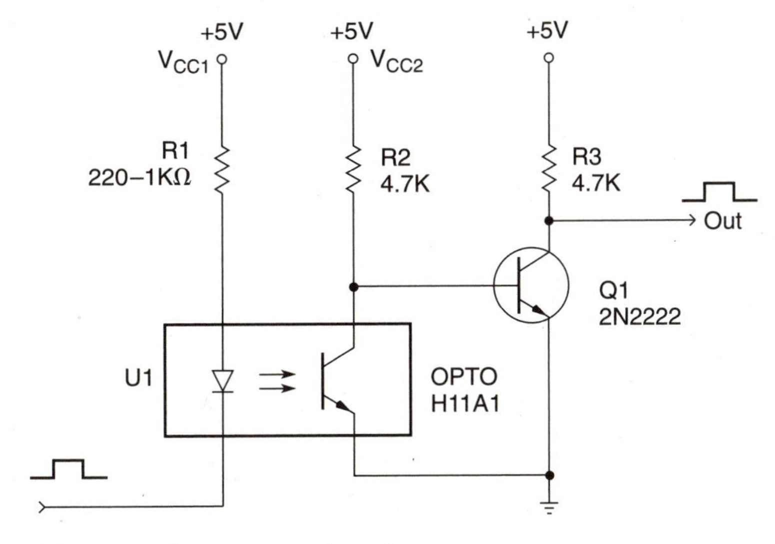

3-9 Inverted output optointerface.

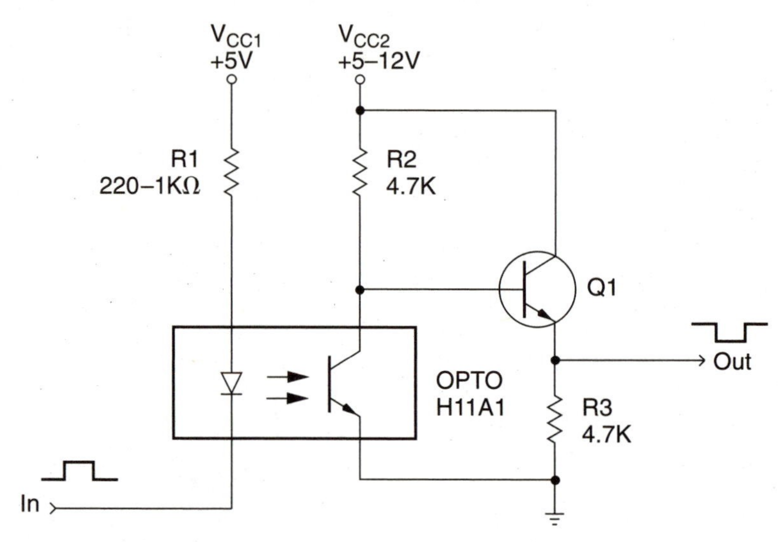

3-10 Noninverted output optoisolator.

any TTL device such as a 7400 Nand gate or 7404 Hex inverter. The LED activates the optocoupler's transistor, and a low-output signal at the collector of the H11A1 is used to drive a second TTL gate at the secondary or isolated output circuit stage. Power to the LED can be from any 5-volt input source. Power from VCC2 is also 5 volts since the secondary circuit is used to drive TTL circuitry.

TTL-to-CMOS optoisolator

Figure 3-12 shows a TTL-to-CMOS optoisolator interface in which a TTL device drives the LED in the first stage of the interface. The LED "turns on" the optocou-

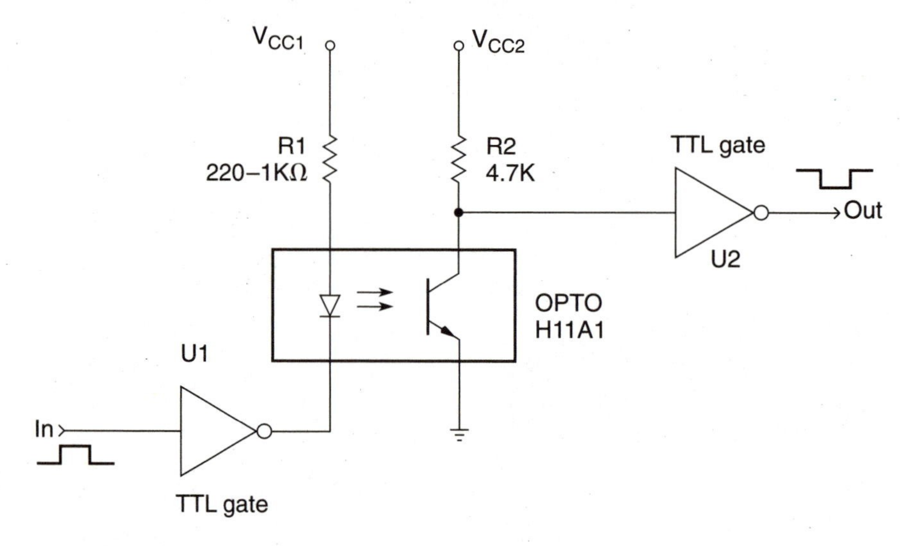

3-11 TTL-to-TTL optoisolator.

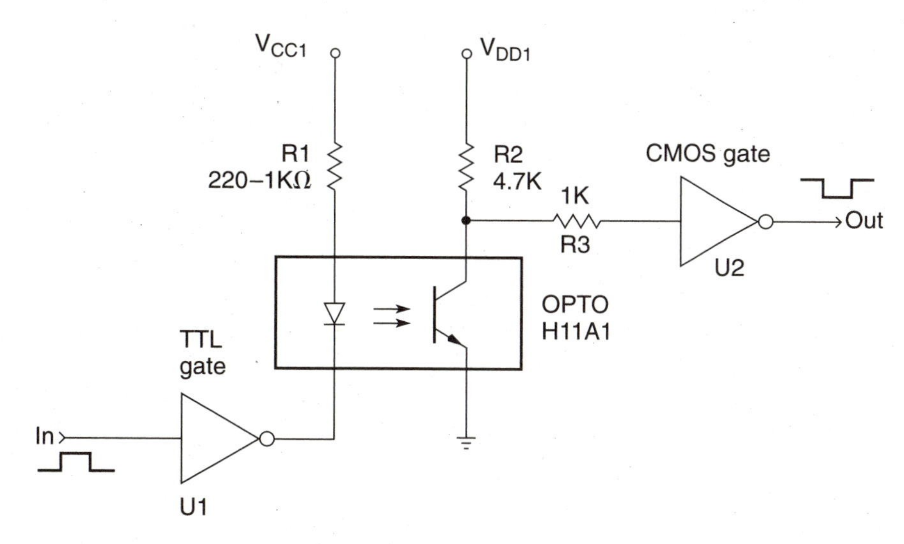

3-12 TTL-to-CMOS optoisolator.

pler's internal transistor. A low output at the transistor's collector is coupled through R3, typically a 1-kilohm resistor, to a CMOS device at U2. A high input signal to the LED through R1 activates the optocoupler, and in turn a high or 5-volt signal is produced at the output of U2. Power from VCC1 is 5 volts, while power to the isolated secondary stage at V_{CC2} can be from 5 to 12 volts since the secondary stage drives CMOS circuitry.

LED drive circuits

The three circuits shown in Figures 3-13, 3-14, and 3-15 illustrate NPN, PNP, and CMOS LED drive circuitry, which can be used to drive most optocouplers. The NPN transistor accepts a "high" input of 5 volts at Q1. When Q1 is turned on, it activates the LED in the optocoupler. The PNP drive circuit takes a high input signal at R1 and turns on transistor Q1, which in turn activates the LED in the optocoupler. In the CMOS drive circuit, a CMOS gate turns on Q1, which in turn activates the LED in the optocoupler. Power to all three circuits at VCC is 5 volts.

Often you will need to drive an LED or optocoupler via a TTL gate, as shown in Figure 3-16. When the enable input is low, as shown, the LED is turned off irrespective of the logic level at the TTL input. When the enable input is high, the LED is forward biased and the logic level at the gate input to pin 1 of the 7400 is high. Conversely, when the logic level input is low, the LED is turned off. The series resistor at R1 is a 330 ohms, and power to the 7400 driver is 5 volts dc.

Optocoupler sensors

So far, this discussion has focused on the drive circuits at the source side of the optocoupler interface. Now let's take a look at the sensor side. The Schmitt trigger is an effective method of coupling an optocoupler to TTL circuitry. The isolated signal from a TIL-102 optocoupler is fed directly to the inputs of a 7413 dual Schmitt

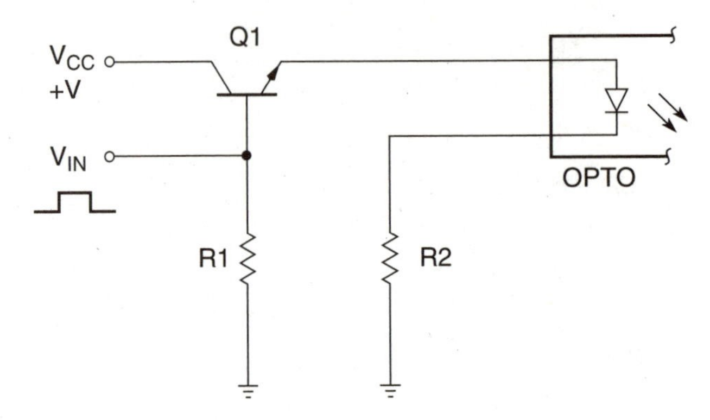

3-13 NPN transistor LED driver.

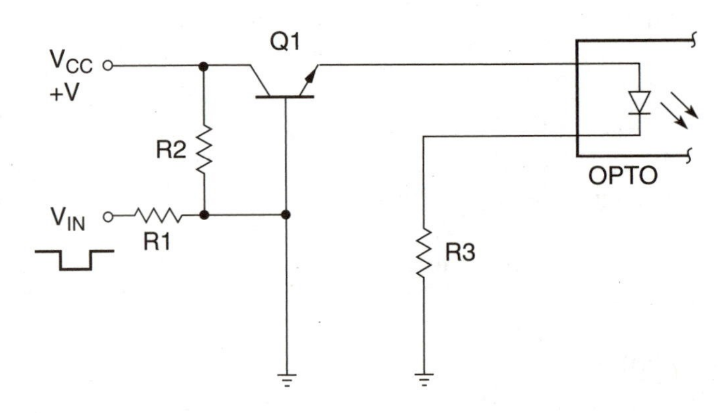

3-14 PNP transistor LED driver.

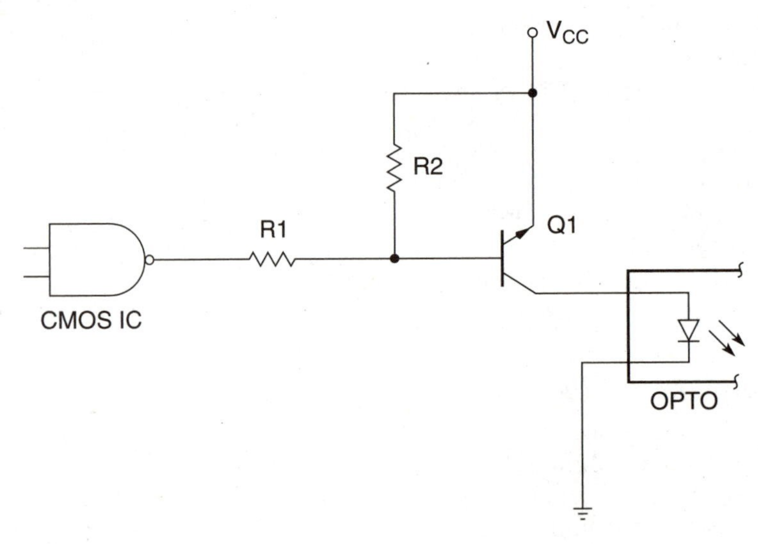

3-15 CMOS IC optocoupler driver.

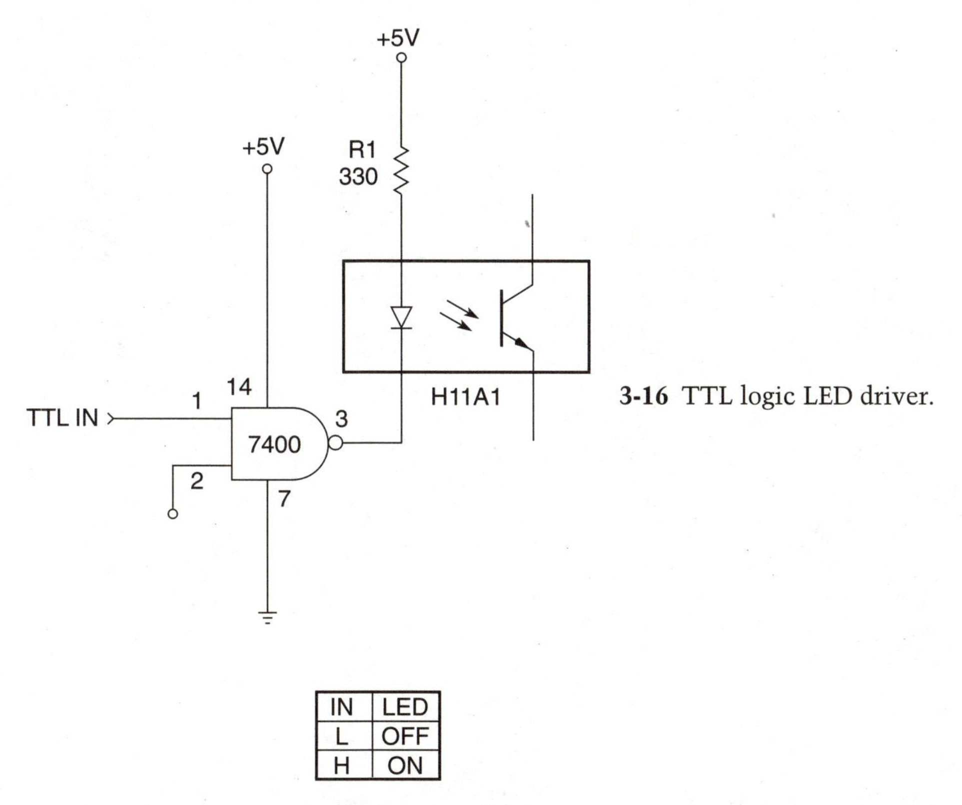

IN	LED
L	OFF
H	ON

3-16 TTL logic LED driver.

trigger IC. The 7413 IC provides excellent noise immunity due to the threshold action of the Schmitt trigger circuitry. Both the inverting and noninverting Schmitt trigger interfaces shown in Figures 3-17 and 3-18 are powered from a 5-volt dc source to ensure compatibility with TTL circuits. Note that the difference between these two Schmitt trigger interface circuits lies in the configuration of the 1-kilohm resistors on the input to the 7413 IC.

Optoisolator relay driver

Often optocouplers are used to drive relays, as shown in Figure 3-19. The optocoupler provides isolation and level shifting. The input resistor at R1 can be tailored to specific input voltages as required. Values between 270 ohms and 1 kilohm are suitable. The output from the optocoupler's internal transistor is coupled to a 2N2222 transistor to provide additional drive current for the 500-ohm, 6-volt relay. The secondary or relay drive stage can be powered from a 9-volt source.

Optoisolator relay driver parts list

R1	270-ohm to 1-kilohm ¼-watt resistor
R2	4.7 kilohm ¼-watt resistor
Q1	2N2222 transistor
D1	1N914 silicon diode
OPTO	TIL-102/103 optoisolator
RLY	500-ohm, 6-volt relay (RS 275-005)

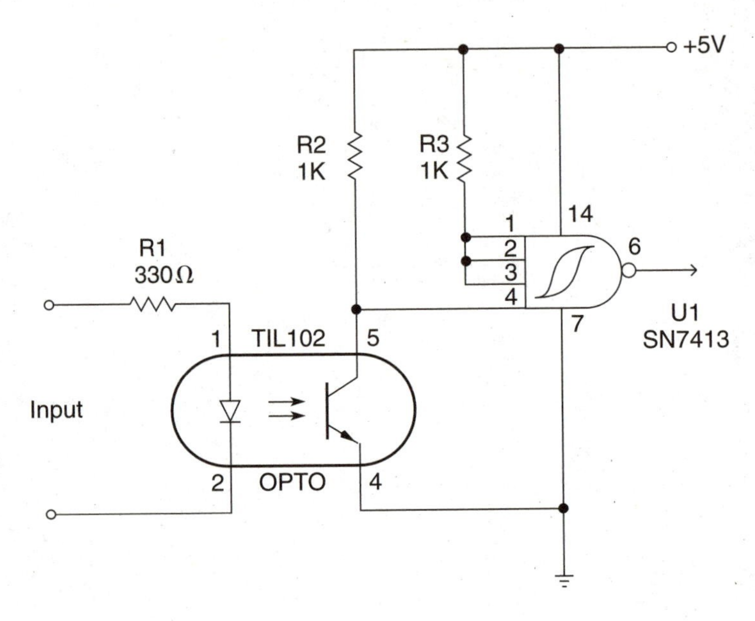

3-17 Noninverting Schmitt trigger output circuit.

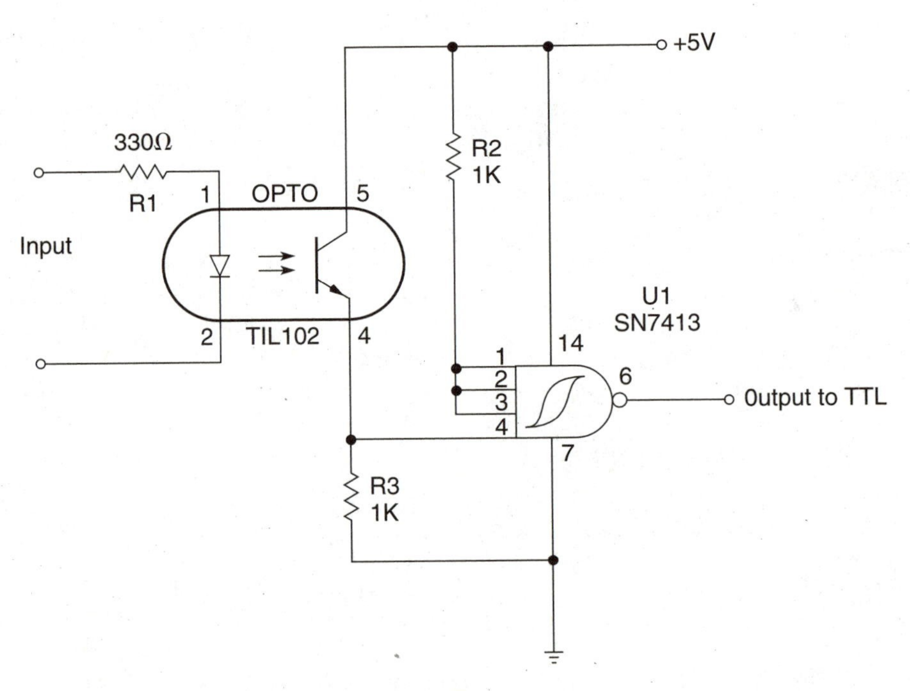

3-18 Inverting Schmitt trigger output circuit.

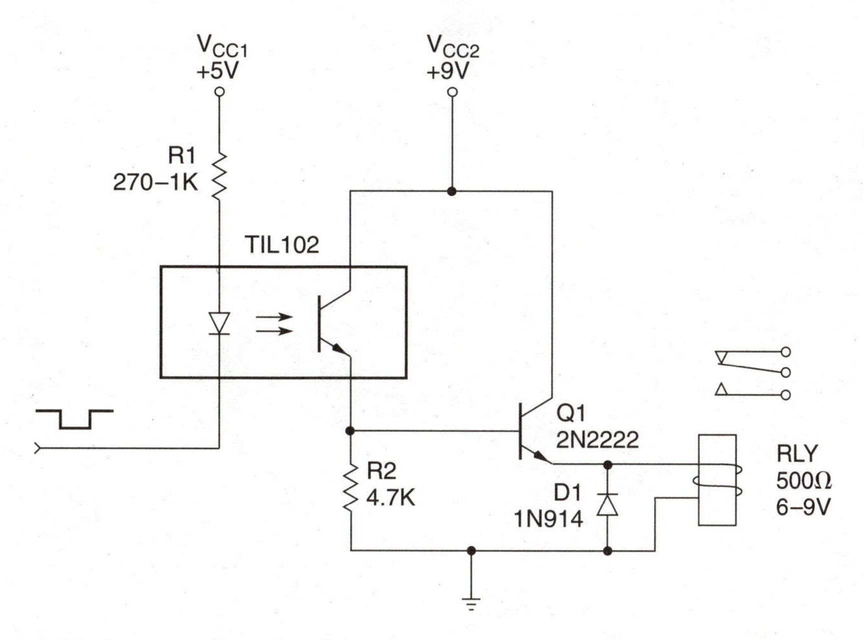

3-19 Optocoupler relay driver.

Optocoupler SCR latch

A simple isolated SCR latch circuit is illustrated in Figure 3-20. The output of the Texas Instruments TIL-120 optoisolator is used to fire the SCR in order to control an external load device. The load device should be resistive. You could also use the SCR latch to drive a relay to allow high current load control, if desired. In order to turn off the load device, remove power from the V_{CC2} by using a normally closed push-button switch. Drive to the optocoupler's LED can be provided from a 5- to 12-volt dc source via a suitable current-limiting resistor at R1.

Optocoupler SCR latch parts list

R1	200-ohm to 1-kilohm ¼-watt series resistor
R2	1-kilohm ¼-watt resistor
OPTO	TIL-120 optoisolator, Texas Instruments
D1	C44 SCR, Texas Instruments Load Resistive device within the current specs

Photodiode amplifier light probe

The next optocoupler IC is the integrated photodiode and transimpedance amplifier light probe, shown in Figure 3-21. This chip eliminates commonly encountered problems found when using discrete detectors and amplifiers, such as leakage current errors, noise pickup, and gain peaking due to stray capacitance.

The Burr-Brown OPT-101 chip is ideal for applications in medical and laboratory instrumentation, smoke detectors, position and proximity detectors, background light controllers, currency changers, and photographic analyzers. The single supply

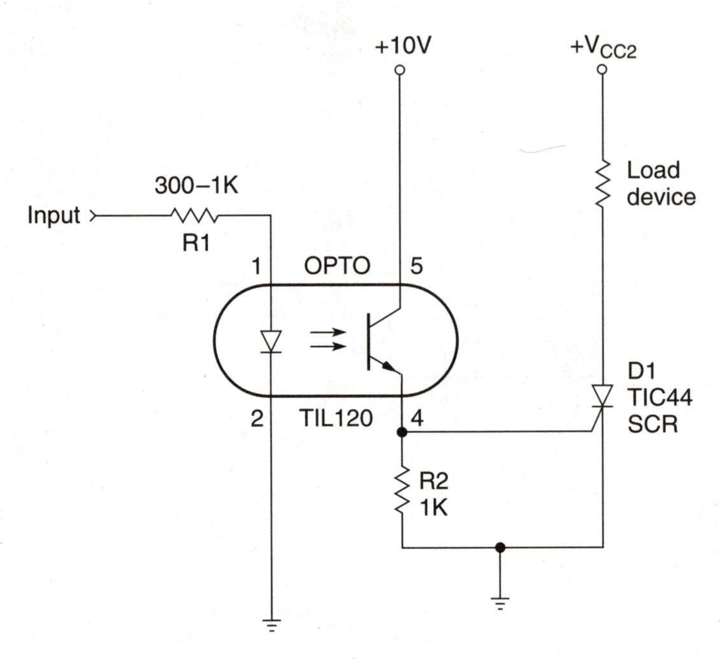

3-20 Optocoupler SCR latch.

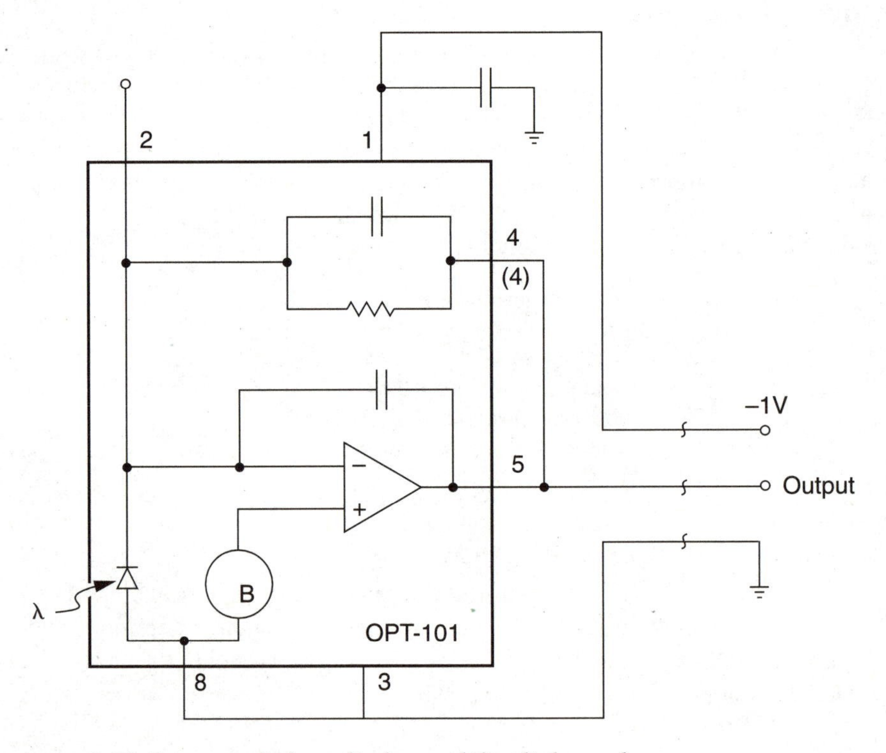

3-21 Integrated photodiode amplifier light probe. Copyright 1989-1994 Burr-Brown Corporation. Reprinted, in whole or in part, with the permission of Burr-Brown Corporation.

photodiode amplifier chip can be operated from 2.7 to 36 volts and consumes only 120 µA of current while offering high responsiveness and a 14-kHz bandwidth. The spectral sensitivity of the linear photodiode amplifier peaks at 750 nm on the low end of the infrared spectrum. Light falling on the internal photodiode is directed to the transimpedance amplifier, which produces an output voltage proportional to the incoming light amplitude. The internal 1-megohm resistor forms the overall gain path for the transimpedance amplifier. To increase the responsiveness of the chip, an external 1-megohm resistor can be added between pins 4 and 5. An external 50-pF capacitor is then placed across the external resistor. The OPT-101 is used in a three-wire remote light sensor probe, shown in Figure 3-21. A number of these probes could be used to remotely monitor room lighting or process control circuitry.

The OPT-101 chip provides a very linear response when uniformly illuminated. A narrowly focused light beam falling on the photodiode element alone provides improved settling time, compared to light falling on the entire chip. You can also power the OPT-101 chip from a bipolar supply, if desired, by supplying a minus voltage to pin 3 of the chip.

A wideband sister chip to the OPT-101 is the new OPT-211, which provides response to 50 kHz, and up to 150 kHz with an external bootstrap buffer. The OPT-211 is a bipolar device that requires two power supplies. If you are interested in these two devices, contact Burr-Brown for their free data book.

Practical optoisolator interface circuits

The remaining portion of this chapter is devoted to practical optoisolator interface circuits that can transfer digital data. First I will describe a two-way optocoupler interface, then move to a current loop interface, and finish with a TTL-to-RS-232 interface and two wireless interfaces.

Two-way optoisolator interface

The diagram in Figure 3-22 presents a simple yet useful two-way optoisolator interface. It is not generally known, but GaAs, GaAs:Si, and AlGaAs:Si near-infrared emitters work better as detectors than do most visible light LEDs. The heart of this novel two-way optoisolator interface is composed of two LEDs and two conventional optocouplers, as shown. The twist here is that the two LEDs at the bottom of the diagram act as both emitters and detectors. They are shown face to face separated by a short piece of heat-shrink tubing. For isolation or long-range applications, the two LEDs can be interfaced via optical fibers.

In operation, a low or high bit at the control-line input forward-biases the LED in one of the two conventional optoisolators. First, let's assume that the control bit is high. This condition causes the LED in optoisolator 1 to be forward- biased, which in turn activates its phototransistor. Any digital signal present at input 1 and at the phototransistor's collector will now forward-bias LED 1 in the two-way LED-LED optoisolator. Now LED 2 functions as a detector and data is presented at output 2 from input 1. Note that the circuit cannot now receive any signal present at the collector of the phototransistor in optoisolator 2 since that transistor is now turned off. When

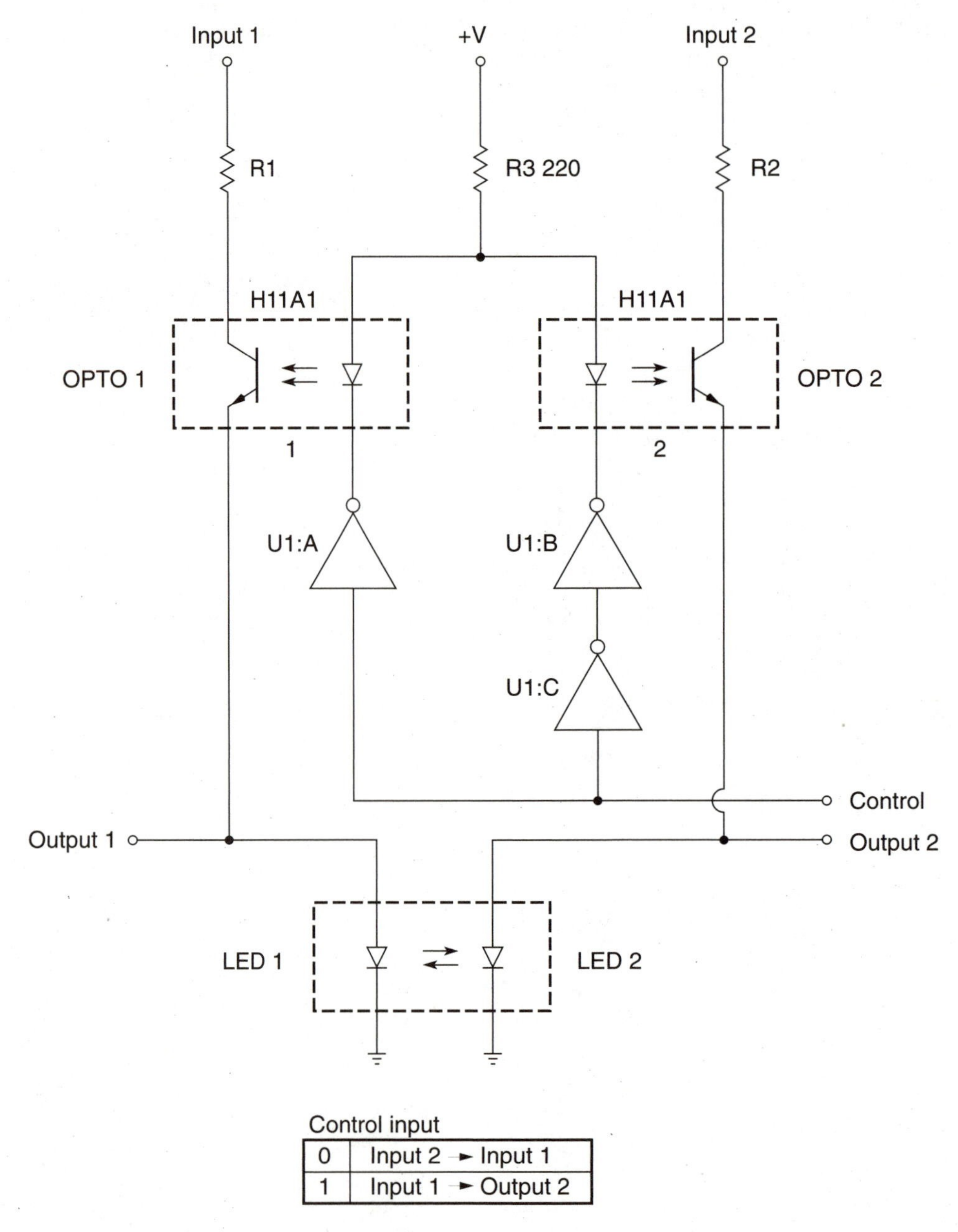

Control input

0	Input 2 ► Input 1
1	Input 1 ► Output 2

3-22 Two-way optoisolator system.

the control bit is changed from high to low, the operating mode is reversed and the LED-to-LED optoisolator transmits in the opposite direction. Now an input signal on input 2 is used to forward-bias LED 2. LED 1 now acts as a detector and the output is present at output 1. This simple yet novel method of transferring digital information can be constructed with a low parts count.

Two-way optoisolator interface parts list

R1,R2	4.7-kilohm ¼-watt resistor
R3	200-ohm ¼-watt resistor
U1	7404 hex inverter
Opto-1,Opto-2	H11A1 optocoupler
LED1, LED2	TRW OP-195 GaAs:Si IR LED

Current loop interface system

Current loop interfacing has become a standard for signal transmission in the process and control industry. Current loops are insensitive to noise and are immune to errors from line impedance. Adding isolation to the 4- to 20-mA current loop protects electrical systems from electrical noise and transients; it also allows transducers to be electrically separated by hundreds of volts in some applications. The current loop system shown in Figures 3-23 and 3-24 illustrates how a CdS photoresistive cell could be used to detect varying light levels and then send the information over twisted wires to a current loop receiver. The voltage output at the current loop receiver is then available to reconstruct the input signal from the CdS cell. You could also use the current loop system to send temperature, pressure, liquid level, gas flow, or position sensor data from one location to another.

The current loop interface consists of transmitter and receiver units separated by a length of twisted-pair wires. The carrier loop transmitter section is based on the

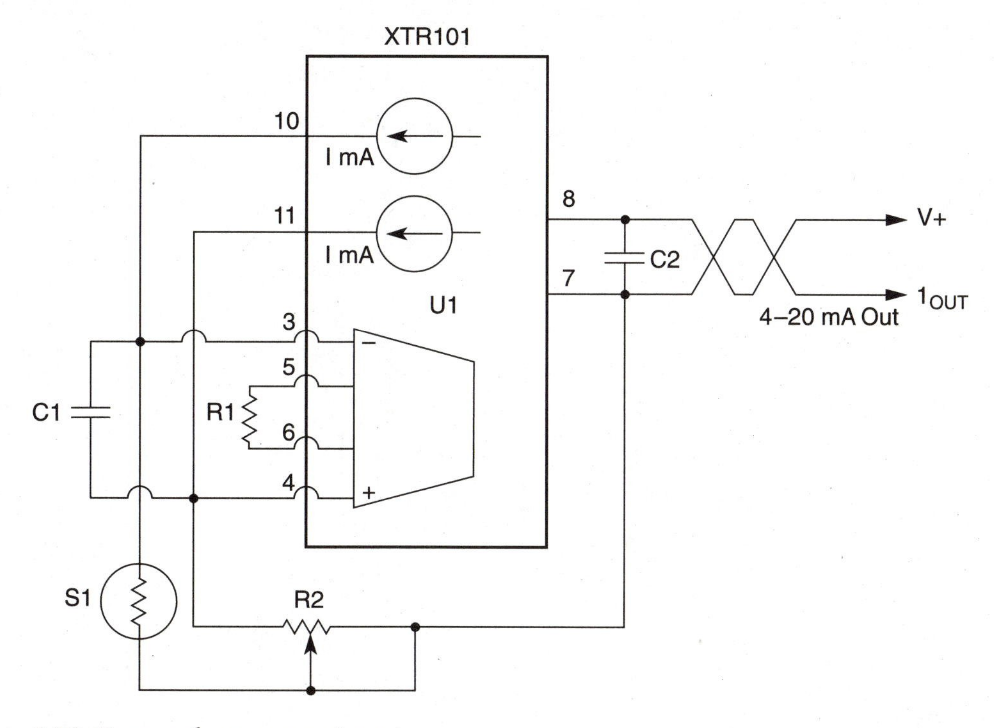

3-23 Current loop transmitter. Copyright 1989-1994 Burr-Brown Corporation. Reprinted, in whole or in part, with the permission of Burr-Brown Corporation.

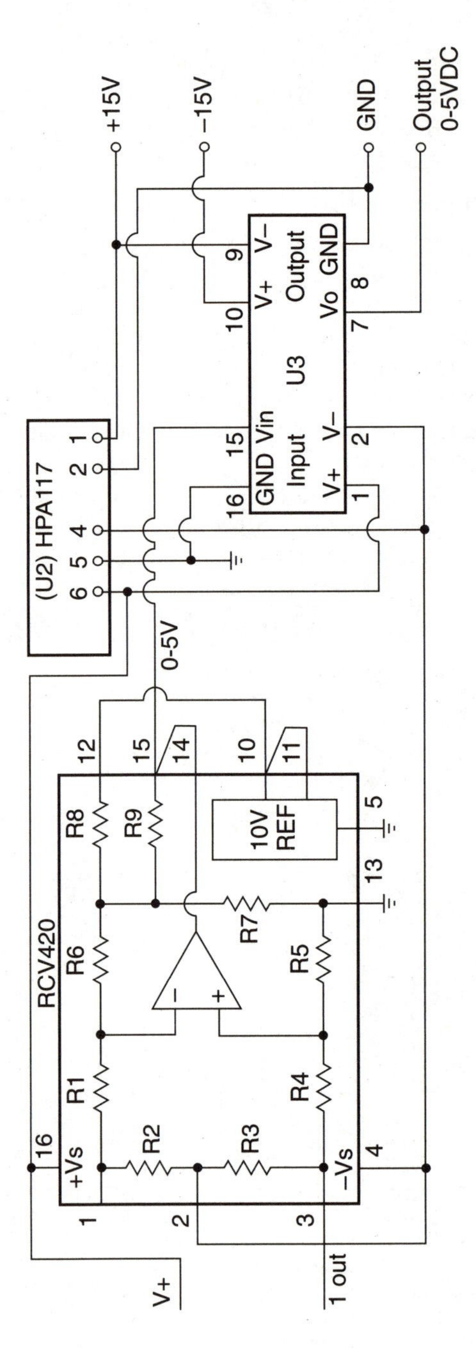

3-24 Current loop receiver. Copyright 1989-1994 Burr-Brown Corporation. Reprinted, in whole or in part, with the permission of Burr-Brown Corporation.

Burr-Brown XTR-101, an eight-pin miniature transmitter chip. The current loop transmitter chip consists of an instrumentation amplifier and two 1-mA current sources for transducer excitation and offset control. Figure 3-23 illustrates the current loop transmitter configured with a CdS photoresistive light sensor as the input device, as shown. Potentiometer R2 is used to zero the transmitter input. Resistor R1 on pins 5 and 6 is used as a "span" setting resistor to get the 20-mA current loop output to full scale based on the sensor used. Try a 50- to 100-kilohm potentiometer and adjust with your particular sensor; then replace it with a fixed resistor.

Only two other components are used in this very straightforward transmitter, two .01-µF capacitors. The output of the current loop transmitter at pins 7 and 8 provides 4 to 20 mA of output. Note that pin 7 is the plus output. The transmitter section is powered by the receiver section. The current loop transmitter is also available in the XTR-103 and XTR-104 models. The XTR-103 contains two instrumentation amplifiers, one for signal and the other for linearization. The model XTR-104 is designed specifically for use with bridge-type sensors. See the Burr-Brown applications handbook for more information on these devices.

The receiver section of the current loop interface is shown in Figure 3-24. The heart of the receiver is the RCV-420 self-contained 4- to 20-mA current loop receiver chip. This chip contains a precision voltage reference, a 75-ohm precision sense resistor, and a ± 40-volt common mode input range difference amplifier. The RCV-420 conditions and offsets the 4- to 20-mA input signals to give a precision 0- to 5-volt output, which can be coupled to an analog meter or to an A/D computer card for data logging. The current loop receiver section consists of three low-cost building blocks. The output of the RCV-420 is coupled to an ISO-122, a low-cost precision isolation amplifier at U3. The input is at pin 15 while the output is available at pin 7. The HPA-117 chip at U2 is a low-cost isolated dc-to-dc converter, which provides a precise ± 15-volt, 30-mA output to the power U1 and U3. The HPA-117 chip receives its power from a ± 15-volt power supply. Both the current loop transmitter and receiver should be mounted in small metal chassis boxes in order to shield them from outside interference.

The current loop interface can be configured with many different types of sensors for detecting temperature, pressure, strain, position, etc. For more information on other sensing configurations, contact Burr-Brown for a copy of their new applications handbook.

Current loop transmitter parts list

R1	See text
R2	5-kilohm potentiometer
C1,C2	.01-µF 25-volt disc capacitor
S1	CdS photoresistive cell sensor
U1	XTR-101 current loop transmitter

Current loop receiver parts list

U1	RCV-420 current loop receiver
U2	HPA-117 isolated dc-to-dc converter
U3	ISO-122P precision op-amp isolation amplifier

TTL-to-RS-232 interface system

The unique optocoupler interface shown in Figure 3-25 provides a TTL-to-RS-232 translation, which is completely isolated and uses the computer's RTS/DTR lines to power the interface. This novel interface can be used to couple a microcomputer to a desktop computer or to interface a TTL sensor to a data logger or laptop.

The optocoupler circuit at the top of the diagram illustrates the RS-232-to-TTL portion of the interface. The overall system supports a 9600-bps transfer rate. The RS-232 TX line is first sent to D1 and D2, which biases the LED in the MET-2 optocoupler. The phototransistor in the optocoupler is fed via the collector at pin 5 to the input of U2, a 74HCTO5 gate that translates the output at R5 to a TTL signal output.

The TTL-to-RS-232 or RX portion of the interface is shown at the bottom diagram in Figure 3-25. In this circuit, a charge pump configuration converts TTL to RS-232. The TTL input signal is first sent to a zener diode at D4 before going to the second 74HCTOS gate at U4. The output of U4 is then sent to the LED in optocoupler U3. The output from the MET-2 at U3 is now sent to Q4 via R14. Transistor Q4 now charges capacitor C2 to about 1 volt less than the RTS voltage, while the TTL line asserts a marking state. As the capacitor is charging, Q1 is biased into saturation and thus provides a negative voltage, with respect to the RS-232 ground, at the RS-232 RXD line output. When a "spacing bit" is driven from the TTL line, Q1 switches off and Q4 switches on. This now biases Q4's emitter up to the RS-232 ground, and the voltage is summed with C2's charge to create an RS-232-compatible spacing signal (approximately 1 volt less than -VRTS). The discharge rate on C2 is limited by R9 to prevent the signal S9 from becoming a problem, at 110 bps. The C3/R10 time constant must be somewhat close or within four times the C2/R9 time constant to ensure that Q1 turns "off" correctly.

Now take a look at the circuit at the extreme right of Figure 3-25. It illustrates the simple scheme that derives the 5-volt potential from the RTS/DTR signal pins. Resistors R19 and R20 and diodes D5 and D6 mix the return current to the RS-232 port so the RTS and DTR drivers split the current drawn by the interface.

This power derivation method can supply 12 mA of current quite easily, even from a laptop. The only drawback to this power scheme is that the TTL device must be isolated from the computer's ground (Earth ground), since the interface treats the RS-232 ground as a positive voltage. If you cannot guarantee this isolation, then you should provide another means to power the interface.

This unique, low-cost interface can provide good isolation and conversion from TTL to RS-232 signals. Note that you could easily adapt this circuit to long, noise-free, optical transmission by using fiber-optic cables and replacing the MET-2 with discrete components.

TTL-to-RS-232 interface parts list

R1	300-ohm, ¼-watt resistor
R2,R3,R16,R18	100-kilohm, ¼-watt resistor
R4	20-kilohm, ¼-watt resistor
R5	1-kilohm, ¼-watt resistor
R7,R11	10-kilohm, ¼-watt resistor

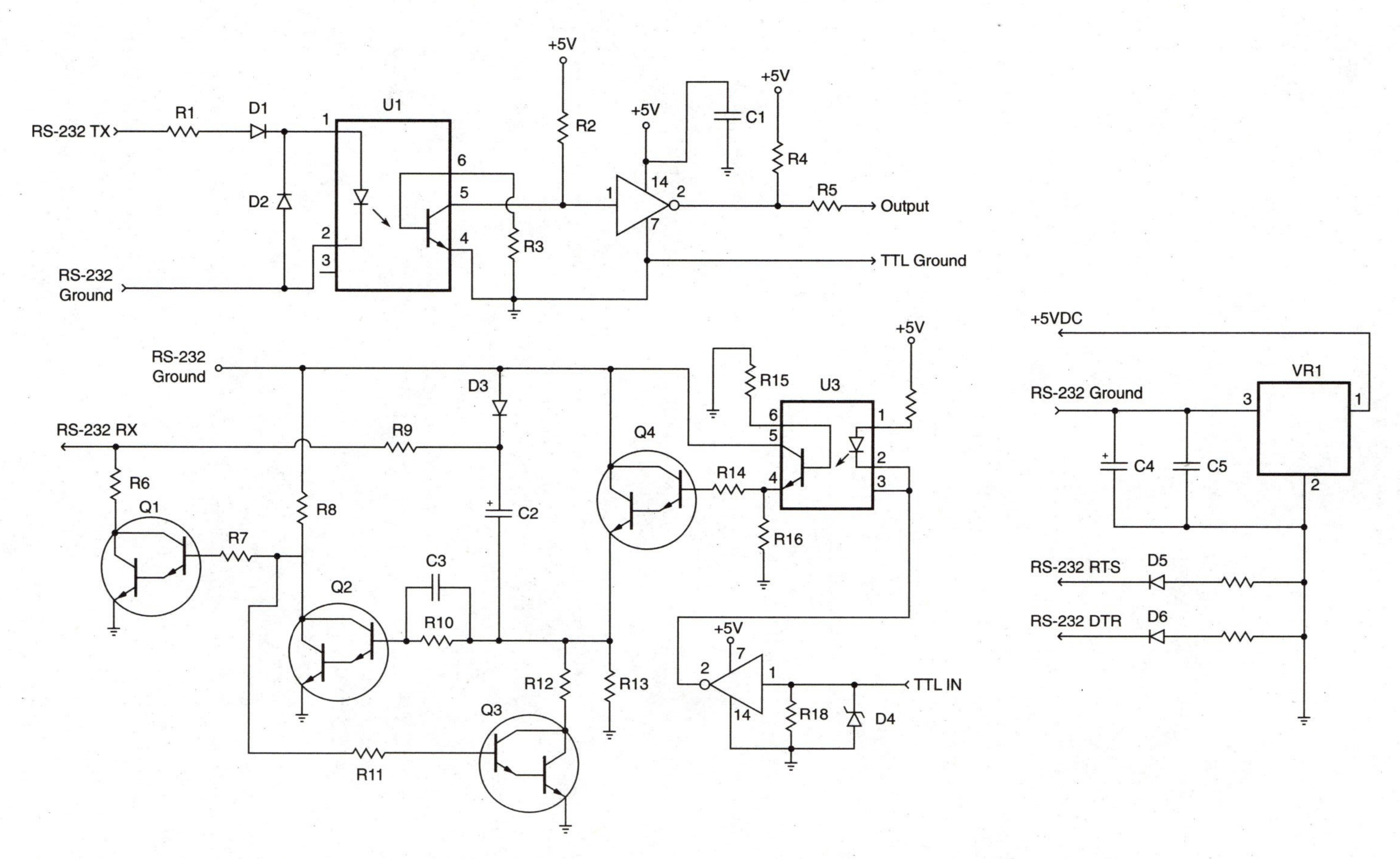

3-25 Optically isolated TTL-to-RS-232 interface.

TTL-to-RS-232 interface parts list continued

R8	24-kilohm, ¼-watt resistor
R10	8.2-kilohm, ¼-watt resistor
R12	100-ohm, ¼-watt resistor
R13	3.9-kilohm, ¼-watt resistor
R15	1-megohm, ¼-watt resistor
R19,R20	43-ohm, ¼-watt resistor
C1,C5	.001-µF, 25-volt ceramic capacitor
C2	3.2-µF, 25-volt electrolytic capacitor
C3	.01-µF, 25-volt ceramic disc capacitor
C4	22-µF, 25-volt electrolytic capacitor
D1,D2,D3,D5,D6	1N914 silicon diode
D4	1N4736 zener diode
U1,U3	MET-2 optocoupler
U2,U4	74HCTO5 IC
VR1	LM7805 5-volt regulator
Q1,Q3,Q4	MPSA13 transistor
Q2	2N6426 transistor

RS-232 infrared data transmission system

Wireless data transmission offers great freedom and mobility by eliminating interconnecting wires between computers and peripherals and between laptops and network connections. Infrared (IR) data communication can be used in many applications where digital data must be transferred from palmtop to desktop computer, from desktop to desktop, or from data logger to laptop.

The Sharp RY5 series IR data communication ICs, shown in Figure 3-26, provide low-power serial data transmission at a 38,400-bps data rate and a distance up to three meters (see Photo 3-1). The RY5 IR data devices are TTL/CMOS-compatible devices that operate from standard 5-volt power supplies.

The RY5AT01 IR data transmitter is a compact three-lead device designed for amplitude shift keying (ASK) modulation. The IR data transmitter is composed of an input buffer, a 500-kHz modulator, and an output IR LED. A positive logic input to the inverting buffer enables the 500-kHz modulator and the IR LED, whereas a logical zero (0) at the input disables the output at the modulator. The RY5AT01 chip can be used in a reliable half-duplex data interface. Full-duplex operation is much more difficult due to local echo at each transceiver end, from reflective objects in short-range line-of-sight applications.

Half-duplex operation can achieve high data transfer rates and low error rates when error checking is used. ASK modulation is referred to as the presence or absence of a modulated carrier. In an actual transmission, a "mark" signal corresponds to a logical (1) while a "space" is denoted by a logical zero (0). The time period of a logical (1) data bit is filled with the 500-kHz modulation frequency. The rising and falling edges of the envelope drive the hysteresis comparator to generate the recovered output signal. Error rates of 10^{-7} can be easily achieved with the RYS IR data modules.

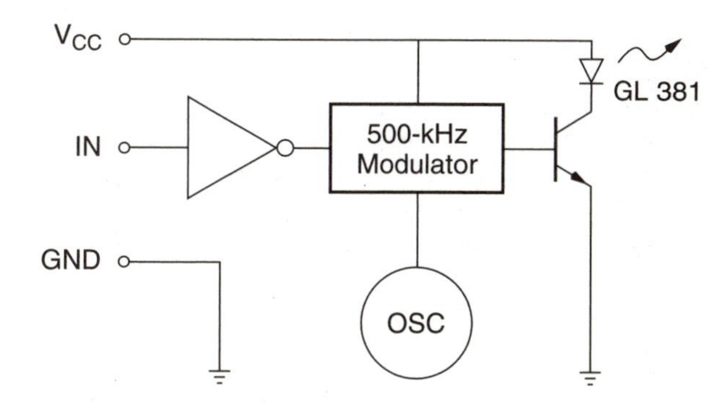

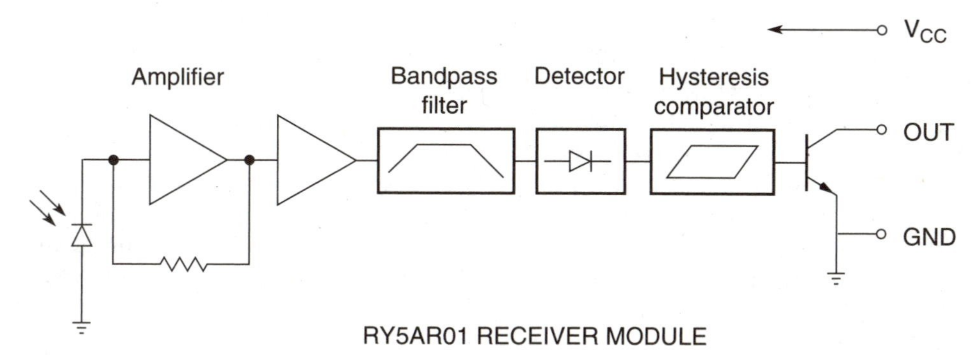

3-26 Infrared transmitter and receiver modules.

Photo 3-1 RS-232 IR data-link transceiver.

The RYSAR01 IR data receiver is a three-lead low-power chip designed for ASK-formatted serial data from the compatible IR data transmitter. The IR data receiver is comprised of a photodiode, amplifier chain, bandpass filter, detector, and hysteresis comparator, as shown in Figure 3-26. The photodiode detector is sensitive to IR energy between 900 and 1050 nm. The recovered data signal at the output of the photodiode is fed to a chain of five internal amplifiers in the RY5AR01 module. An internal AGC and limiter stage is used to prevent saturation of the amplifier chain if a strong signal is present. Conversely, a weak signal input will be amplified to a usable signal.

A bandpass filter is used to select the desired signal source as the primary input recovered by the receiver. The bandpass filter suppresses signals in the 38-kHz modulation band used by many home remote controls and fluorescent lamps. The receiver chip responds only to the 500-kHz transmitter signal modulation and rejects all other input signals. The detector stage is an envelope detector that responds to the presence of a correctly modulated signal. The detector output feeds the hysteresis comparator, which provides a TTL/CMOS-compatible output. A positive signal that exceeds the upper comparator threshold will "set" the internal comparator, and a negative signal that exceeds the lower threshold will reset the internal comparator. Hysteresis minimizes the probability of receiving incorrect information. The comparator output is the logical complement of the received signal. A logical (1) at the transmitter produces a (0) at the output of the receiver module. The IR data receiver has an open collector output that can be used with signals up to 10 volts. A 33-kilohm pull-up resistor from VCC is recommended. The nominal current should not exceed 200 mA through the output circuit.

A practical two-way serial IR data transmission system is shown in Figure 3-27. The IR interface uses two RY5AT01 IR transmitters, two RY5AR01 receivers, and two Maxim MAX233 RS-232 driver/receiver chips. The IR data transfer interface is a bidirectional half-duplex system. This minimum-component interface can be readily used between desktop computers or between a desktop and a laptop. Each half of the IR data interface can be built on a tiny circuit board housed in a small plastic box at each end of the transmission link. A 9-volt battery can be used to power the 5-volt, low-current regulators, as shown. To maximize the transmission distance between the transmitter and receiver modules, you will need to use IR filters at both ends of the IR link. Another alternative to an IR filter is to use two pieces of 100 ASA exposed color print film to reduce broadband noise from stray light sources. A piece of exposed film is placed at both the transmitter and receiver. Once completed, you can easily test the IR interface with conventional communications software, which might appear to demonstrate full duplex when echo canceling is chosen in the setup parameters.

Data transfer protocols using a "packet" format that are not dependent on simultaneous bidirectional handshakes will operate most effectively. Direct communication between the two modules is the first step in setting up a full IR data link.

A second test setup can be configured as follows. Prepare a test file prior to the actual test. After configuring the port setup, select File Transfer. Use a file-transfer protocol without error correction to first demonstrate a one-way file transfer without handshaking. Then gradually test the bidirectional capabilities of the system. You

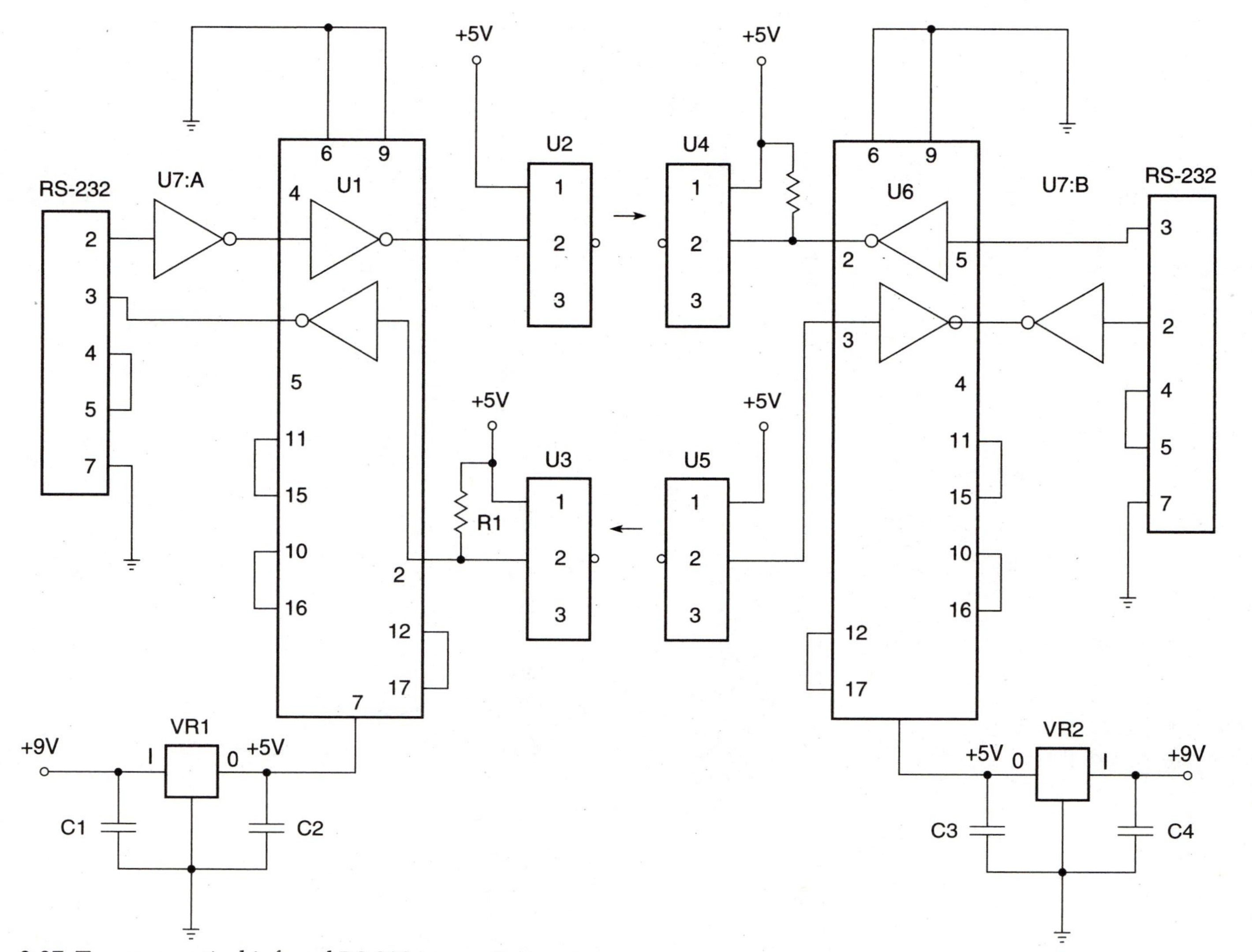

3-27 Two-way optical infrared RS-232 transceiver system. From Forest Mimms' *Circuit Scrapbook* (Howard W. Sams & Co., 1987). Copyright by Forest M. Mimms III. Used with permission.

are now well on your way to creating your own serial IR interface, which can be used in many different applications.

A new RY5AR01 IR data receiver chip will increase the IR link range from 1 to 3 meters. The RY5AR01 1-meter receiver and the RY5AR01 3-meter receiver are pin-compatible and cost about the same.

RS-232 infrared data transmission system parts list

U1	Maxim MAX-223 RS-232 driver/receiver
U2,U5	RY5AT01 transmitter IC
U3,U4	RY5AR01 receiver IC
U6	Maxim MAX-233
U7	SN7404 inverter
R1,R2	33-kilohm ¼-watt resistor
C1,C3	47-μF, 25-volt electrolytic capacitor
C2,C4	1000-pF, 25-volt electrolytic capacitor
VR1,V22	LM785L low-current 5-volt regulator

High-speed RS-232 infrared computer interface

The high-speed infrared RS-232 serial interface circuit shown in Figure 3-28 could well become one of the most used circuits in your computing future. You can readily use the high-speed interface to send data between two computers at rates of up to 115,200 bps. The beauty of this high-speed interface is that it consists of only two integrated circuits and a handful of discrete components. The heart of the high-speed IR interface is the multistandard CS8130 IR transceiver chip from Crystal Semiconductor. The IR transceiver is unique in that it will operate in four remote modes: IrDA, HPSIR, ASK, and TV. The CS8130 chip incorporates a PIN diode preamplifier, threshold detector/decoder, and demodulator block in the receive section and a data/control decoder, baud-rate generator, FIFO buffer, modulator, and two LED drivers in the transmitter section of the chip.

The CS8130 chip was configured in the IrDA mode for this high-speed interface project. The transceiver chip uses a 3.684-MHz crystal for its timing or baud-rate generator. The CS8130 IR transceiver is coupled to a MAX562 serial interface chip, which is ported to a DB9F serial connector. Constructing two of these IR interface circuits would allow you to send programs or data between two computers with high speed and excellent reliability. A BPV23NF silicon PIN photodetector diode is used for receiving IR impulses from the second transceiver unit, which is connected between pins 6 and 7 of the CS8130 chip. Two TSHA5502 IR LEDs are driven from pins 1 and 4 of the CS8130. Power bypass capacitors are used at both the VA+ power lead on pin 8 and the VD+ power lead on pin 12 of the transceiver. The interface circuit can be powered from two penlight or two C cells through a series diode for portable operation, if desired. The interface circuits can be built on a small PC board and duplicated for both ends of the system. Note that CS8130 is a 20-pin SSOP minicomponent and MAX-562 is a conventional DIP chip. The prototype was constructed on a dual-sided PC board measuring 2 × 3 inches.

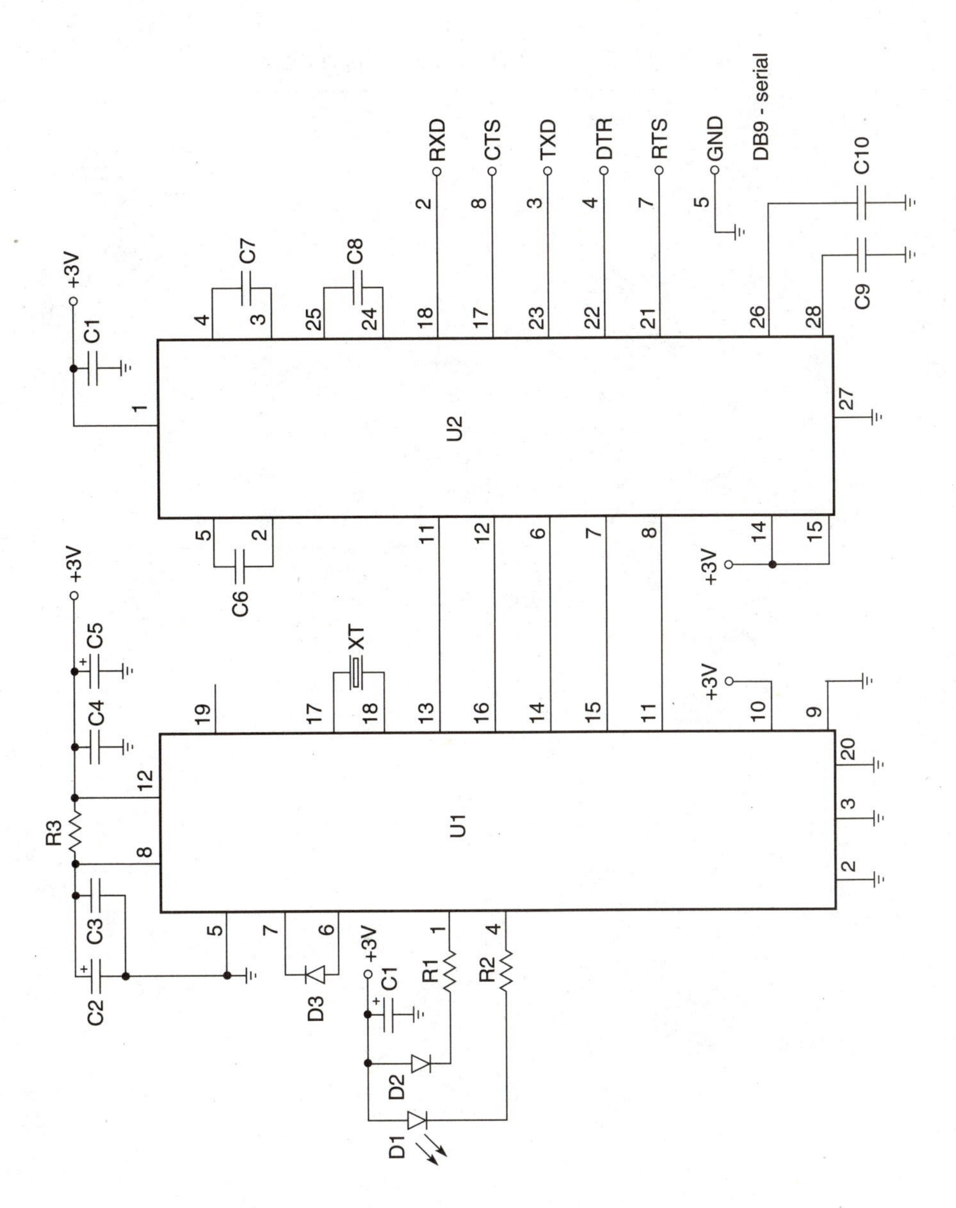

3-28 High-speed IR RS-232 computer interface. Crystal Semiconductor.

The CS8130 chip is addressable through software control, and there are 28 registers that are completely user-programmable. I will not go into the programming details of each register, but a Windows program is available for high-speed data transfer directly from Crystal Semiconductor. Contact Wayne Alvareze at Crystal for more details: 512-445-7222.

High-speed RS-232 IR interface parts list

R1,R2	5.5-ohm, ¼-watt resistor
R3	10-ohm, ¼-watt resistor
C1	47-µF, 25-volt electrolytic capacitor
C2,C5	10-µF, 25-volt electrolytic capacitor
C3,C4	.1-µF, 25-volt disc capacitor
C6,C7,C8,C10,C11	.33-µF, 25-volt disc capacitor
C9	.68-µF, 25-volt disc capacitor
D1,D2	TSHA5502 IR LED diode
D3	BPV23NF photodiode
Xtal	3.864-MHz crystal
U1	CS8130 multistandard IR transceiver chip
U2	MAX-562 serial interface chip
DB-9F	serial connector
Software	Windows drivers (contact Crystal Semiconductor)

4
CHAPTER
Light-meters

Light-meters serve the varied interests of scientists, electronic hobbyists, and photographers in a host of applications, from monitoring relative light and reflective light to gauging moonlight. Scientists monitor infrared and ultraviolet radiation to determine atmospheric transmission and absorption of sunlight. Scientists also measure the amount of UV light and ozone to monitor our safety. As an amateur scientist or electronic hobbyist, you can use the many light-meters presented in this chapter for your own observations.

Figures 4-1 and 4-2 illustrate simple yet sensitive light-meters. In Figure 4-1, a cadmium sulphide (CdS) photoresistive cell is connected in a series loop consisting of a 9-volt transistor radio battery, a 100-kilohm potentiometer, and a 0- to 1-mA meter. This circuit represents a series current loop, since all the components pass current around the loop. The second simple light-meter, shown in Figure 4-2, uses a silicon solar cell as a light detector and power source. The silicon solar cell or photovoltaic cell forms a series loop comprised of a 1-kilohm potentiometer and a 0- to 1-mA meter. Increasing the number of solar cells connected in parallel will produce higher sensitivity.

The last simple light-meter is shown in Figure 4-3. This light-meter uses a phototransistor base and collector junctions to form a photodiode in series with a 1-kilohm potentiometer and 0- to 1-mA current-displaying meter.

Linear light-meter

A linear light-meter is shown in Figure 4-4. This circuit employs a low input-bias op-amp to give a steady dc indication of light level. To reduce circuit sensitivity to light, if desired, you can reduce R1, but not less than 100 kilohms. The capacitor values in this light-meter were chosen to provide a time constant sufficient to filter high-frequency light variations from extraneous sources such as fluorescent lamps. The output of the light-meter can be fed to a voltmeter, multimeter, or A/D card inside a personal computer. The light-meter can be powered from a 9-volt source.

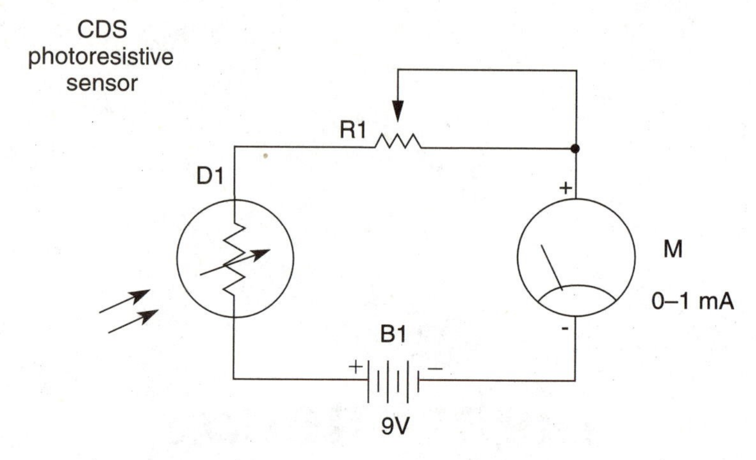

4-1 Photoresistive light-meter.

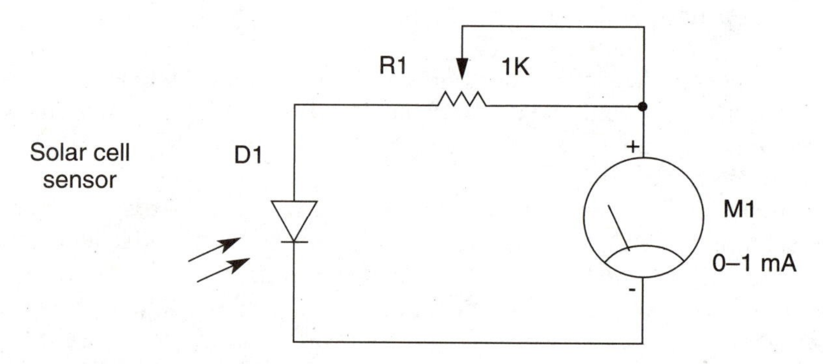

4-2 Solar-cell light-meter.

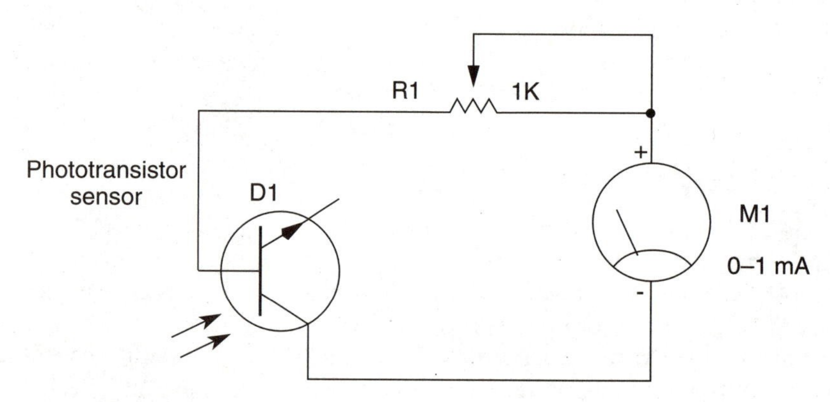

4-3 Phototransistor light-meter.

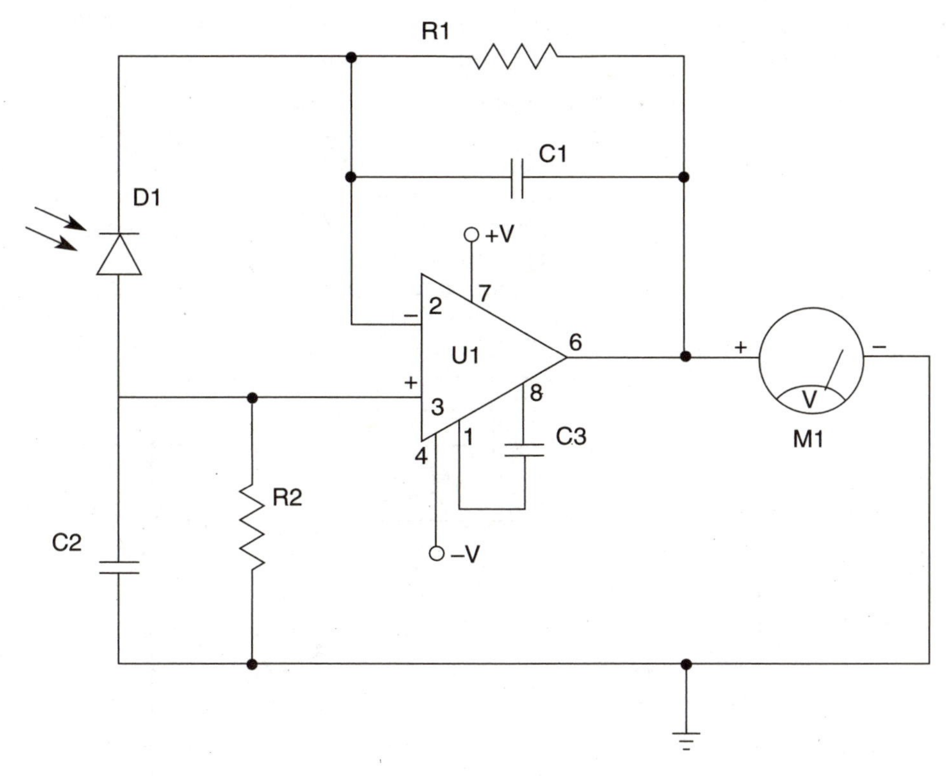

4-4 Linear light-meter.

Linear light-meter parts list

R1,R2	10-megohm, ¼-watt resistor
C1,C2	1,000-pF, 25-volt Mylar capacitor
C3	30-pF, 25-volt disc capacitor
U1	LM308 op-amp
M1	0- to 10-volt meter
D1	0SD1-0 photodiode

Logarithmic light-meter

The light-meter illustrated in Figure 4-5 provides a meter reading that is directly proportional to the logarithm of the input light power. The logarithmic circuit behavior arises from the nonlinear diode PN junction voltage relationship. The diode in the op-amp output circuit prevents the output voltage from becoming negative, which could happen at low light levels due to amplifier bias currents. Potentiometer R2 is used to adjust the meter to full-scale deflection in order to calibrate the log light-meter circuit. The light-meter is powered from a single 9-volt battery. The log light-meter's output is in series with a current-displaying meter.

Logarithmic light-meter parts list

R1	5.6-kilohm, ¼-watt resistor
R2	5.6-kilohm potentiometer
R3	10-kilohm, ¼-watt resistor
C1	30-pF, 25-volt disc capacitor
D1	0SD1-0 photodiode sensor
D2	1N4148 silicon diode
M1	0- to 100-mA meter
B1	9-volt battery

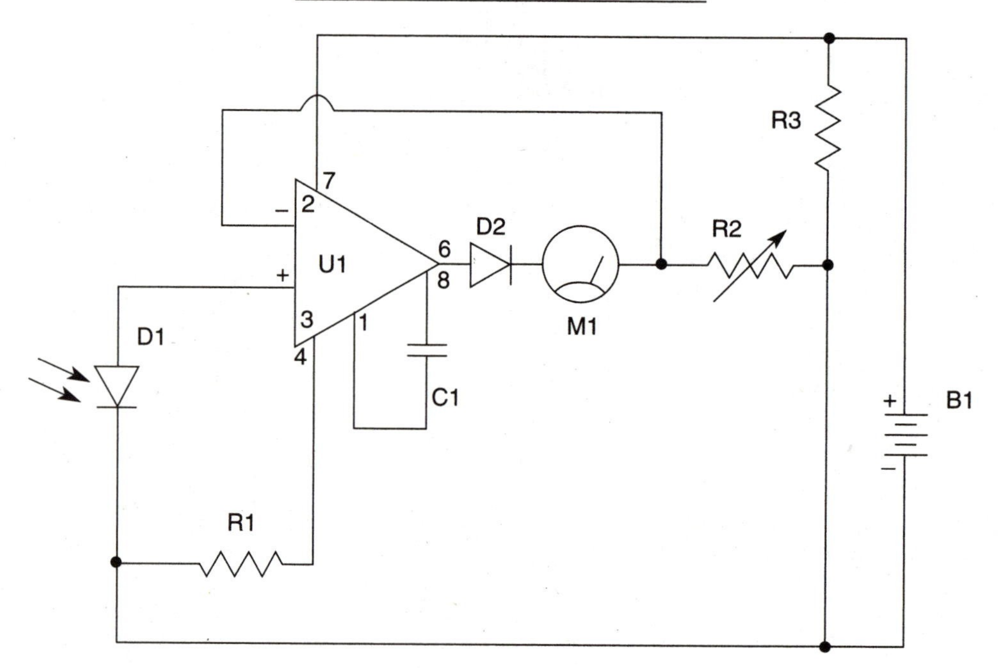

4-5 Logarithmic light-meter.

Programmable light-meter

The next light-meter, shown in Figure 4-6, is programmable. A cadmium or CdS cell is used to detect light in this circuit. The resultant resistive change to varying light on the CdS cell is coupled to pin 7 of an LM339 comparator op-amp. Potentiometers R1 and R3 are used to control the input levels to each 339 comparator section. In essence, the programmable light-meter consists of two comparator circuits, which act as a window detector circuit.

In operation, you would adjust R1 and R3 so the LED at the output glows when light at the detector is above or below the desired set points of R1 and R3. The outputs from both LM339 comparators are tied together and fed to Q1, which is used to

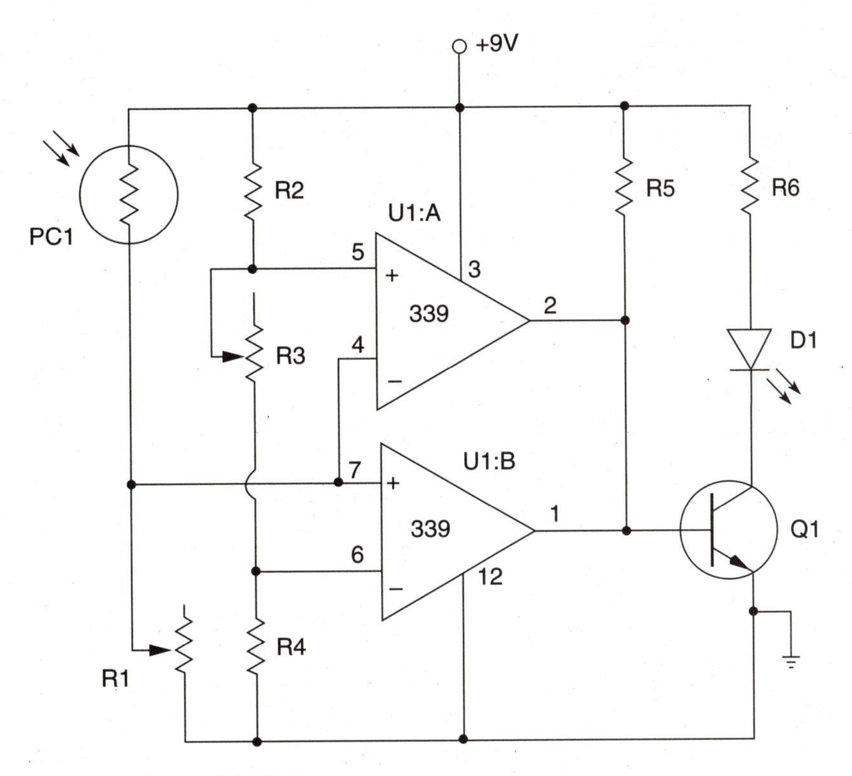

4-6 Programmable light-meter.

drive the LED. Note that a small relay could be substituted for the LED in order to use the output for control applications. The programmable light-meter can be operated from a 9-volt transistor radio battery.

Programmable light-meter parts list

R1,R3	1-megohm potentiometer
R2	15-kilohm, ¼-watt resistor
R4	5-kilohm, ¼-watt resistor
R5	10-kilohm, ¼-watt resistor
R6	1-kilohm, ¼-watt resistor
D1	LED or low-current relay
PC1	CdS photoresistive cell
Q1	RS-2009 transistor or equivalent
B1	9-volt transistor radio battery

Bargraph light-meter

A bargraph light-meter is featured in Figure 4-7. In this light-meter, a photodiode or IR-sensitive silicon cell can be a light sensor to the input of an LM308 op-amp. A 1-megohm, 2-percent potentiometer or range switch with individual gain resistors at R2 is used to select exposure to a ½ f-stop resolution. A 1000:1 f-stop range is possible with this circuit. The bargraph light-meter uses two LM3915 bargraph-driver

ICs connected together in order to drive 20 LEDs. You can also use this circuit as an infrared light-meter by substituting an IR-sensitive silicon diode with an IR filter ahead of the detector diode. The bargraph light-meter can be made pocket-sized, if desired, since it can be easily powered from a 9-volt battery.

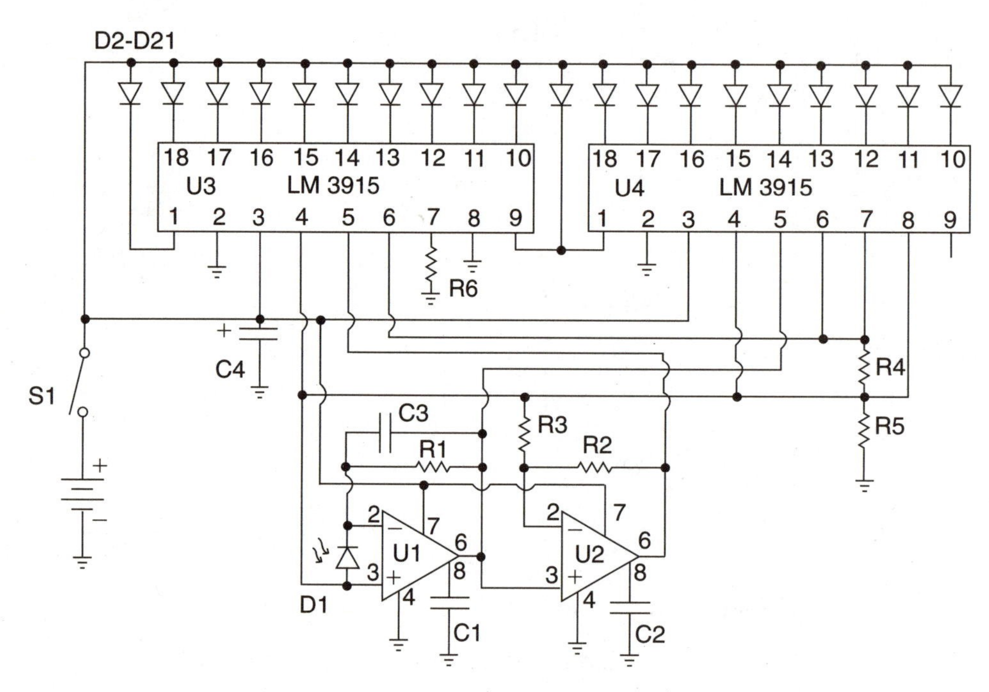

4-7 Bargraph light-meter.

Bargraph light-meter parts list

R1	1-megohm potentiometer
R2	120-kilohm, 2-percent, ¼-watt resistor
R3	15-kilohm, 5-percent, ¼-watt resistor
R5,R6	6.2-kilohm, 2-percent, ¼-watt resistor
C1,C2	100-pF, 25-volt silver mica capacitor
C3	10-pF disc capacitor
C4	2.2-µF, 25-volt electrolytic capacitor
U1,U2	LM308A op-amp
U3,U4	LM3915 LED driver
D1	Honeywell SD3421 silicon light detector diode
D2,D21	LEDs
S1	SPST switch
B1	9-volt battery

Ultrasensitive light-meter

An ultrasensitive light-meter that can be used as a relative light-meter is shown in Figure 4-8. A silicon solar cell is used to detect light in this circuit. Current flow is set up between pin 2 and 3 of an LM741 op-amp. Sensitivity is set by the potentiometer at R1, while R2 functions to zero the circuit. A range switch function is configured with a rotary switch at S1. Five different resistor/capacitor networks are used across pins 2 and 6 in the gain path of the op-amp. Position 1 of S1 is the low-sensitivity position, while position 5 is the highest gain setting. Note that this circuit is extremely sensitive, so set it to the lowest gain range when initially powering the circuit. The ultrasensitive light-meter is powered from two 9-volt batteries. A plus voltage is applied to pin 7 of the 741, while a minus voltage is connected to pin 4. In operation, first set S1 to position 1 and apply power. Then carefully adjust R1 to set the meter to zero, and adjust R2 with the detector dark in order to zero the op-amp itself. The ultrasensitive light-meter is extremely sensitive to normal sunlight, which might damage the meter movement.

Ultrasensitive light-meter parts list

R1	5-kilohm potentiometer
R2	10-kilohm potentiometer
R3	10-megohm, $\frac{1}{4}$-watt resistor
R4	1-megohm, $\frac{1}{4}$-watt resistor
R5	100-kilohm, $\frac{1}{4}$-watt resistor
R6	10-kilohm, $\frac{1}{4}$-watt resistor
R7	1-kilohm, $\frac{1}{4}$-watt resistor
C1	.002-µF, 25-volt disc capacitor
C2	.02-µF, 25-volt disc capacitor
C3	.2-µF, 25-volt disc capacitor
C4	2.2-µF, 25-volt electrolytic capacitor
C5	22-µF, 25-volt electrolytic capacitor
D1	Silicon solar cell
S1	Five-position rotary switch
U1	LM-741 op-amp
M1	0- to 1-mA meter
B1,B2	9-volt battery

Sunlight passing through the atmosphere is scattered or absorbed by dust, haze, smoke, and molecules of many gasses, as shown in Figure 4-9. Debris from major volcanic eruptions can often remain in the atmosphere for up to five years. Atmospheric transmission directly affects the Earth's temperature. Scientists regularly measure atmospheric light transmission from sunlight passing through the atmosphere. As an amateur scientist, you can observe and record many different aspects of atmospheric transmission, such as cloud height, UV radiation, ozone, sun angles, and the optical thickness of the atmosphere, all from your own backyard.

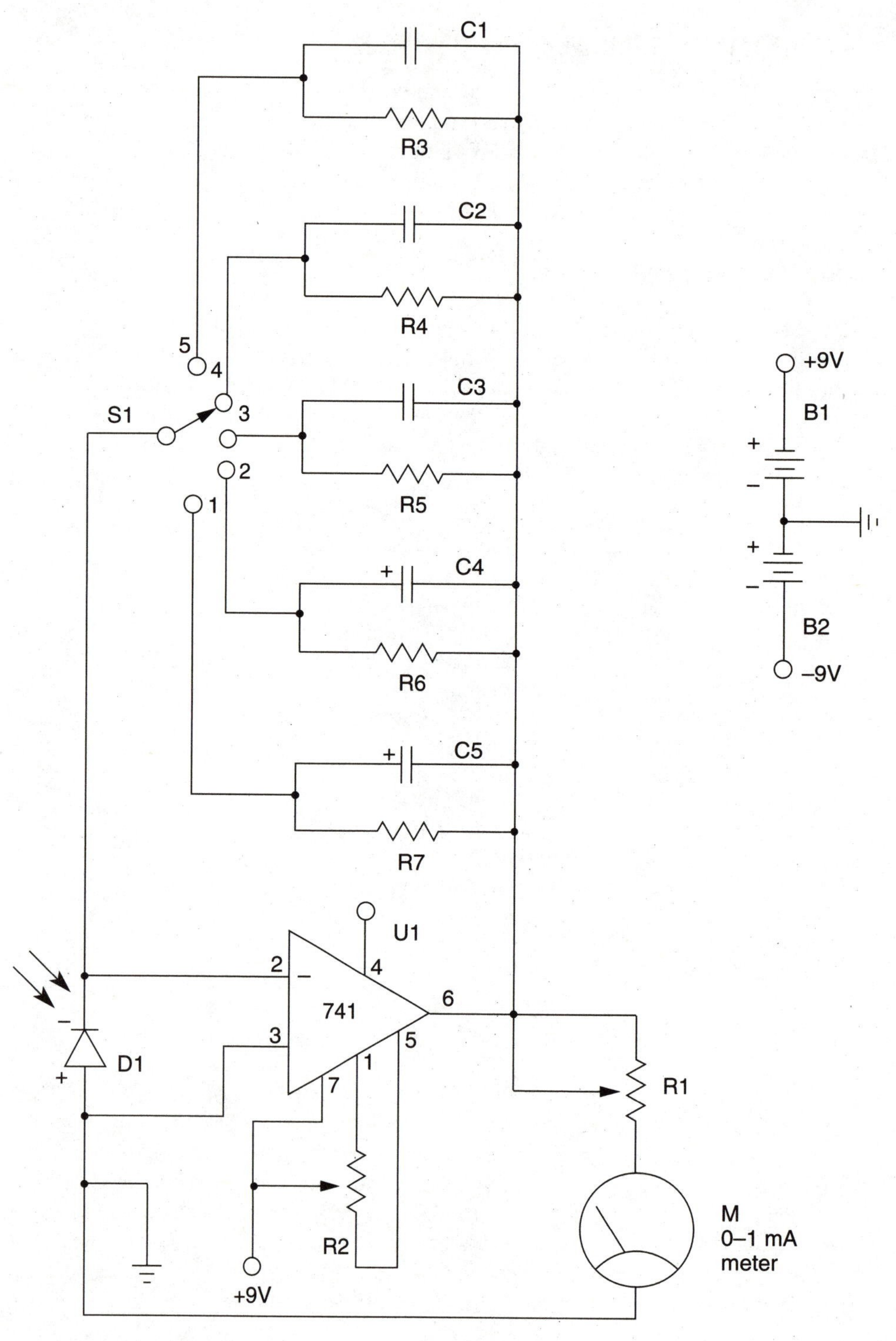

4-8 Ultrasensitive light-meter.

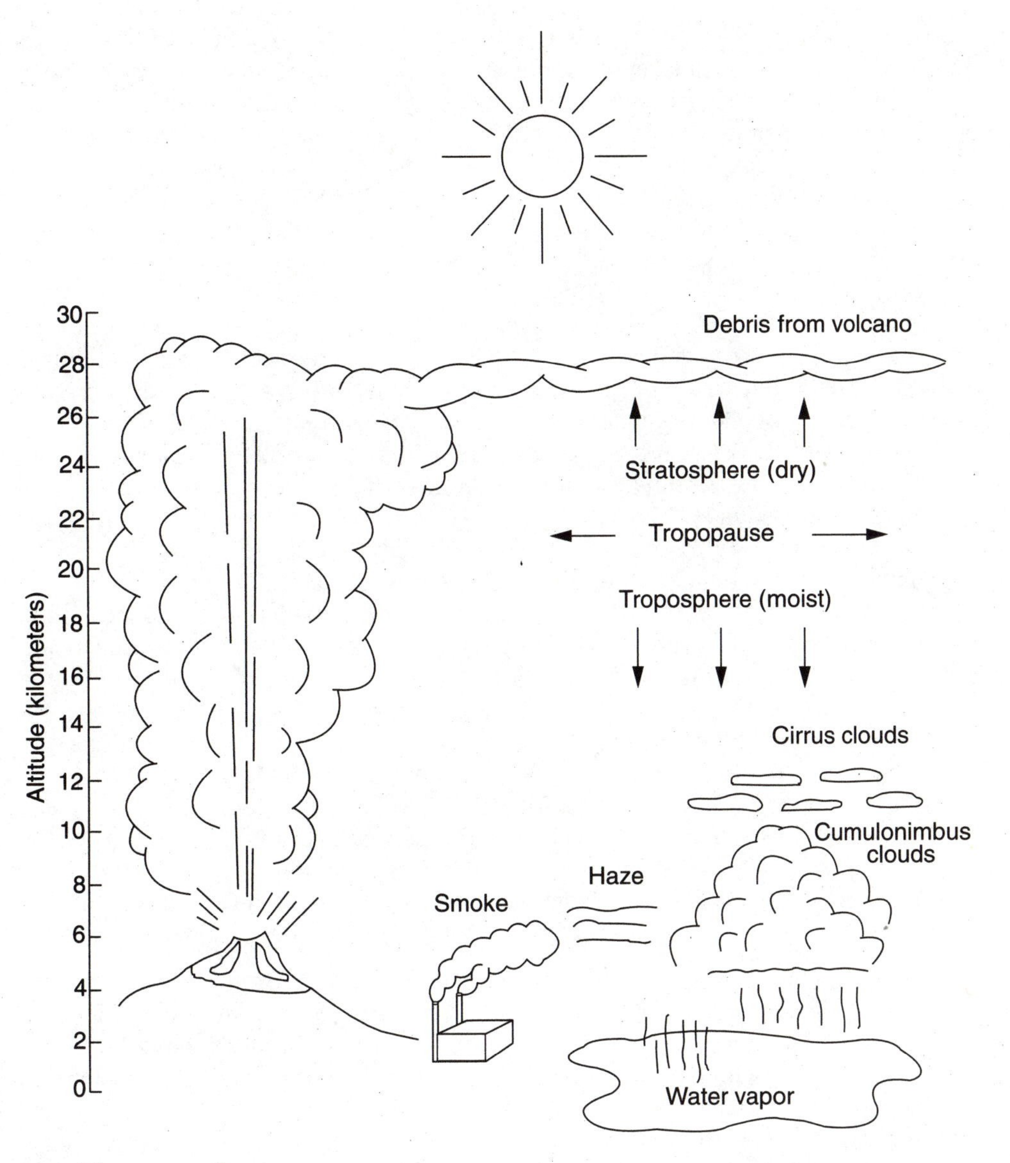

4-9 The atmosphere.

The tropopause layer of the atmosphere is tucked in between the very dry stratosphere and the moist air of the troposphere. The height of the tropopause varies with the latitude and weather systems, and it usually resides at about 17 kilometers at the equator and at about 7 kilometers at the North and South Poles. Note that volcanic haze and meteor dust travel relatively unimpeded in the stratosphere. The ozone layer includes about 90 percent of the total ozone, and the remaining 10 percent can be found in the troposphere. The ozone layer absorbs most of the sun's ultraviolet radiation. Volcanic haze and both natural and man-made (anthropogenic) gasses can destroy the ozone layer. The atmosphere contains 78 percent nitrogen, 21 percent oxygen, and 1 percent argon; other gasses such as ozone, carbon dioxide, methane, carbon monoxide, and sulphur dioxide; and smoke, dust, and water vapor.

Natural haze is caused by smoke from forest fires, water vapor, fog, dust, sea salt, cirrus and stratus clouds, and the photochemical reactions of sunlight and various gasses emitted by plants. Anthropogenic or man-made haze is caused by emissions from coal-fired power plants, fireplaces, and wood stoves, as well as the contrails of high-altitude aircraft and gasses from internal-combustion engines. Man-made haze is especially concentrated over the eastern United States and eastern Europe.

Haze significantly reduces direct radiation from the sun and increases diffuse radiation, while slightly reducing the total radiation from the sun. Haze also scatters some solar radiation back into space, causing a cooling effect. Haze greatly increases diffuse radiation on plants and animals, which in effect are shaded from sunlight.

When the sky is free from haze, the sun appears as a brilliant disk in the deep blue sky and the clouds stand out in high contrast. If some haze is present, the sun is generally surrounded by a bright glow and clouds near the horizon are difficult to resolve. Considerable haze causes the sun to appear dim, the entire sky becomes a pale milky blue, and the clouds blend into haze and are difficult to see.

The temperature of the Earth is regulated partly by clouds, as mentioned earlier. Warm air can contain more water vapor, hence more clouds. The clouds reflect sunlight back into space, thus cooling the Earth. Recording a fraction of the sky covered by clouds can provide important information about the effect of clouds on the climate. The fraction of the sky covered by clouds is measured in tenths or eighths (octas). One way to measure cloudiness is to photograph the clouds using a film camera on a tripod. The camera is pointed downward towards a hemispherical or wide-angle mirror placed on the ground. To estimate the cloudiness in each quadrant of the compass, average the four estimates to get the overall cloudiness.

You can readily measure the height of clouds by using the following formula:

$$\text{height in feet} = 227\ (T - \text{dew point})$$

Cumulus clouds form when warm, humid air rises to where the air temperature falls below the dew point. Knowing that air temperature falls at about 2.77°C or 5.5°F per .3 kilometer, you can calculate the height of the base of a cumulus cloud. Note that T is the ground temperature in degrees Fahrenheit and the dew point is in degrees Fahrenheit.

You can measure the dew point by making a wet/dry relative-humidity instrument, or *hygrometer*. You can use a hygrometer directly or you can use two thermometers, one dry thermometer and a second wet bulb thermometer. A wet bulb thermometer consists of a thermometer with wet gauze or fabric over the thermometer, with air from a fan blowing on the sensor. A sling hygrometer could also be used to determine the dew point. A sling hygrometer consists of two thermometers, one wet and one dry, placed in a housing that can rotate on a handle. To get a reading, you quickly rotate the thermometers around. To calculate the dew point, use the following formula:

$$\text{dew point} = (5T\ \text{wet} - 2T\ \text{dry})\ /\ 3 \quad \text{in degrees Centigrade}$$

Note that T = temperature. You can measure the approximate intensity of sunlight outside the atmosphere by using a light-meter known as a radiometer to make a series of measurements of the intensity of sunlight over half a day. You could then plot the results on a graph. One axis of the graph represents the logarithm of the sun's intensity. The other axis represents the thickness of the atmosphere through which the measurements are made. If the atmosphere remains stable over the course of the measurement, then the points on the graph should fall in a straight line. Extending this line to an atmospheric thickness of zero gives the intensity of sunlight outside the atmosphere. By using various types of filters, you could observe the sun's output at different wavelengths.

A simple solar-cell radiometer is shown in Figure 4-10. Silicon solar cells or photovoltaic cells respond well to visible and near-infrared radiation from the sun. You can use a solar cell to track daily variations in sunlight quite easily. Connect the red or positive lead from the solar cell to the red or plus lead of a digital multimeter. Then connect the black or minus lead of the solar cell to the minus input of the multimeter. Switch your multimeter to measure current. If the solar cell output in full sunlight exceeds the range of the multimeter, block parts of the cell or switch the meter to a higher current range. Be sure to place the solar cell in the same location every day in order to make accurate and reliable readings.

You can convert a solar-cell radiometer into a sun photometer by using a collimator tube painted flat black inside and out, placed over the solar cell to restrict the

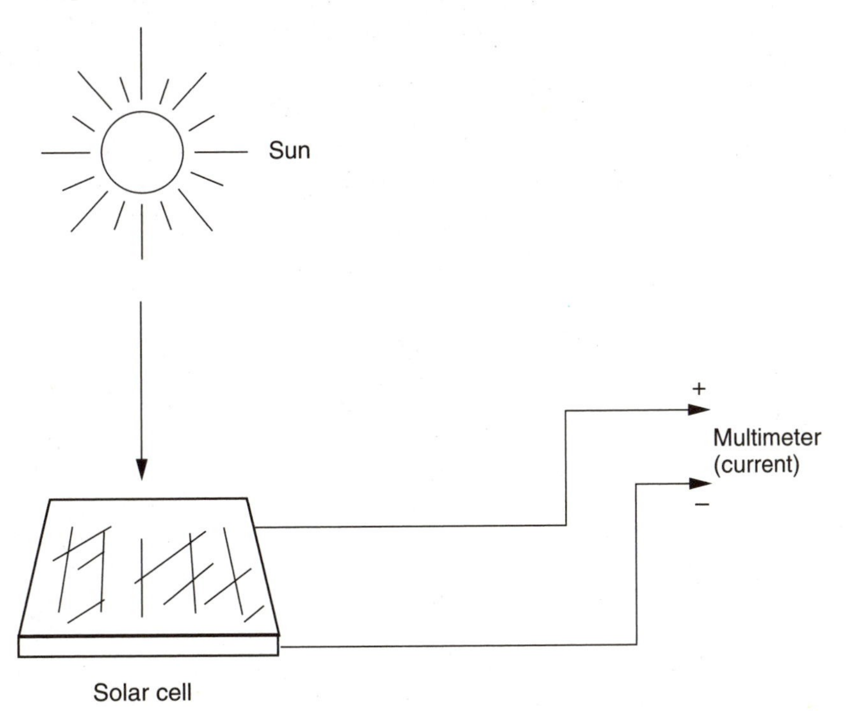

4-10 Simple solar-cell radiometer.

field of view of the sensor, as shown in Figure 4-11. A plastic or color gel filter is placed between the collimator and the solar cell to select the wavelength of interest. You can use a sun photometer to measure the optical thickness of the atmosphere. Atmospheric optical thickness (Aot) is a measure of the clarity of the air in a vertical column through the atmosphere. Aot indicates the amount of haze, smog, smoke, dust, and volcanic aerosols in the atmosphere. A small Aot indicates a clean atmosphere. You can measure the atmospheric optical thickness with a sun photometer and a calculator with a natural logarithm key (Ln). A simplified formula for measuring atmospheric optical thickness is:

$$\text{Aot} = (\text{Ln Io} / \text{Ln I}) / M$$

The extraterrestrial constant (ET) is defined as Io and is the signal the sun photometer would measure above the atmosphere (see Figure 4-12). The signal measured at a specific sun observation period is defined as I. Air mass determined during the observation is defined as M.

The ET constant can be obtained by first measuring I for half a day every 30 minutes near noon. This reading should be done more often at lower sun angles. Then plot the logarithm (Ln) of I versus M at each observation, and draw a straight line through the points. The Ln of the ET constant is where the line intercepts the vertical axis where $M = 0$. You can use the linear regression feature of your calculator or your computer and spreadsheet program to find the intercept at $M = 0$. To compute

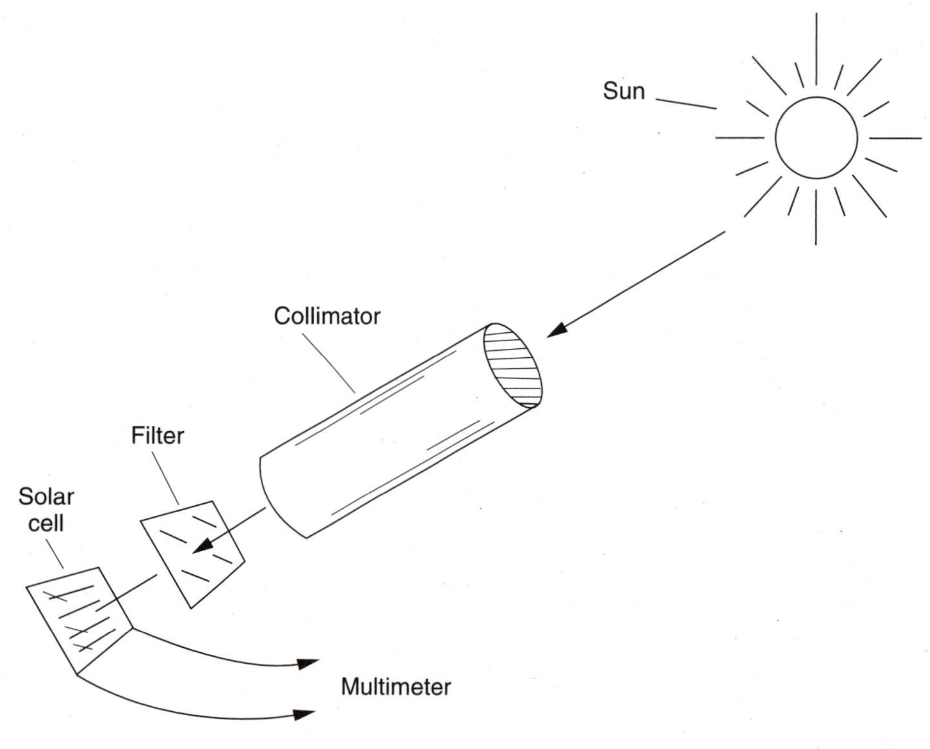

4-11 Sun photometer.

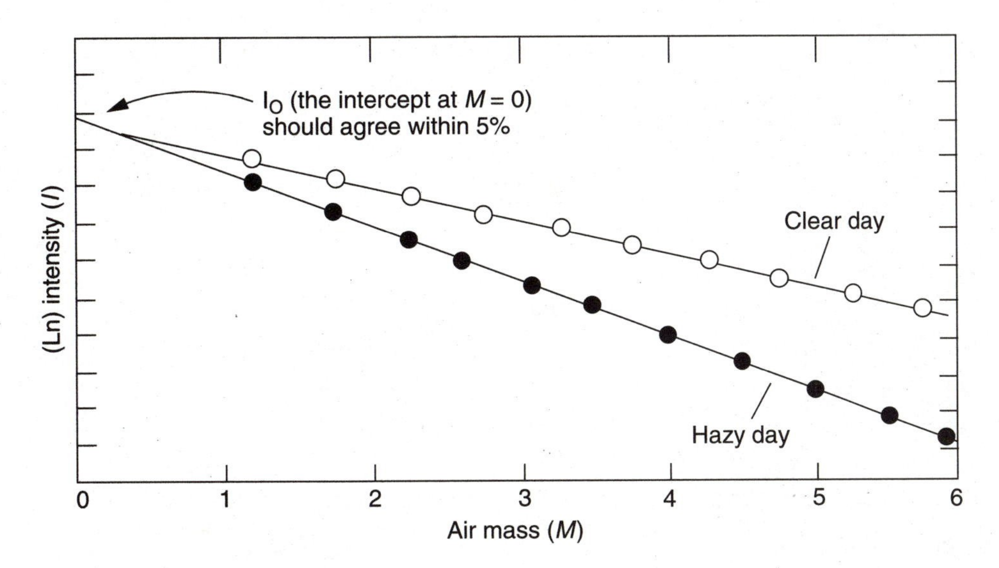

4-12 Extraterrestrial constant.

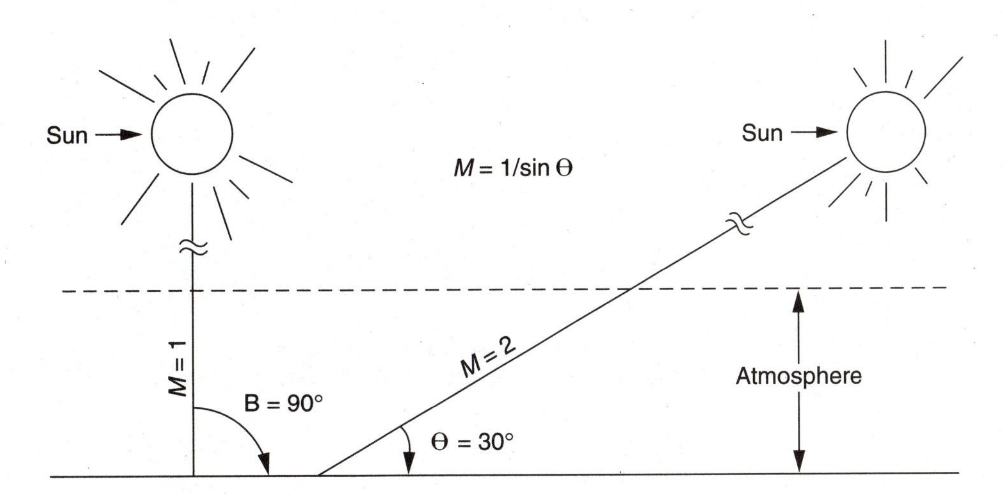

4-13 Air mass.

the air mass (*M*), see Figure 4-13. The air mass is defined as 1/sin θ, where sin θ is the angle of the sun above the horizon.

Simple sunlight photometer

The self-contained sun photometer in Figure 4-14 is a simple variation on the sun photometer theme. This circuit consists of only three electronic components: a silicon solar cell, a resistor, and an analog meter. Locate a light-tight plastic or

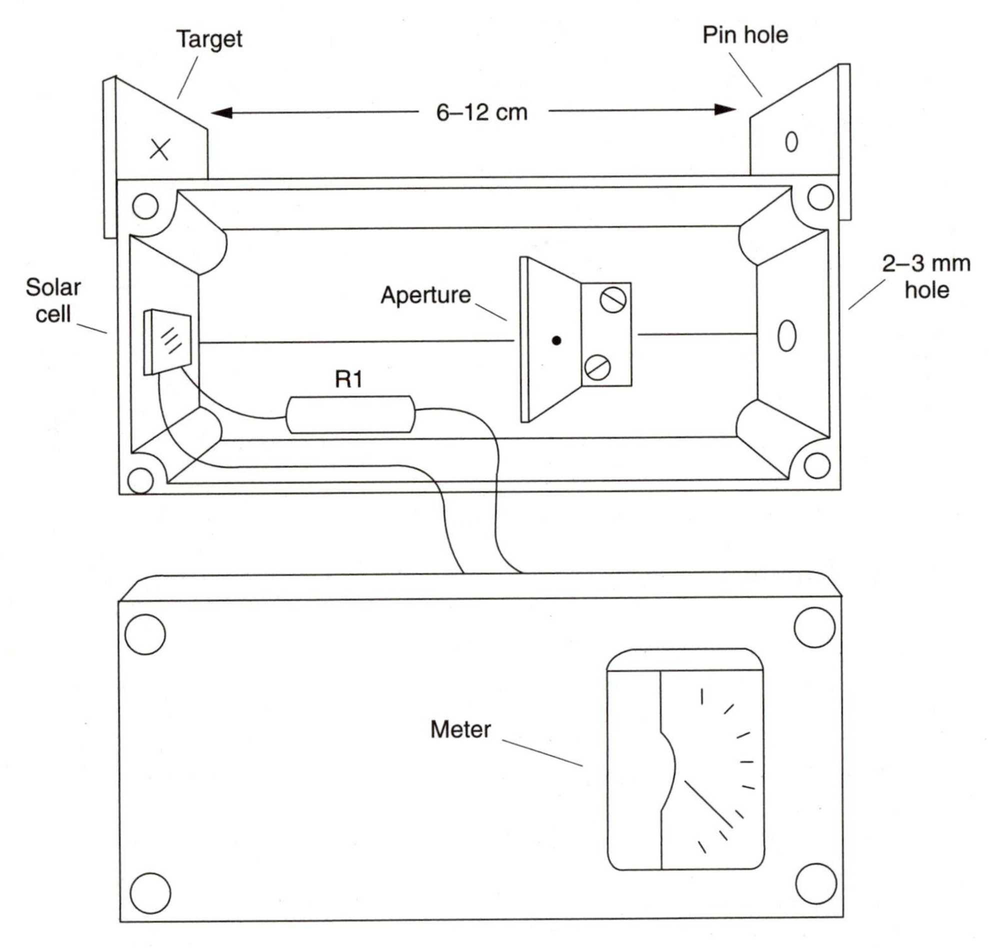

4-14 Self-contained sun photometer.

Simple sunlight photometer parts list

R1	Resistor (see text)
S	Silicon solar cell
M	0- to 100-mA meter
Misc	Aluminum chassis box, plastic vane, and plastic vane w/aperture

aluminum case and drill a 2- to 3-cm hole at one end of the enclosure, as shown. Next, mount a small plastic vane at either side of the chassis box with epoxy. Drill a pin hole at the light input side of the box, and mark the vane at the solar-cell end with equal divisions from the center point of the vane. Then create an aperture about a third of the way from the light-input end of the enclosure. Cement a silicon solar cell to the rear of the box and connect it to a potentiometer in series with a 0- to 100-mA meter to reduce the current from the solar cell and avoid damaging the meter. Use a 500-ohm to 1-kilohm potentiometer for R1; once an appropriate setting has been established, you can substitute a fixed resistor. Adjust the potentiometer for 80 to 100 degrees of full scale.

Advanced sun photometer

An advanced sun photometer is depicted in Figures 4-15 and 4-16 and in Photo 4-1. The heart of an advanced sun photometer is the Honeywell SD3421-002 photodiode and filter assembly, shown in Figure 4-15. The photodiode is soldered to an ⅛-inch phone plug behind the end cap, as shown. The end cap is then coupled to a ⅜-inch brass union. A 12.5-mm diameter interface filter that transmits 500, 850, or 1000 nm is sandwiched between two O rings. The conical cap is fitted to the ⅜-inch brass union. A collimator is fashioned from an aluminum tube painted both inside and out with flat black paint. The 45-mm collimator is then pressed into the conical cap. All of these fittings should be available at a hardware store. The photodiode is available from Newark Electronics and the filter is available from Edmund Scientific (see the Appendix).

The electronic parts for the advanced sun photometer are shown in Figure 4-16. The heart of this photometer is the Texas Instruments TLC-271 IC. The photodiode sensor is inputted to the op-amp via J1 to pin 2. You then adjust the gain of the op-amp circuit by using a rotary switch to select from three gain resistors: R1, R2, and R3. Resistor R4, a 100-kilohm potentiometer, is used to zero the op-amp. A voltage divider is set up between R5 and R6, between pins 4 and 7 of the op-amp. A single 9-volt battery is used to power the advanced sun photometer circuit. The output from the op-amp can be fed to any digital voltmeter or multimeter for accurate and repeatable readings.

The collimator tube limits the photodetector's sky view to 2 degrees; therefore, make sure to point the detector at the same sun angle at the same time of day to ensure repeatability of measurements. You can use a camera tripod to hold the advanced sun photometer.

Advanced sun photometer parts list

R1	10-kilohm, ¼-watt resistor
R2	100-kilohm, ¼-watt resistor
R3	1-megohm, ¼-watt resistor
R4	100-kilohm potentiometer
R5,R6	1-megohm, ¼-watt resistor
C1	.01-µF, 25- volt disc capacitor
U1	TLC-271 op-amp (Texas Instruments)
S1	Three-position rotary switch
B1	9-volt battery
D1	Honeywell SD3421-002 photodiode
F1	500-, 850-, or 1000-nm, 12.5-mm interference filter
J1	⅛-inch phone jack
P1	⅛-inch phone plug

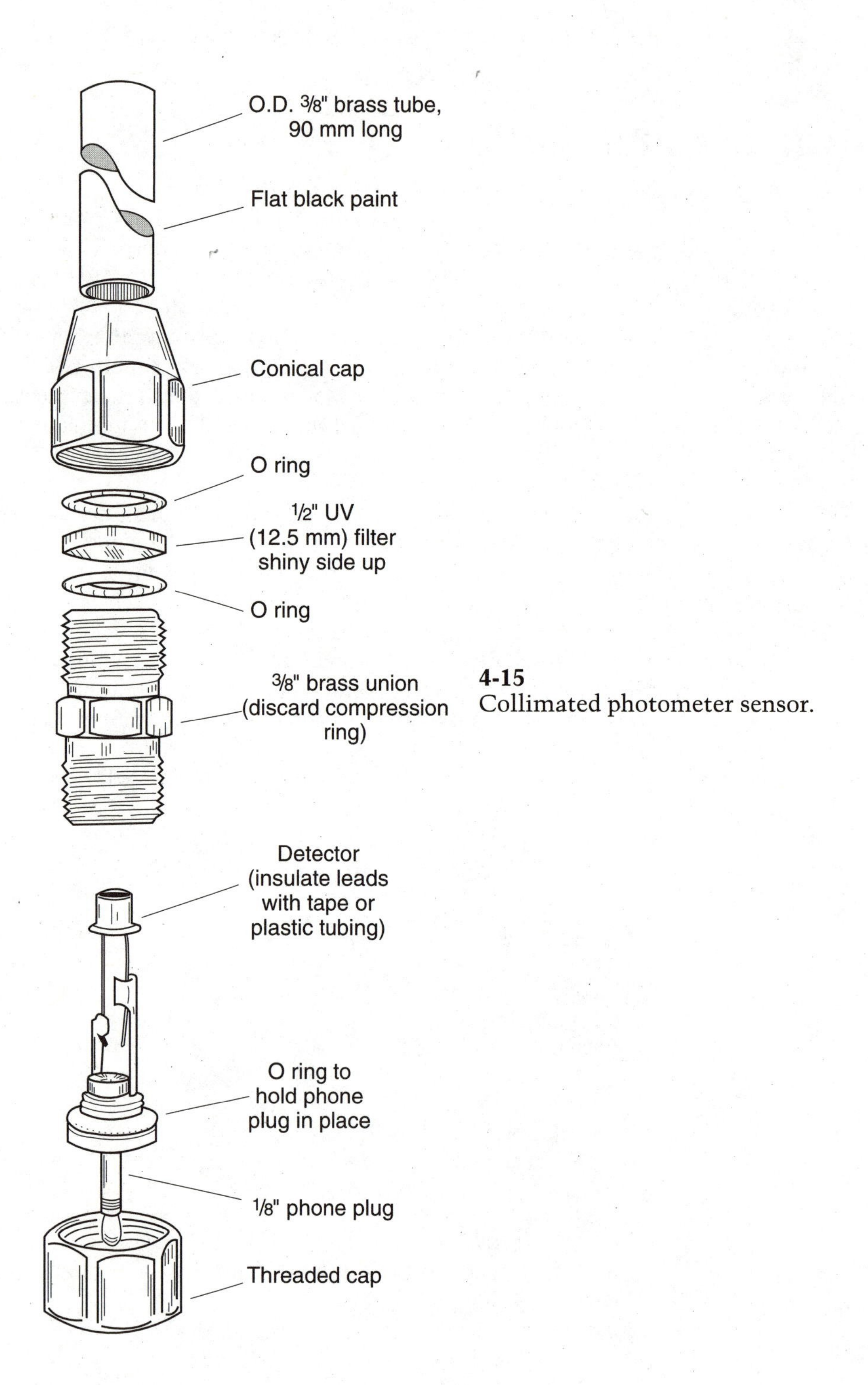

4-15
Collimated photometer sensor.

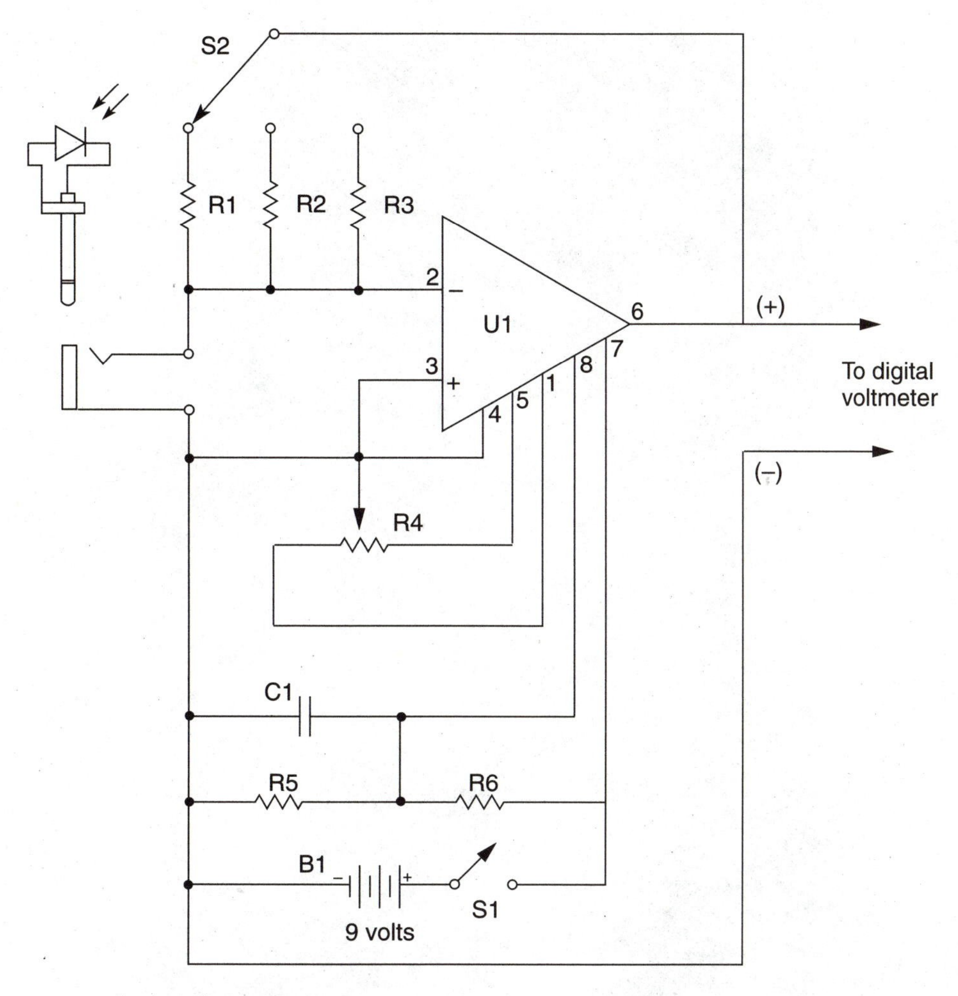

4-16 Advanced sun photometer.

Digital UV photometer

Fortunately, most ultraviolet radiation from the sun never reaches the Earth's surface due to the blanket of ozone. Without the ozone barrier, ultraviolet radiation would be so strong in intensity that most plants and small animals would soon perish. Volcanic eruptions as well as other natural phenomena and human foibles have all contributed to altering the composition of the atmosphere. Many people around the world have become concerned about ozone holes that have appeared at the South Pole in recent years. As a result, many ground-monitoring stations have been set up around the world to measure ultraviolet radiation.

Too much ultraviolet radiation can cause erythema (reddening of the skin) and sunburn. Plastics can also decompose prematurely when exposed to radiation near 300 nm. The ultraviolet spectrum falls between 280 and 320 nm.

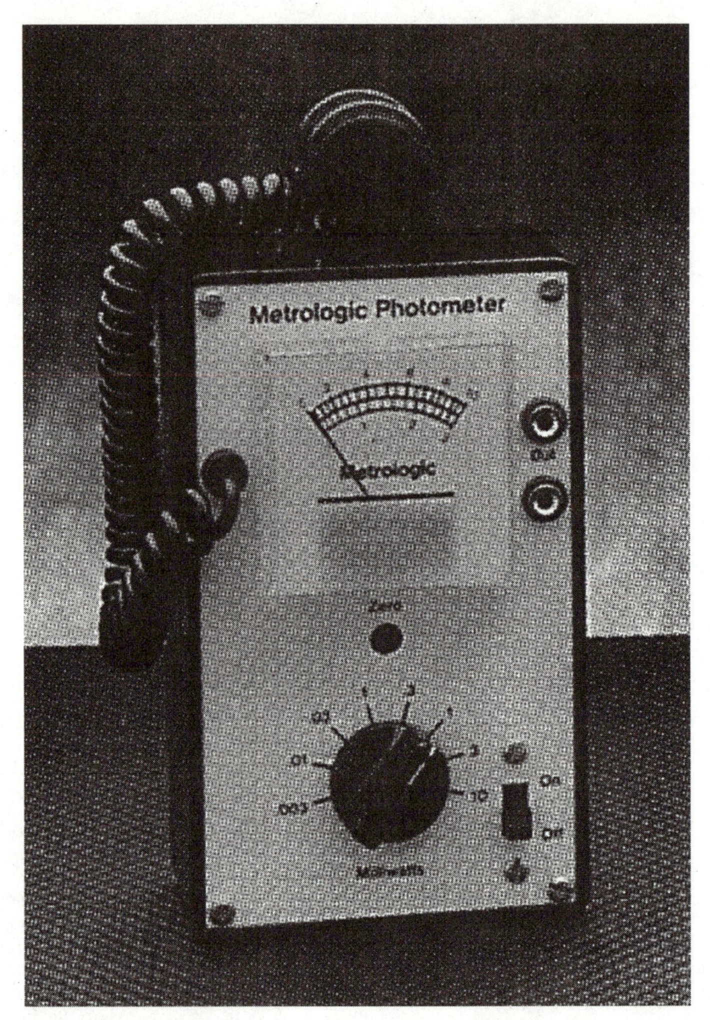

Photo 4-1 Photometer.

As an amateur scientist, you can collect ultraviolet light readings via your own digital ultraviolet-B photometer, shown in Figure 4-17. The heart of the ultraviolet-B photometer is the gallium phosphide (GaP) photodiode, which responds only to UV radiation, unlike most UV-sensitive silicon diodes, which also respond to near-infrared radiation. The GaP photodiode is placed into a detector assembly. The photodiode is then inserted into a transistor socket and soldered to a $\frac{1}{8}$-inch phone plug behind an end cap screwed to a $\frac{3}{8}$-inch brass union, as shown. A UV filter is then placed between two O rings and inserted between the $\frac{3}{8}$-inch brass union and a conical end cap. A brass or aluminum collimator tube painted flat black inside and out is fitted into the opposite end of the tapered conical cap.

The output of the detector assembly of the digital UV meter is plugged into the electronic measuring circuit illustrated in Figure 4-17. A Texas Instruments TLC-271 op-amp is at the center of the ultraviolet-B photometer. A rotary switch positioned

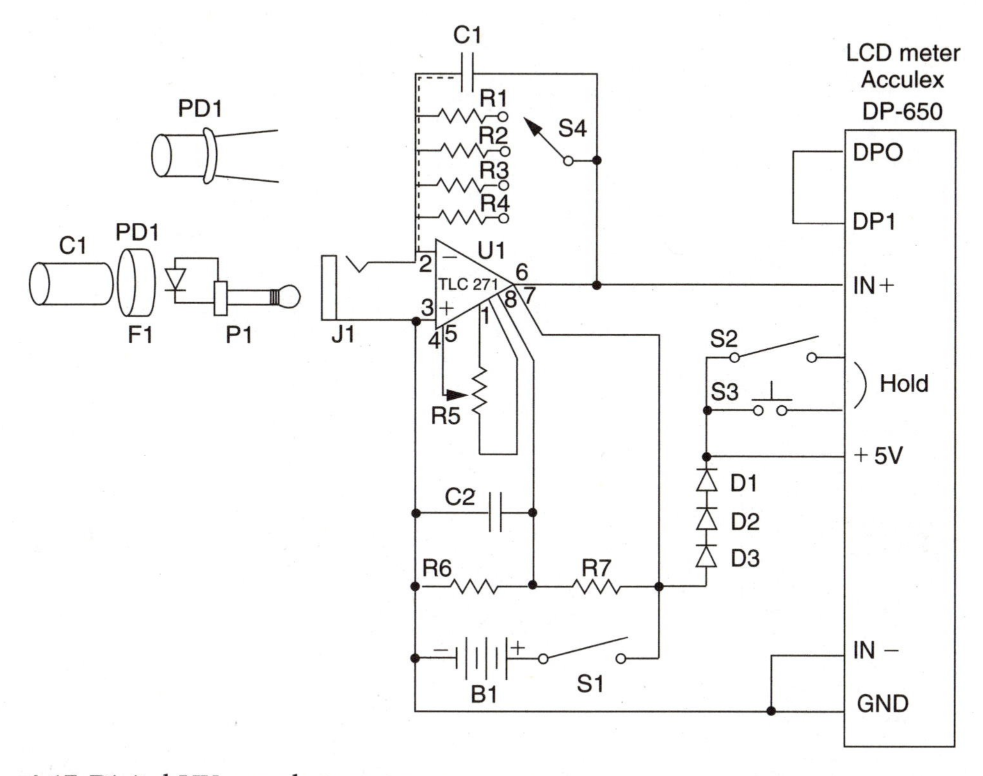

4-17 Digital UV sun photometer. From *Science Probe* Magazine (Gernsback Specialty Press, Nov. 1992) Forrest Mimms III. Used with permission.

close to the op-amp is used to switch between gain resistors R1 through R4. Note the capacitor filter across the gain resistors. Potentiometer R5 is used to electronically zero the op-amp. A voltage divider is set up between pins 3 and 7 of U1 using two 1-megohm resistors. Power switch S1 is used to power the ultraviolet-B photometer via four AA cells. The output of the TLC271 op-amp is fed to the input of an Acculex DP-650 LCD digital voltmeter. The digital voltmeter is powered from the 6-volt AA-cell battery pack via three 1N914 diodes, which are used to drop 1 volt to power the 5-volt display unit. Switches S2 and S3 are hold buttons, switch S3 is a momentary hold button, and S2 is a "close to hold" toggle switch. If you want, you can substitute other suitable LCD digital voltmeters for your UV-B photometer. By selecting a 12.5-mm, 300-nm UV filter ahead of the GaP photodiode, you now have an ultraviolet-B monitoring instrument.

If you wanted to monitor ozone in the atmosphere, you would need to measure a column of atmospheric ozone. You can determine the total amount of ozone in a column through the atmosphere by measuring two wavelengths of ultraviolet radiation at the same time. Ozone strongly absorbs ultraviolet radiation from the sun with a wavelength of about 330 nm. This absorption is so efficient that, under normal conditions, practically no radiation below 295 nm reaches the ground. Ozone absorbs shorter wavelengths more efficiently than longer wavelengths; therefore, you can measure the amount of ozone by using dual-wavelength absorption spectroscopy.

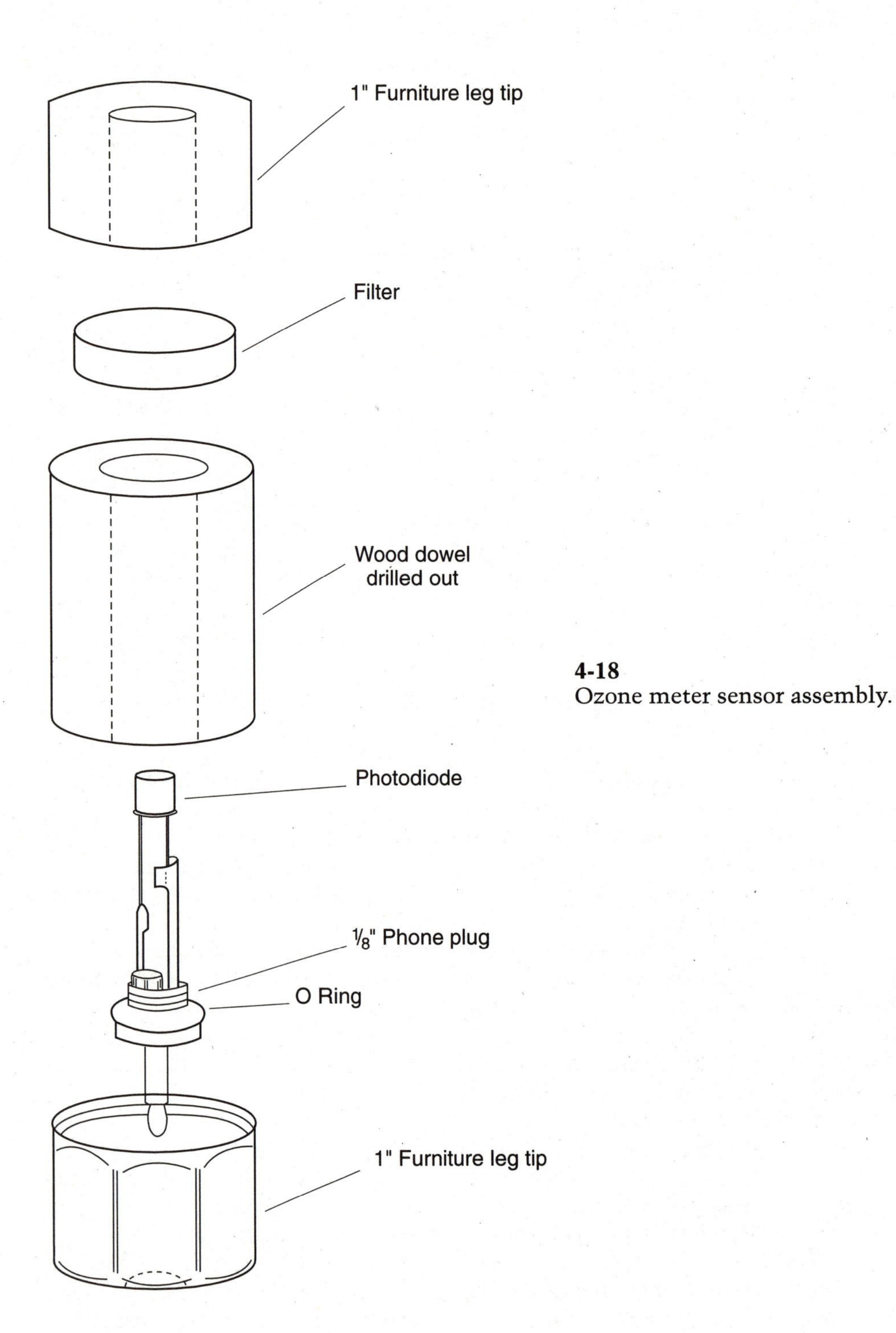

4-18
Ozone meter sensor assembly.

As you might imagine, the most important components in constructing an ozone-monitoring instrument are the two selective filters. In order to create your own ozone detector, you need to build a second UV-B photometer because you will need two identical detectors, amplifiers, and LCD display units. The only difference between the two UV-B detectors is that each unit uses a different filter (see the ozone sensor assembly diagram in Figure 4-18).

For locations around and above 35 degrees north latitude, you will need to obtain one filter in the 305- to 310-nm range and a second filter in the 325- to 330-nm range. The smaller-diameter UV filters (12.5 mm) are easier to work with and much cheaper than 25-mm filters.

To construct your own ozone monitor, you will need to locate a chassis box in which to house both UV-B photometers. Squeezing all the components of the two radiometers into a single chassis box requires careful planning.

After installing both UV-B photometers into a single enclosure, carefully check all your wiring, install the batteries, and turn on the power switch. Both displays should come to life. Block both photodiodes and adjust the R5 zero-trimmer potentiometers until both readouts indicate zero. Using the ozone meter is somewhat complicated.

First go outside and point the ozone meter towards the sun while watching the shadow that the upper alignment vane casts on the lower vane, until you see the spot of sunlight on the lower vane (see Figure 4-19). Align the ozone meter until the spot of sunlight is centered over the alignment mark.

A camera tripod is ideal to secure the ozone meter while making measurements. Hold the ozone meter securely and take a reading; then press Hold to save the reading for both channels. Ozone measurements are usually made at mid-morning and mid- afternoon. Try and make at least three observations per measuring session and record them in a notebook so you can compare them with calibrated data from the Total Ozone Mapping Data (TOMS) satellite. You can obtain reprints of the satellite data from NASA, at the Goddard Data Center in Greenbelt, MD 20771 (code 933.4).

To actually calculate the amount of ozone, you first have to determine the sun angle used to take the measurements. With that data in hand, you can compute an ozone column in the atmosphere with the following formula (not for the faint of heart):

$$O^3 = \frac{\text{Log } L1\text{^}/L2\text{^} - \text{Log } (L1/L2) - (b1 - b2) \times (P \times M/1013)}{(a1 - a2) \times M}$$

The following are ozone formula descriptions you will need to calculate a column of ozone in the atmosphere with the preceding formula:

- L1^ and L2^ are the intensities of the two wavelengths outside the atmosphere, known as the *extraterrestrial constant*. You can obtain this constant from the recent *Handbook of Chemistry and Physics,* available from CRC press. Look under satellite measurements of sunlight.
- L1 and L2 are the intensities of the two wavelengths during actual ground measurements.
- a1 and a2 are the absorption coefficients for ozone at the two wavelengths.

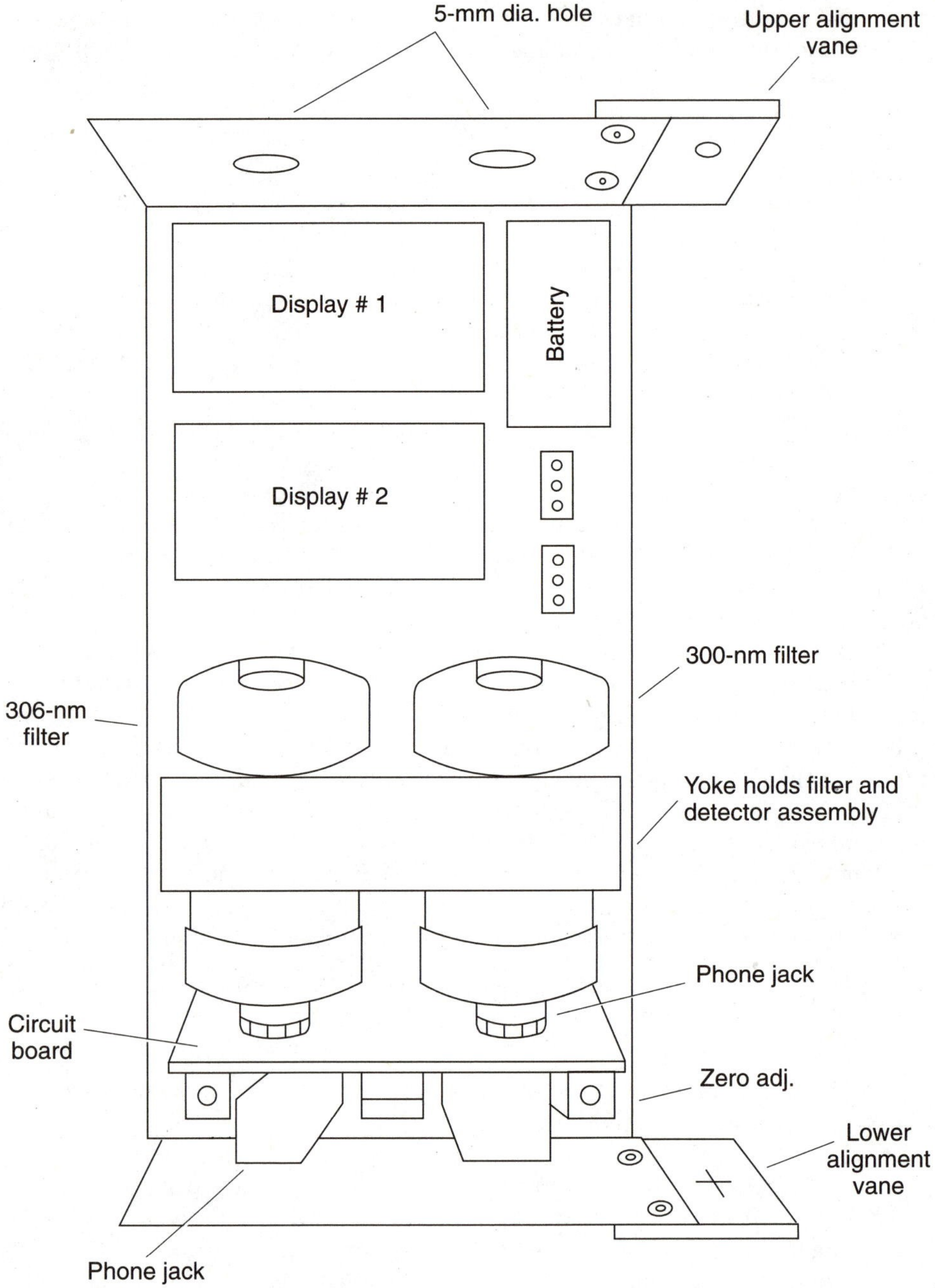

4-19 Digital ozone meter. From *Science Probe Magazine* (Gernsback Specialty Press, Nov. 1992). Forrest Mimms III. Used with permission.

This information for ozone can be found in published tables. See "Absolute Absorption Cross-Sections of Ozone in the 185-350nm Wavelength Range" by L.J. Molina and M.J. Molina in the *Journal of Geophysical Research*, vol. 91, no. D13, December 20 1986, pages 14,501 to 14,508.

- b1 and b2 are the Rayleigh scattering coefficients for air at the two wavelengths for Rayleigh scattering. See the second column of Table 3 in "Tables of Refractive Index from Standard Air and the Rayleigh Scattering Coefficient" by Rudolf Penndorf in the *Journal of the Optical Society of America*, vol. 47, no. 2, February 1957, pages 176 to 181.
- *M* is the air mass (approximately $1/\sin c$, where *c* is the angle of the sun above the horizon. Mount a bubble level on your ozone meter and, when the bubble is centered, measure the length of the shadow cast by the upper vane on the instrument. The tangent of the sun's angle above the horizon is the length of the upper vane's shadow. Whether you measure the sun's angle directly or not, it's very important to record the date and exact time of observation, which will allow you to calculate the sun's angle by computer.
- *P* is the mean barometric pressure of the observation site in mb (inches of mercury × 33.864 = pressure in millibars).

For further reading, please refer to the following articles:

"How to monitor ultra-violet radiation from the sun." Forrest Mimms III. *Scientific American*, August 1990.

"How to measure the ozone layer." Forrest Mimms III. *Probe* magazine, November 1992.

"The effect of bandwidth on filter instrument total ozone accuracy." *Journal of Applied Meteorology*, vol. 16, August 1977.

"Problems in the use of interference filters for spectrophotometric determination of total ozone." *Journal of Applied Meteorology*, vol. 16, August 1997.

Digital UV photometer parts list

PD1	Gallium phosphide G1961 (Hamamatsu Electronics)
R1	1-megohm, ¼-watt resistor
R2	10-megohm, ¼-watt resistor
R3	50-megohm, ¼-watt resistor
R4	100-megohm, ¼-watt resistor
R5	100-kilohm potentiometer
R6,R7	1-megohm, ¼-watt resistors
C1	100-pF, 25-volt mica capacitor
C2	.01-µF, 25-volt disc capacitor
D1,D2,D3	1N914 silicon diodes
S1,S2	SPST toggle switch
S3	Normally open push-button
B1	6-volt battery (4 AA cells)
DISP	Acculex DP-650 LCD voltmeter (+/- 200-mV input)

Digital UV photometer parts list continued

P1	⅛-inch phone plug
J1	⅛-inch phone jack
Filters	For locations between 29 and 35 degrees, use 300- and 306-nm filters; above 35 degrees latitude, use 305 to 310 nm for the short wavelength and 325 to 330 nm for the long wavelength.

Note that UV interface filters are available from Twardy Technology Inc. or MicroCoatings (see the Appendix).

5
CHAPTER
Optical sensors

This chapter will present many different types of optical sensors and sensing systems. I will describe simple optical sensors, from position sensors to optical counters, optical rotary encoders, and a phototachometer. One of my favorite sensors is the position-sensitive detector, which can be used in a host of interesting and useful applications, from laser tracking and laser "spot" following to triangulation sensors. Also presented in this chapter are a turbidity meter and a light-to-frequency converter and computer interface. Later in this chapter I will discuss more complex projects, such as an IR voltage-to-frequency and frequency-to-voltage data converter link and a light-wave temperature link. Then you can construct the Hunter's Companion, a tracking device for locating large downed animals. Finally, a long-range heat/cold detector and an experimental lightning detector close the chapter.

A *sensor* is any device that converts some physical attribute into an electrical or electronic signal. An optical sensor uses solar cells, cadmium-resistive cells, photoresistors, or optical diodes to detect a physical attribute via infrared, visible, or ultraviolet light. A number of optical sensors are shown in Photo 5-1. Sensing systems usually consist of three components. The first is the actual sensor. This chapter examines light sensors, but there are a variety of different kinds of sensors for detecting information such as heat, pressure, and acceleration. The second component of a sensing system is an amplification stage, which amplifies and/or conditions an often low-level sensing signal. The third component is the display or control portion of the system. This could be a meter (analog or digital) or an analog-to-digital board in a personal computer. In sensing systems, a sensor is often used to control devices such as lights, motors, or fans and the control can be either a discrete control circuit or a computer.

Optical slot window/door sensor

The door/window intrusion sensor depicted in Figure 5-1 can be used in a variety of ways. It could be used to detect the opening of windows, doors, or drawers, or as an intrusion detector in an alarm system. You could also modify the circuit to act

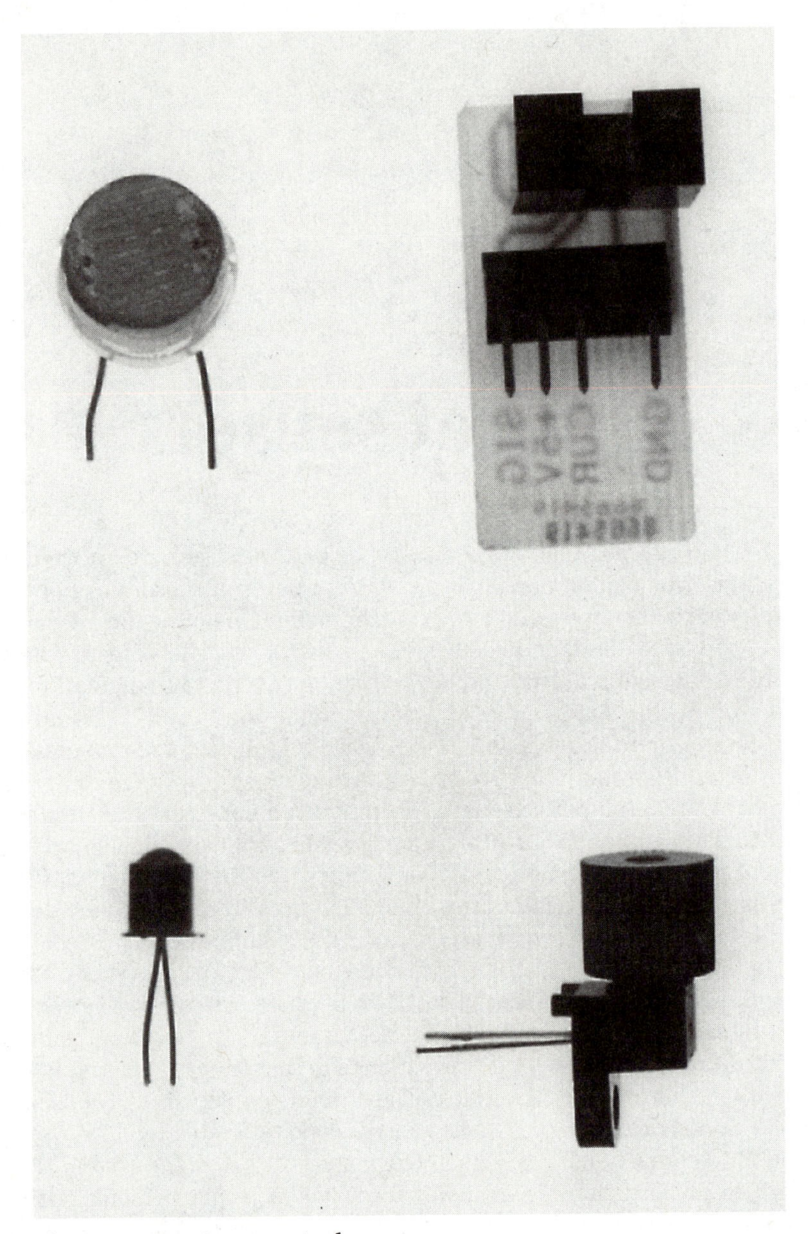

Photo 5-1 Various optical sensors.

as a rotary encoder. The heart of the door/window sensor circuit is two 555 IC timer chips. The first timer/oscillator at U1 consists of a 555 configured to operate as a free-running oscillator that drives the LED, as shown. The second 555 timer is connected as a monostable multivibrator that functions as a missing pulse detector. When the slot in the optoisolator is blocked, pulses from the LED cannot reach the phototransistor, and the output from the missing pulse detector U2 is low. When the slot is open or when light from the LED is allowed to strike Q1, the slot begins its timing cycle and its output goes high until the timing cycle is completed. When the output of U2 is high, the piezo sounder is activated. The timing cycle of the missing pulse detector at U2 is controlled by R5 and C2.

If the timing cycle is longer than the interval between incoming pulses from the LED, the output from U2 will stay high until a new pulse arrives. Therefore, the piezo sounder will emit a continuous alarm tone. If, on the other hand, the timing cycle is shorter than the interval between pulses, the missing pulse detector will complete its cycle before the next pulse arrives. This will cause the piezo sounder to emit a pulsating tone. The intrusion detector uses an opaque flag attached to a bracket on a door/window or desk drawer. When the door or drawer is opened, the flag moves from the optoisolator and the alarm sounds. You could use a number of these optical sensors around your home or office to protect it from intruders. You could also replace the piezo buzzer with a small relay, if desired. The intrusion detector can be powered with any 5- to 15-volt power source.

Optical slot door/window sensor parts list

R1	47-kilohm, ¼-watt resistor
R2	1-kilohm, ¼-watt resistor
R3	220-ohm, ¼-watt resistor
R4	10-kilohm, ¼-watt resistor
R5	1-megohm, ¼-watt resistor
C1	4.7-µF, 25-volt electrolytic capacitor

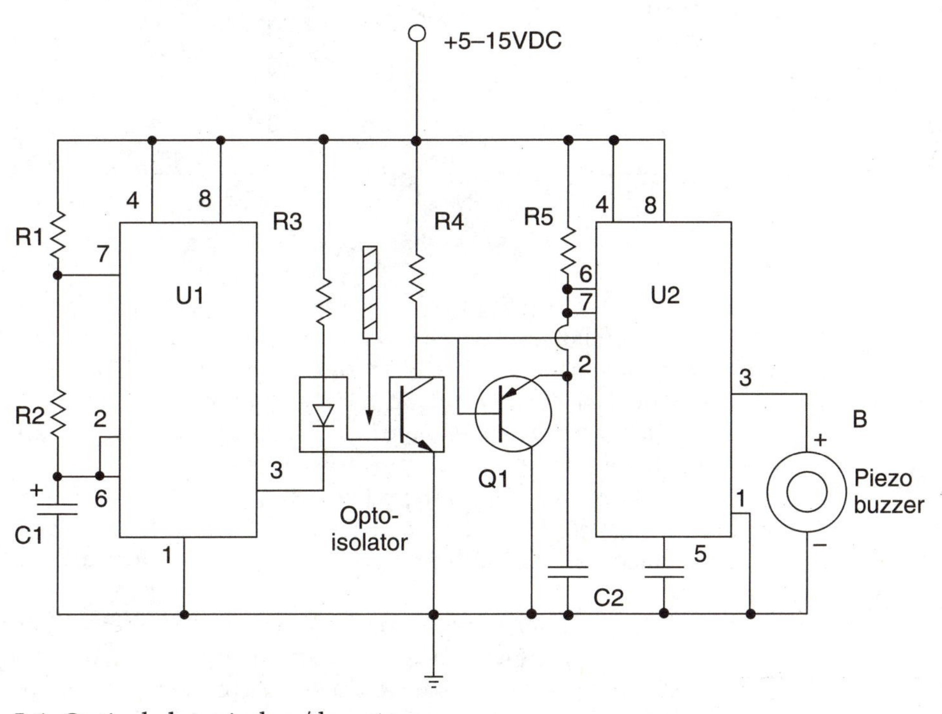

5-1 Optical slot window/door sensor. From Forrest Mimms' *Circuit Scrapbook* (Howard W. Sams & Co., 1987). Copyright by Forrest M. Mimms III. Used with permission.

Optical slot door/window sensor parts list continued

C2	.047-µF, 25-volt disc capacitor
C3	.01-µF, 25-volt disc capacitor
Q1	2N3906 transistor
U1,U2	LM555 IC timer
Opto	13B1 optoisolator (GE)

Film-strip position sensor

The film-strip position sensor illustrated in Figures 5-2, 5-3, and 5-4 illustrates a low-cost sensor you can use to make measurements on devices that will not tolerate loading or drag. Most sensors, such as linear variable differential transformers (LVDTs) and resistive encoders, inherently cause loading. The film-strip position sensor is small enough to be placed in very tiny places where other types of sensors would be simply too large. Possible applications are micropositioners and microscope stages.

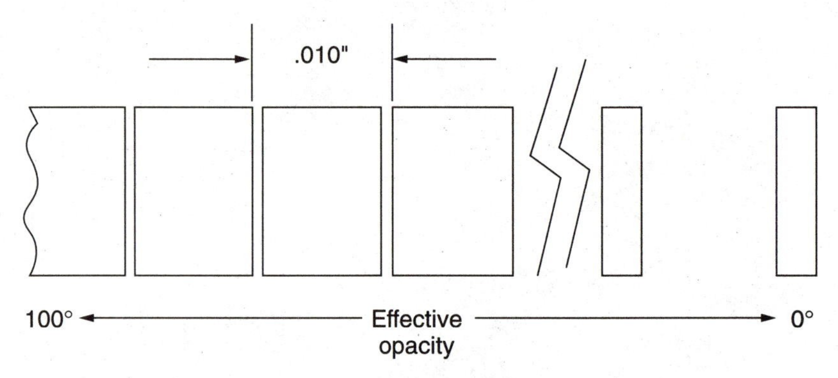

5-2 Opacity film strip.

The principle behind the film-strip position sensor is quite simple. A film strip, which varies linearly in opacity along its length, is attached to a moving part and positioned so it modulates a beam of light in a slotted optical switch or optoisolator. Generally, it is quite difficult to construct a film strip with precise linearly varying density. A simple solution to this problem consists of a pattern of opaque bars spaced on .01-inch centers (see Figure 5-2). Varying the width of the spacing from 0 to 100 degrees from one end of the strip to the other essentially creates an optical pulse-width modulated output. The .01-inch spacing appears to be adequate for this type of detector since the LED and phototransistor die or substrata are about .05 inch apart.

The film-strip pattern can be either manually drawn or computer-generated on paper or Mylar, and then reduced photographically. A high-quality image is not required to operated this sensor. Figure 5-3 illustrates placement of the film strip in a typical optical-switch assembly. The width of the slot is .01 inch, so mechanical tol-

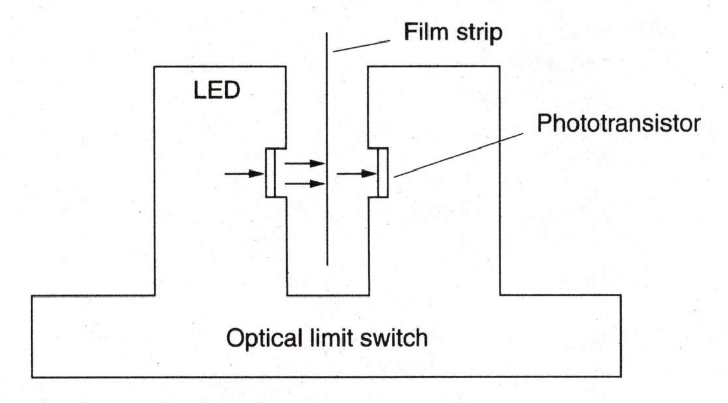

5-3 Optical film-strip switch.

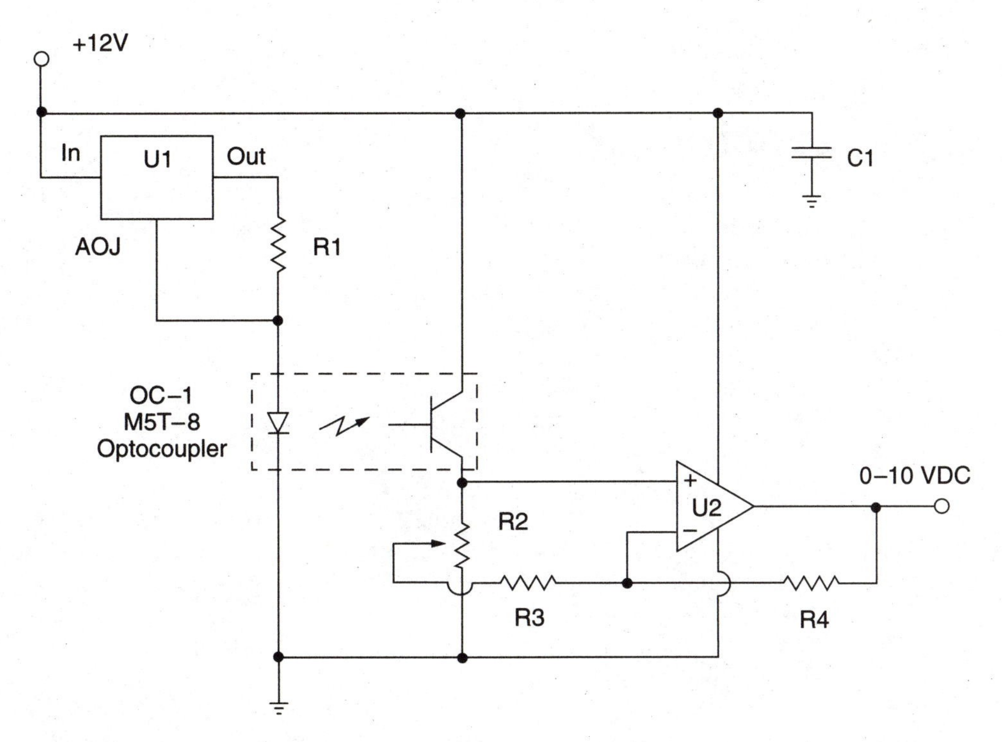

5-4 Film-strip interface.

erance is not crucial. The electronic interface for the film-strip position sensor is shown in Figure 5-4.

The LM317L voltage regulator provides a constant 20 mA of current to the LED in the optical switch. At this level, the full-scale current of the phototransistor might vary from .2 to 1 mA. The 5-kilohm potentiometer is adjusted to provide a 1-volt full-scale output. To obtain the best linearity in the system, keep the voltage across the

5-kilohm potentiometer to a low value. The voltage from the 5-kilohm potentiometer is amplified by the op-amp to provide a full 0 to 10 volts of dc output. The gain of the op-amp is set by the ratio of Ra and Rb. The output of the film-strip position sensor can be fed to a digital panel meter, an A/D card in a personal computer, or a dedicated microcontroller.

Film-strip position sensor parts list

R1	62-ohm, ¼-watt resistor
R2	5-kilohm potentiometer (trimmer)
Ra	1-kilohm, ¼-watt resistor
Rb	9.1-kilohm, ¼-watt resistor
U1	LM317L voltage regulator
U2	LM358 op-amp
C1	.1-µF, 25-volt disc capacitor
OC-1	M5T-8 optocoupler
FS	Film strip (see text)

Optical rotary encoder

The optical rotary encoder is ideal for digital panel controls or in position-sensing applications where long life, reliability, high resolution, and precise linearity are crucial. Optical rotary encoders are often used in computer- aided design and computer-aided manufacturing (CAD/CAM), counting circuits, machine tools, speed controls, and other positioning applications.

The Bourns EN series optical encoder, shown in Figure 5-5, is a 1-inch-square device that converts rotary input into 5-volt electrical pulses or signals of both counts and direction of rotation. The electrical pulses can be further processed via a microprocessor or a counting card placed in a personal computer.

Optical rotary encoders (OREs) combine an LED light source, a photointerruptor disk, and an IC light receiver consisting of a phototransistor or photodiode and an amplifier/signal conditioner. Light passing through the slots in the interrupter is sensed and translated into digital pulses by the phototransistor detector inside the ORE. The ORE has no contacts or wipers that can wear out or create electrical noise, and they can be operated at high speeds if desired. The EN series optical encoders use a standard ⅛-inch shaft for easy implementation. Rotary encoders generally provide a *quadrature*, or two-channel output, as shown in Figure 5-6. ORE outputs are compatible with both CMOS and TTL systems. Up to 256 quadrature output cycles per shaft revolution are provided with the EN series encoders. Quadrature output allows the ORE to determine count pulses and direction information, with the addition of the electronic circuit shown in Figure 5-7. Photo 5-2 illustrates a speed/direction counter circuit, which could be used with both optical rotary encoders or with the bifurcated light pipes shown.

The outputs from the optical rotary encoder are 5-volt dc signals, coupled to two Schmitt triggers that are followed by three Nand gates. Nand gates U2A and U2B feed

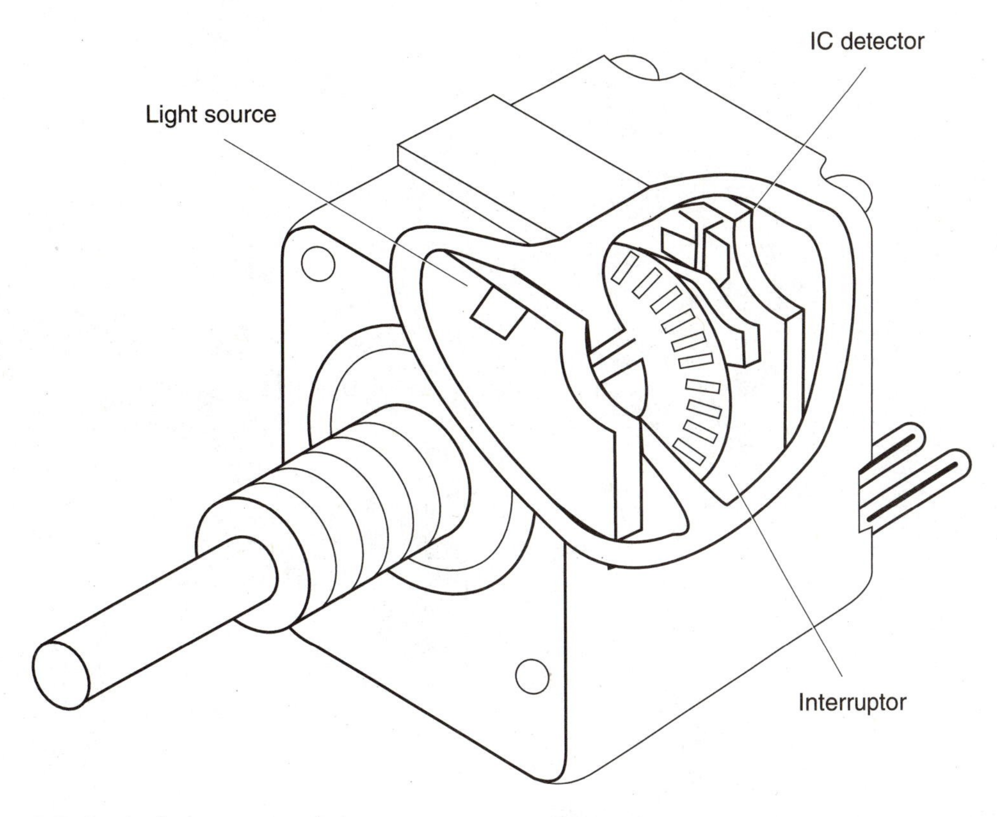

5-5 Optical rotary encoder. Bourns, Inc.

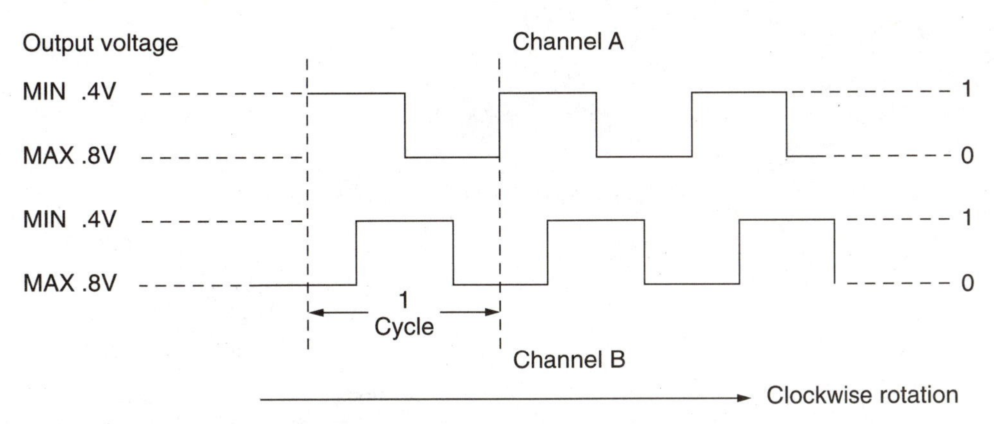

5-6 Quadratic output logic diagram. Bourns, Inc.

a 7474 flip-flop, which provides directional output. Nand gate U2C of the steering logic circuit is coupled to a Schmitt inverter at U1:A and then on to a series of inverters that provides a short-term delay to the counting pulse output. The output from the 74L504 is then available to provide counting pulses. The resulting 5-volt pulses for direction and speed can be used directly to drive an I/O card in a computer.

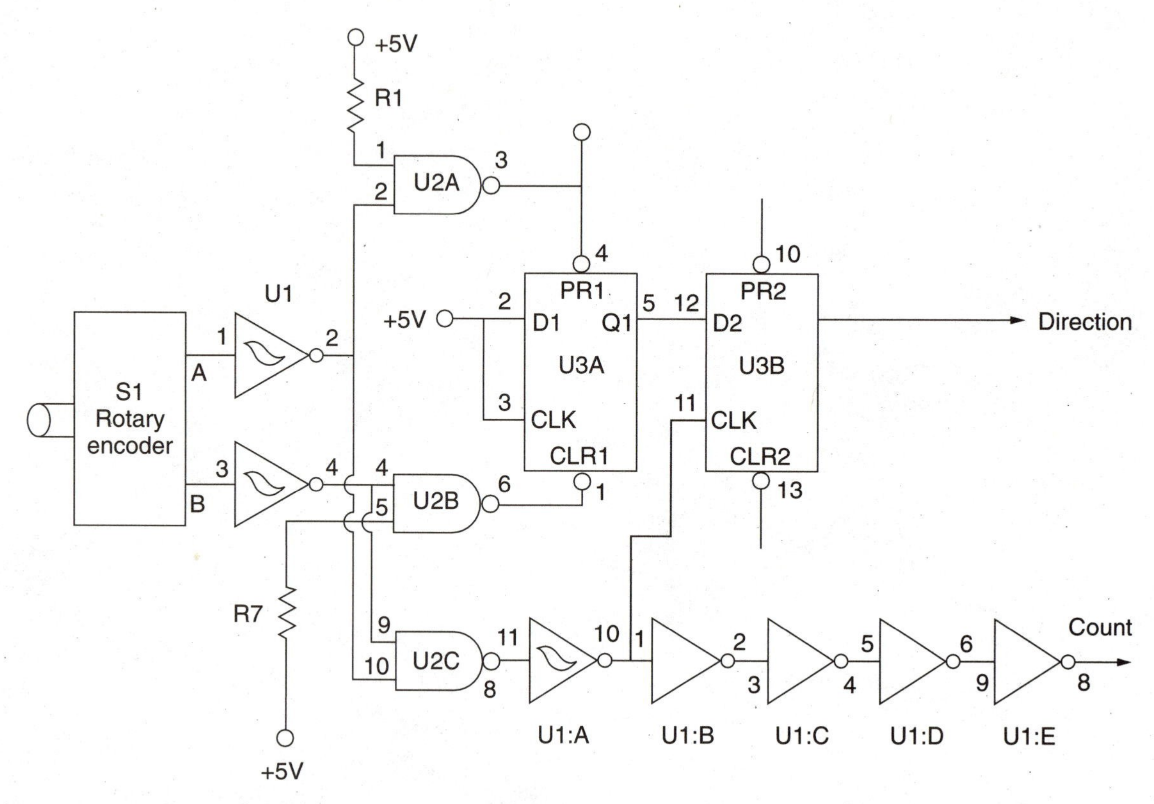

5-7 Speed/direction interface. Bourns, Inc.

Special counting cards are available that directly accept the outputs of the rotary encoder and process the speed and direction of multiple encoders via a personal computer, thus eliminating the interface. Optical rotary encoders are available in various mounting configurations and output resolutions.

Optical rotary encoder parts list

R1,R2	4.7-kilohm, ¼-watt resistor
U1	7414 Schmitt trigger IC
U2	7400 quadrature Nand gate
U3	7474 flip-flop
U4	7404 hex inverter
S1	Optical rotary encoder

Bifurcated light-pipe interface

You can also implement counting and position sensing by using a bifurcated fiber-optic light pipe, shown in Figure 5-8 and Photo 5-3. Two polished fiber-optic light pipes are fused together at a single junction of two light pipes. A single emerging light pipe is then used as a sensing head, which can count very small parts, cali-

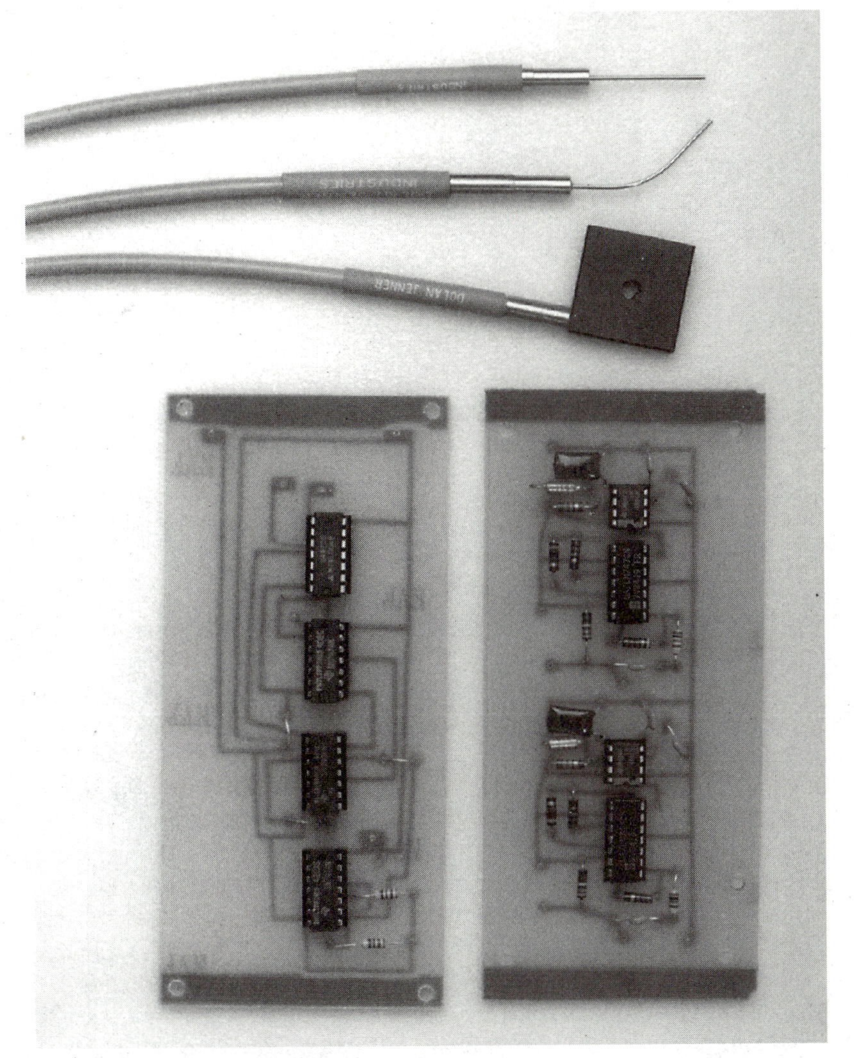

Photo 5-2 Optical counting circuit.

brated markings, or rulings, such as microscope gradations. You could also use a bifurcated sensor to monitor both linear and rotary stages or tables.

An infrared LED is placed at one end of the two light pipes, as shown. An IR phototransistor is placed at the opposite end of the Y-shaped light pipe. The sensing end of a bifurcated light is available in straight or curved sensing heads, which can fit into the tiniest of places. The sensing head is available in many forms and diameters.

The output of the phototransistor is coupled to a light-detector interface, shown in Figure 5-9. The output of the phototransistor feeds an LM741 op-amp, which amplifies the phototransistor's output signal. The output of the op-amp is then passed onto an LM555 timer IC, which provides a precise 5-volt conditioned pulse. The output

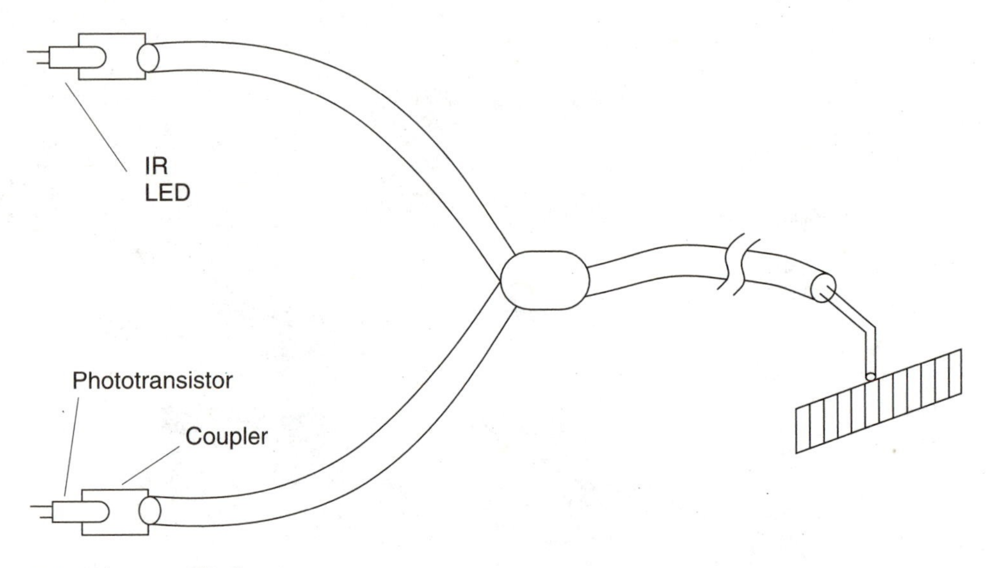

5-8 Bifurcated light pipe.

of this pulse shaper can then be sent to an I/O card in a personal computer or to a dedicated microprocessor.

So far you have a means to provide counts, but a second light-pipe assembly is required to provide direction. The two light assemblies should be placed next to each other, as shown in Figure 5-10.

One sensing head should be placed or aligned on an opaque ruled line, while the second light-pipe sensing head assembly should be aligned precisely in between the next ruling or marking, as shown. By using two light-pipe assemblies, you have in essence created a discrete optical rotary encoder, which can sense direction and provide counts. The outputs from the two pulse-shaper circuits can then be fed to the steering logic circuit, shown earlier, to achieve both direction and count information. The bifurcated light-pipe counter/direction-sensing heads can be placed in very tight places and are ideal for micropositioning or optical microscope stages. See the Appendix for information on obtaining bifurcated light-pipe assemblies.

Bifurcated light-pipe interface parts list

R1,R2	100-kilohm, ¼-watt resistor
R3	4.7-kilohm, ¼-watt resistor
C1	.1-µF, 25-volt disc capacitor
C2	470-pF, 25-volt disc capacitor
C3	.01-µF, 25-volt disc capacitor
Q1	MRD300 phototransistor (Motorola)
U1	LM741 op-amp
U2	LM555 IC timer

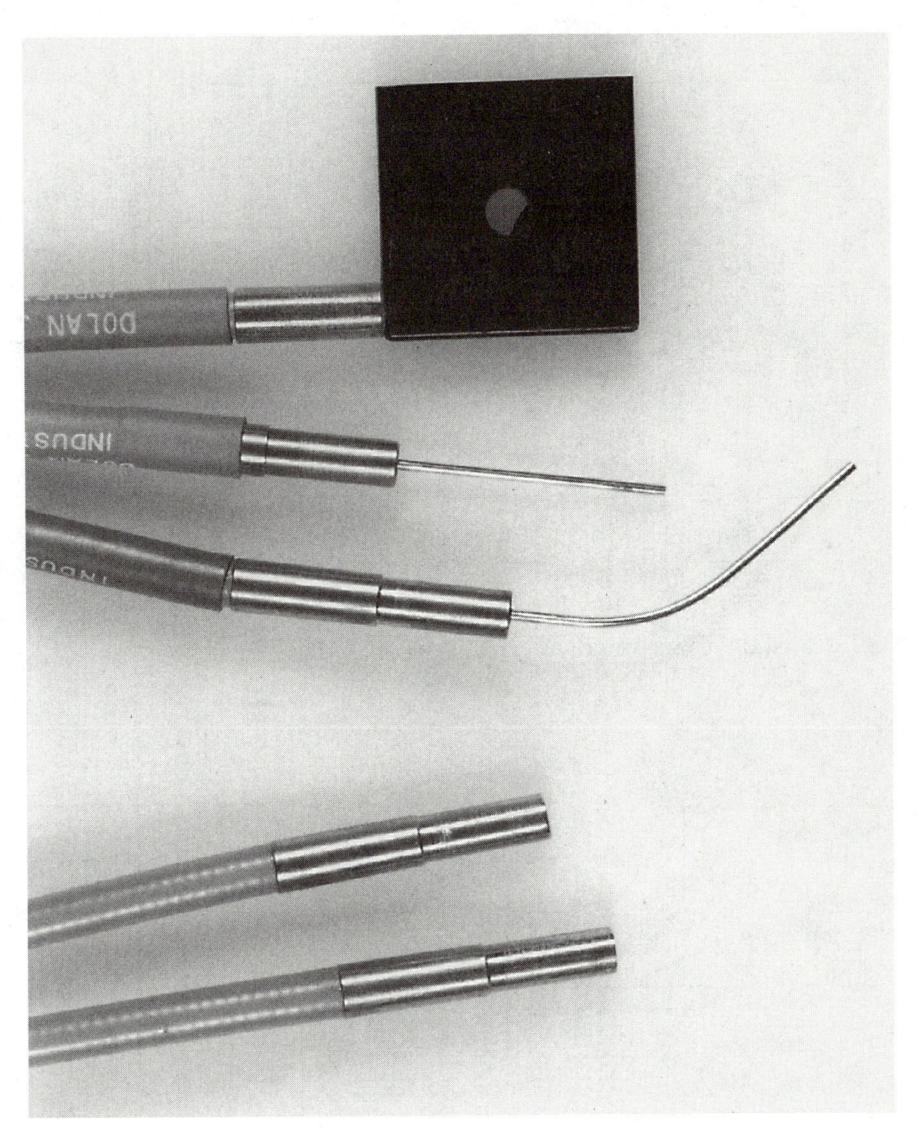

Photo 5-3 Bifurcated light-pipe heads.

Position-sensitive detector

The silicon position-sensitive detector, or PSD, is a relatively new optoelectronic sensor that can provide continuous position data of a light spot traveling over its light-sensitive surface. The PSD is one of my favorite optical sensing devices; it can be used for optical position-sensing devices, angle sensing, laser spot following, remote control, feedback, and cosmic particle detectors. Position-sensitive detectors have also been used for displacement sensing and auto range-finding applications.

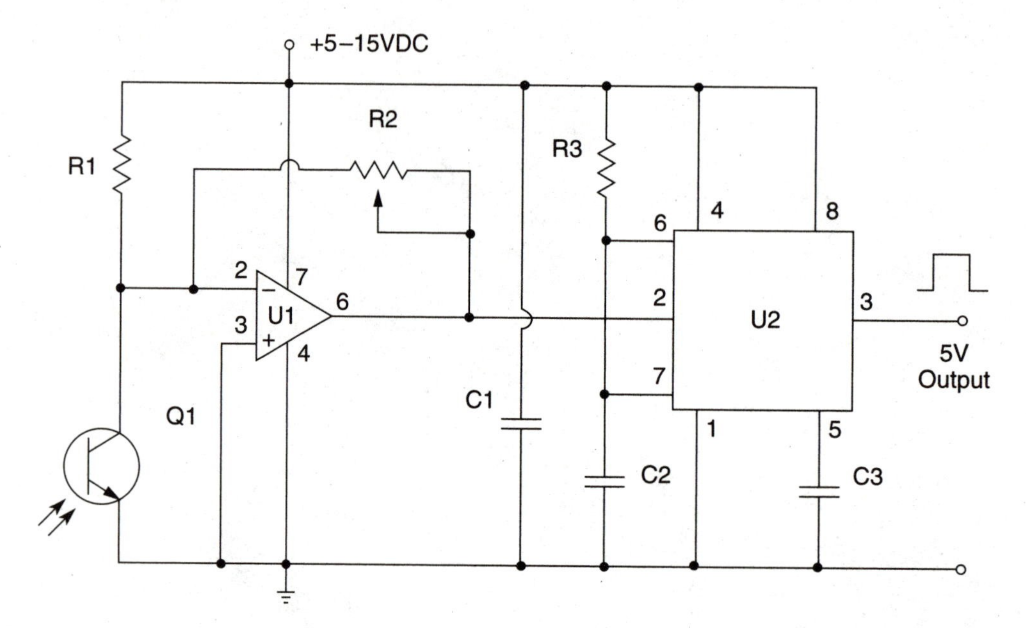

5-9 Bifurcated light-pipe interface.

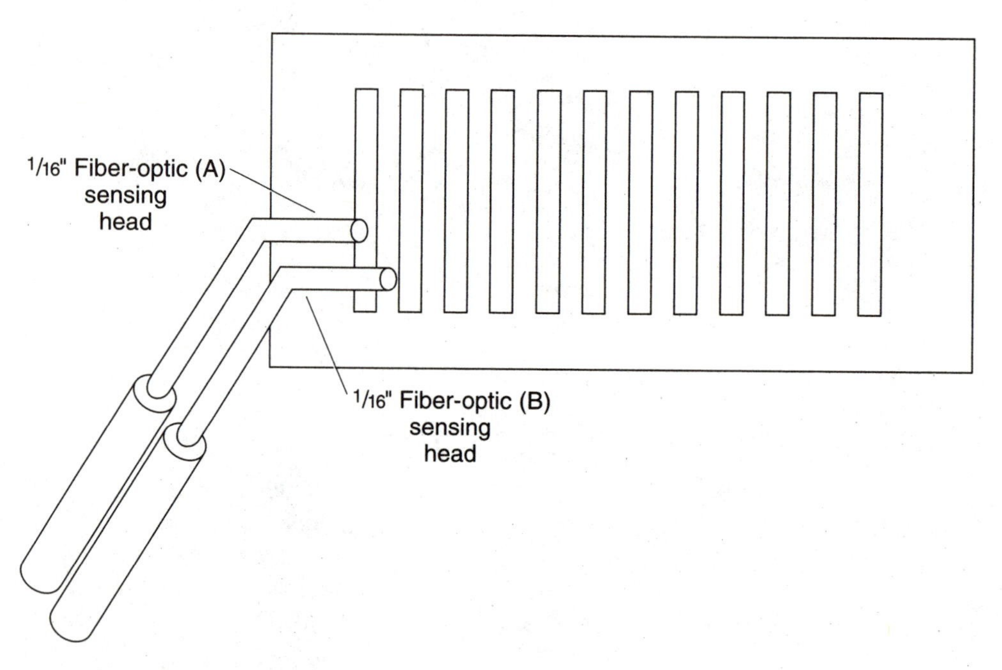

5-10 Dual light-pipe sensing.

Major features of the PSD include high position resolution, wide spectral response, high-speed operation, and good reliability.

The three-lead silicon PSD consists of monolithic photo PIN diodes with either one or two uniform resistive surfaces. PSDs come in many different configurations and sizes, and are available in one- and two-dimensional outputs for precise x and y position indication. Photo 5-4 displays various types of position-sensitive detectors.

The diagram in Figure 5-11 illustrates the internal structure of a typical PSD device. The PSD consists of three separate layers; the first layer, the P layer, is placed on top of the stack, with an I or current layer sandwiched in between the P layer and the lower N layer. Light striking the PSD is converted photoelectrically and detected by two electrodes at the P layer. When a spot of light energy falls on the PSD, an electrical charge proportional to the light energy is generated at the incident position. The electrical charge is driven through the resistive P layer and collected by the electrodes. Because the resistive portion of the P layer is uniform, the photocurrent collected by an electrode is inversely proportional to the distance between the light-spot position and the electrode.

The PSD shown is a one-dimensional device, but two-dimensional, two-layer PSDs are available that will display both x and y position information. Essentially, the x-y, two-dimension PSD is a two-layer device with the addition of a dc restoration stage between the final op-amp U5 and the analog divider.

The electronics interface for the PSD is shown in Figure 5-12 and Photo 5-5. A bias voltage is applied to the center lead of the PSD or N-layer electrode. The two active electrodes are directed to an an LF351 op-amp. Resistors R1 through R9 are 10-kilohm, 1-percent resistors. Resistors R1 and R2 on the op-amp inputs are used for

Photo 5-4 Position-sensitive detectors.

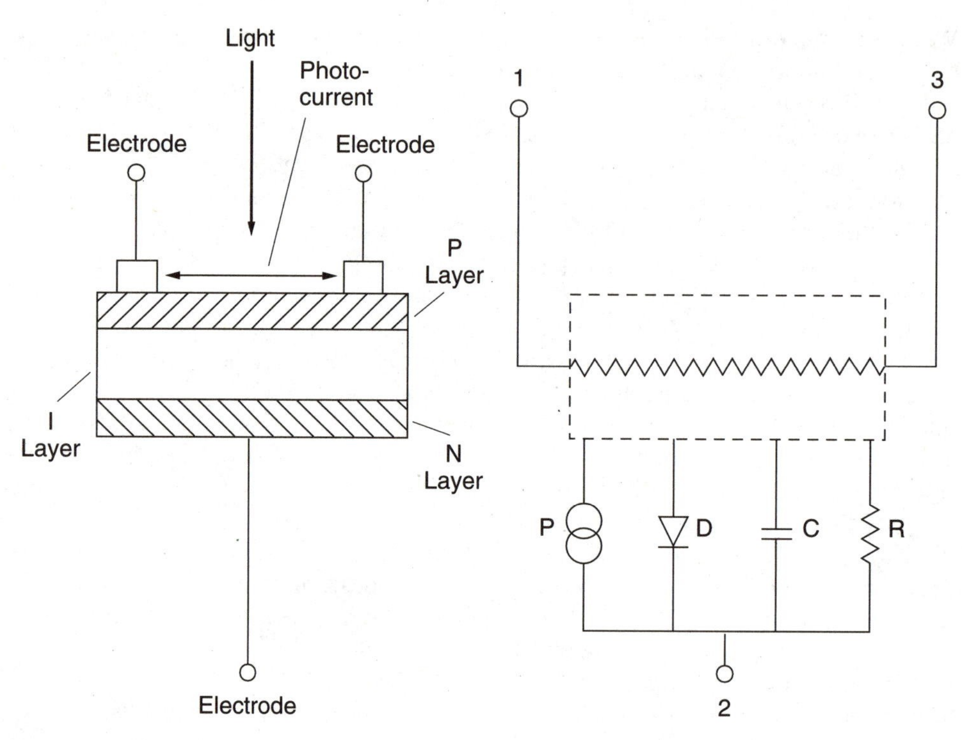

5-11 Position-sensitive detector (PSD).

range or level adjustment. The outputs of the differential amplifiers at U1A and U1B are fed into stages U2A and U2B. The output of U2A is coupled to U3, which in turn is fed to the summation input of an AD532, an analog divider chip. The output from U2B is connected to the current input on pin 13 of the AD532. Offset potentiometers are connected to pins 9, 12, and 11 of the AD532.

The final output stage of the analog divider is presented on pin 2 of the AD532. This output voltage varies from 0 to 10 volts dc, relative to the position of the light spot on the surface of the PSD sensor.

The position voltage can be coupled to an A/D card in a personal desktop computer as a single-channel input; if a multichannel A/D card is used, you could monitor a number of position sensors all at once. Another approach would be to eliminate the voltage divider chip and separately connect both the ΣI and ΔI inputs directly into two separate A/D card inputs. Some simple software would have to be written to accomplish the division. The only drawback to this second approach is that it occupies two channels of an A/D card for a single PSD device. Using two-dimensional *x-y* PSD detectors and a computer, you can keep track of a number of *x-y* positioners at once for manufacturing applications.

PSD parts list

R1,R2	10- to 50-kilohm, input-level, 1-percent, ¼-watt resistor
R3,R4,R5,R6	10-kilohm, 1-percent, ¼-watt resistor
R7,R8,R9,R10,R11	10-kilohm, 1-percent, ¼-watt resistor

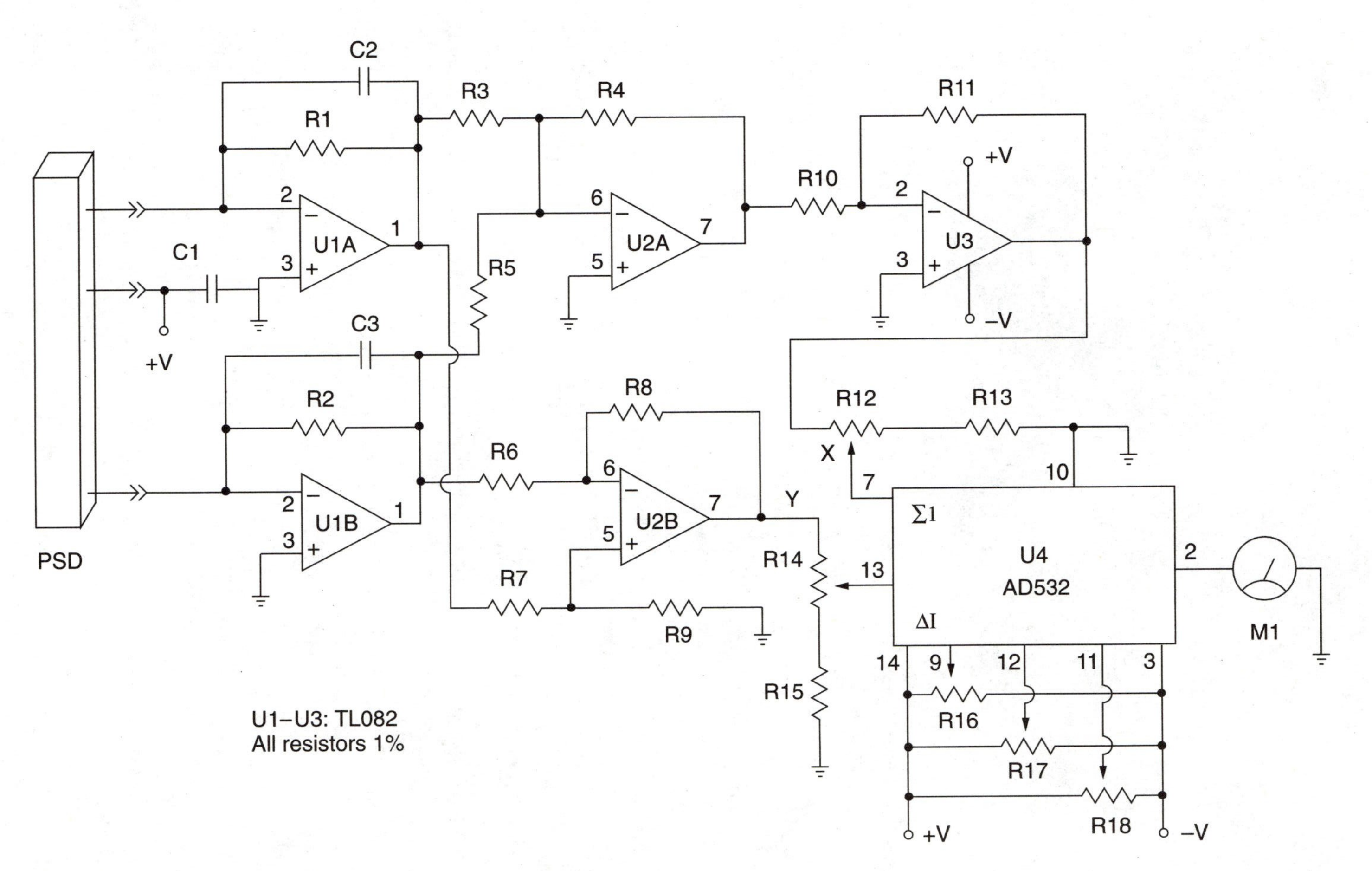

5-12 Position-sensitive detector electronics interface. Hamamatsu, Inc.

Photo 5-5 Position-sensitive detector and circuit.

PSD parts list continued

R12,R14	5-kilohm trim potentiometer
R13,R15	7.5-kilohm, ¼-watt resistor
R16,R17,R18	20-kilohm trim potentiometer
C1	4.7-µF, 25-volt electrolytic capacitor
C2,C3	50-pF, 25-volt disc capacitor
U1,U2,U3	TL-082 op-amp (Texas Instruments)
U4	AD532 (Analog Devices)
PSD	Position-sensitive detector (Hamamatsu)

Optical tachometer

You can determine the speed of almost any rotating object quite easily by using the noncontact or reflective tachometer shown in Figure 5-13. Rotating shafts, fan blades, and tires can be quickly scanned to determine actual rotating speeds. You can use this straightforward optical or phototachometer to observe many rotating speeds, from 2500 to 50,000 rpm.

Light pulses striking the phototransistor Q1 produce voltage pulses at the input of U1, a 741 op-amp. Op-amp U1 is connected as a Schmitt trigger device. The output pulses on pin 6 are first differentiated by C6 and R6. The resultant voltage spikes are then coupled to U2, an LM555 timer IC. The 555 is configured as a one-shot, passing its output on pin 3 to diode D1. The output from D1 feeds Q2 and energizes the FET. The FET provides a constant current source to produce pulses with constant amplitude at R7. The averaged output pulses are then coupled through R18 and then onto a 50-mA meter. Capacitor C11 is added into the circuit at low rpms to dampen the meter movement. Speed determination or calibration is performed at pin 6 of the LM555. A five-position range switch inserts one of five resistors that, when combined with C7, determine U2's speed calibration.

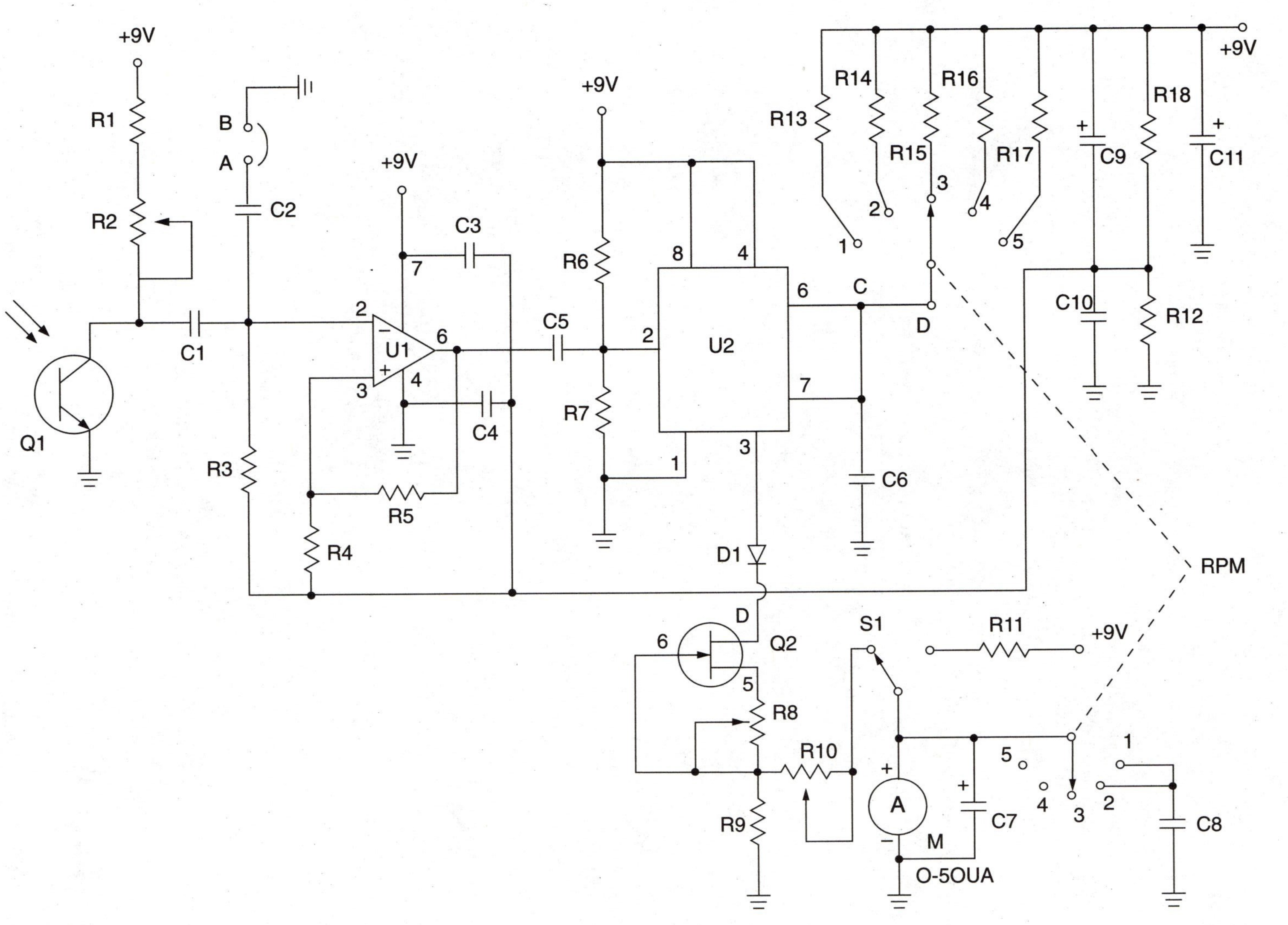

5-13 Optical tachometer.

To calibrate the phototachometer, R17 and R18 are first adjusted to mid-position and the range switch set to 2500 rpm. After disconnecting the wire between points C and D in the schematic, R17 is adjusted so the voltmeter reads 1 volt. The C-D wire is now reconnected and the range switch set to 10,000 rpm. A 3-volt peak 120-Hz sine wave is now applied to points A and B, equal to a 7200-rpm count.

Finally, you must check for low-level 120-Hz modulation rejection of incandescent light sources by aiming the phototransistor at a 50- to 75-watt bulb while varying the sensitivity control R16 over its range. If the meter does not remain set at zero during this test, then increase the input hysteresis by changing R3 to 10 kilohms.

The phototachometer circuit can be powered by a 9-volt transistor radio battery, which will last for a long time with average use. The phototachometer is a very useful item for any optical experimenter.

Optical tachometer parts list

R1,R12,R18	3.9-kilohm, ¼-watt resistor
R2	100-kilohm potentiometer (trim)
R3	150-kilohm, ¼-watt resistor
R4	5.1-kilohm, ¼-watt resistor
R5,R13	100-kilohm, ¼-watt resistor
R6,R7	47-kilohm, ¼-watt resistor
R8	5-kilohm potentiometer (trim)
R9	1-kilohm, ¼-watt resistor
R10	10-kilohm potentiometer
R11	200-kilohm, ¼-watt resistor
R14	50-kilohm, ¼-watt resistor
R15	25- kilohm, ¼-watt resistor
R16	10-kilohm, ¼-watt resistor
R17	5-kilohm, ¼-watt resistor
C1	.002-µF, 25-volt disc capacitor
C2	.05-µF, 25-volt disc capacitor
C3,C4	.1-µF, 25-volt disc capacitor
C5	.001-µF, 25- volt disc capacitor
C6	.068-µF, 25-volt disc capacitor
C7,C9,C10	20-µF, 25-volt electrolytic capacitor
C8,C11	100-µF, 25-volt electrolytic capacitor
Q1	HEP-P0001 NPN phototransistor or equivalent
Q2	HEP-P0010 FET or equivalent
U1	LM741CN op-amp
U2	LM555 timer
D1	1N91 diode
S1	SPDT switch
S2	Dual-deck, five-position rotary switch
M	0- to 50-mA meter

Turbidity meter

An application often arises when you need to measure the relative opacity of a liquid such as water. The clever IR receiver/display circuit shown in Figures 5-14 and

5-15 is being used as a turbidity meter. A turbidity meter measures the opacity of a medium such as water through a conduit or container.

The diagram in Figure 5-14 depicts the turbidity meter circuit. In operation, an LM567 tone decoder at U1 generates a modulated signal that drives LED1. The frequency of the modulation produced at IR LED1 is determined by potentiometer R6. Transistor Q3 responds to the modulated IR beam from LED1. Transistor Q1 amplifies the ac component of the infrared beam from LED1, while transistor Q2 is used to drive the 0- to 10-volt dc meter, which displays the relative strength of the light beam sent through a medium such as water, as shown in Figure 5-15. Light from the LED is allowed to pass through the container, as shown. At the opposite side of the liquid container, the phototransistor is aimed to accept light from the LED. For best results, the IR LED and the phototransistor should be collimated using a darkened tube. A strong beam of light passed through the medium will produce a lower meter reading, while a lower level of light through the medium will produce a higher meter reading. If you anticipate longer-range applications, then use a lens arrangement at both the LED and phototransistor. The turbidity meter can be powered with a 9-volt battery or a dc wall-cube power supply.

The turbidity meter unit can be readily adapted to a host of other applications where a continuous voltage display is required. You could also use the output of the turbidity meter to drive an analog-to-digital converter card placed in a personal computer by removing the meter and loading the output with a resistor at the emitter of Q2.

Turbidity meter parts list

R1	47-kilohm, ¼-watt resistor
R2	2.2-kilohm, ¼-watt resistor
R3	220-kilohm, ¼-watt resistor
R4,R6	1-kilohm, ¼-watt resistor
R5	25-kilohm potentiometer
C1,C5,C6,C7	.01-µF, 25- volt disc capacitor
C2	.02-µF, 25-volt disc capacitor
C3	680-pF, 25-volt Mylar capacitor
C4	.047-µF, 25-volt disc capacitor
C8	47-µF, 25-volt electrolytic capacitor
Q1	IR phototransistor (RS 276-142)
Q2,Q3	2N3904 transistor
U1	LM567CN tone decoder
L1	3- to 10-MHz coil
M1	0- to 10-volt dc meter
B1	9-volt battery

Light-to-frequency converter

Most light-sensing elements convert light to an analog signal in the form of a current or voltage, and must be further amplified and converted to a digital signal in order to perform sampling functions. Important considerations in this conversion process are dynamic range resolution, linearity, and noise.

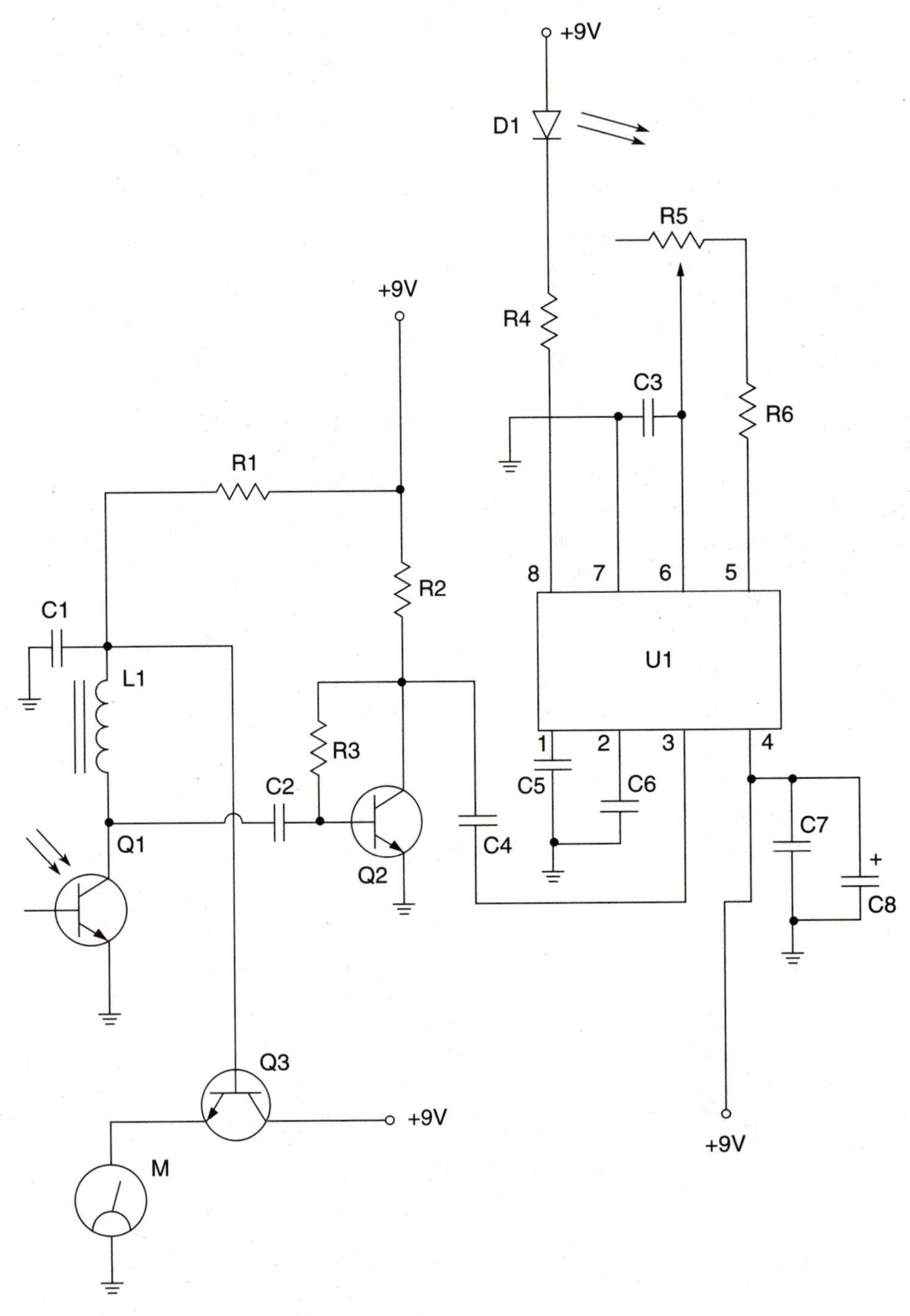

5-14 Turbidity meter circuit.

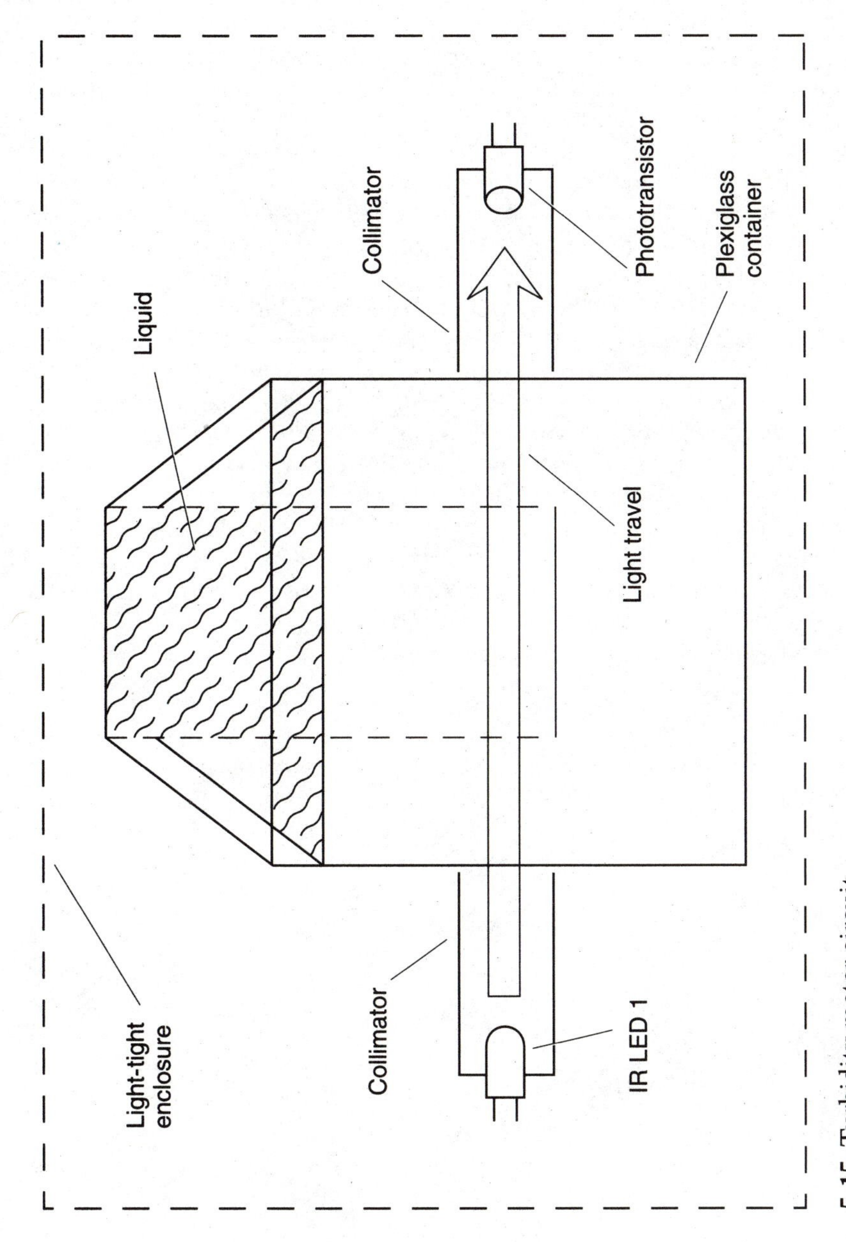

5-15 Turbidity meter circuit.

A modern-day solution to the problem of light-intensity conversion and measurement that provides many additional benefits over discrete conversion techniques is converting light intensity to frequency directly. Light intensity can vary over many orders of magnitude, which further complicates the problem of maintaining resolution and signal-to-noise ratio over a wide input range. Converting from light intensity to frequency all on one chip overcomes many limitations imposed on dynamic range, and noise and analog-to-digital resolution problems are solved by using the frequency conversion method.

Enter the TSL-230 integrated light-to-frequency converter shown in Figure 5-16. Since the conversion of light to frequency is performed "on chip," the effects of external interference leakage currents and noise are greatly minimized and the frequency output from the chip can be easily transmitted to a remote location. Since the frequency output is in the form of data, the interface requirements can be minimized to a single microcontroller port, counter input, or interrupt line.

The TSL-230 light-to-frequency converter chip consists of a 10 × 10 photodiode matrix and a current-to-frequency converter. The photodiode elements produce a photocurrent proportional to the incident or incoming light. Light-sensitivity input can be controlled by S0 and S1, which allows for 1, 10, and 100 ×, thus providing two decades of adjustment. The current-to-frequency converter uses a unique switched capacitor charge metering circuit to convert the photocurrent to a frequency output. The output of the current-to-frequency converter is a train of pulses that feeds the input of an adjustable output scaler. The output scaling can be set via control lines S2 and S3 (see Table 5-1). The control lines allow a direct output in the divide-by-one mode as well as division of the converter frequency by 2, 10, or 100, which results in a 50-percent duty cycle square wave output.

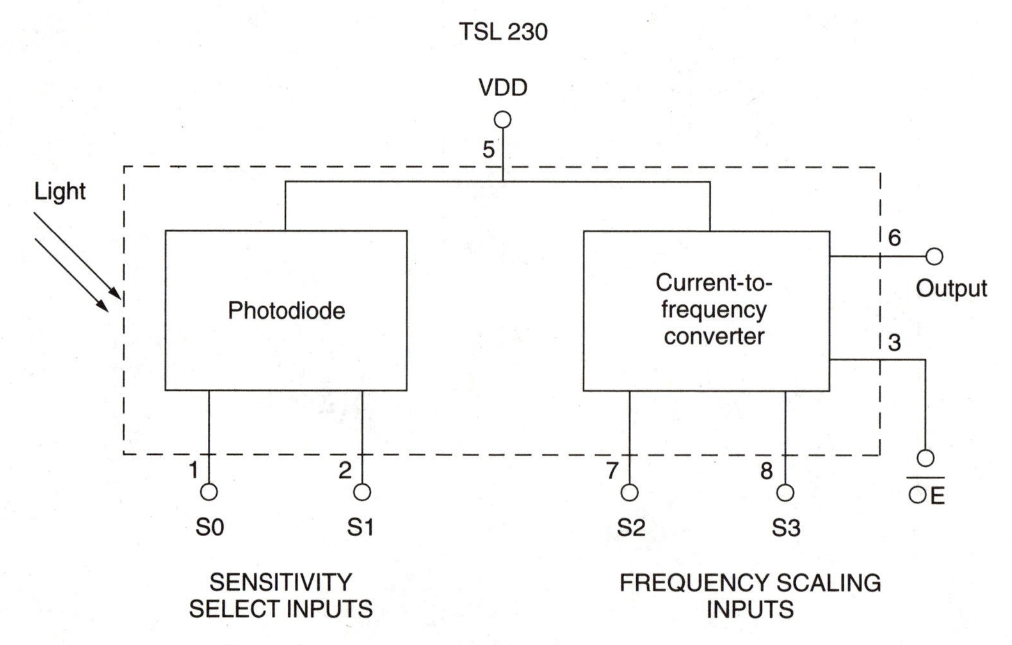

5-16 TSL-230 light-to-frequency converter. Reprinted by permission from Texas Instruments.

Table 5-1 TSL-230 Light-to-frequency converter (sensitivity and scaling).

S1	S0	Sensitivity
L	L	power down
L	H	1 ×
H	L	10 ×
H	H	100 ×
S3	**S2**	**FO scaling**
L	L	1
L	H	2
H	L	10
H	H	100

The TSL-230 light-to-frequency converter is designed to directly interface to logic level inputs. Circuitry at the output stage has been added to limit pulse rise and fall times, which helps to lower electromagnetic radiation. A buffer or line driver is recommended for driving output lines longer than several inches from the TSL-230 chip. An active output enable (OE) line is provided on the chip. When the OE line is high, the output is in a high-impedance state. The OE is useful when several TSL-230 devices are sharing a common output line.

The flexibility of the TSL-230 means it can be used in a variety of system configurations. Although simple light-measurement systems can be built using the TSL-230 with timer/controller circuits, you can achieve maximum versatility by using a microprocessor. The TSL-230 can also be interfaced to a personal computer through the parallel or serial port, or through an interrupt line. The TSL-230 can also be interfaced directly to a dedicated microprocessor, as shown in Figure 5-17. Figure 5-17 illustrates a PIC16C54 microcontroller interface that uses a general-purpose I/O port for frequency input. A hardware timer is not present in this microprocessor. Fast clock speed and rapid instruction execution allow accurate timing functions to be implemented in software. Period measurement is accomplished with a tight software loop; for simplicity of operation, the period is obtained by measuring the high going pulse width and multiplying by 2. The following program shows the 8-bit period measurement routine for a PIC16C54:

```
;8-bit period measurement routine for PIC 16C54
;assumes input signal is 50% duty cycle (divide by 2)
;does not account for overflow
          movlw    ffh         ;FF is hex for 255
          movwf    pcnt        ;initalize period counter
Loop 1    btfsc    porta, 0    ;check port
          goto     loop 1      ;wait for low level
Loop 2    btfss    porta, 0    ;check port
          goto     loop 2      ;wait for high level
Loop 3    incf     pcnt        ;begin counting
          btfsc    porta, 0    ;check port
          goto     loop 3      ;count while high
;value in percent (8 bit) represents period/2
PIC16C54
;Resolution: 8-bit LSB = .80 s
```

```
;Minimum freqency: 4.8 kHz
;Maximum freqency: 1 MHz
;Measurement time: 1 to 210 µs
;Range: 460 B
```

Generally, the choice of interface and measurement technique depends on the desired resolution and data acquisition rate. For a maximum data acquisition rate, period measurement techniques are used, as shown in with the PIC16C54 interface.

Using the divide-by-two output, you can collect data at a rate twice that of the output frequency, or one data point every microsecond for full-scale output. Period measurement requires using a fast reference clock with resolution directly related to the reference clock rate. Output scaling is used to increase the resolution for a given clock rate or maximize the resolution as the light input changes. Period measurement is used to measure rapidly varying light levels or to make very fast measurement of a constant light source.

You can obtain maximum resolution and accuracy by using frequency measurement, pulse accumulation, or integration techniques. Frequency measurements provide the added benefit of averaging out random or high-frequency variations (jitter) resulting from noise in the light signal. Resolution is limited mainly by available counter registers and allowable measurement time. Frequency measurement is well suited for slowly varying or constant light levels and for reading average light levels over short periods of time. Integration can be used to measure exposure, i.e., the amount of light present in an area over a given period of time.

The Texas Instruments TSL-230 light-to-frequency converter is a versatile eight-pin DIP device that can readily perform high-resolution light measurements with a minimum of components.

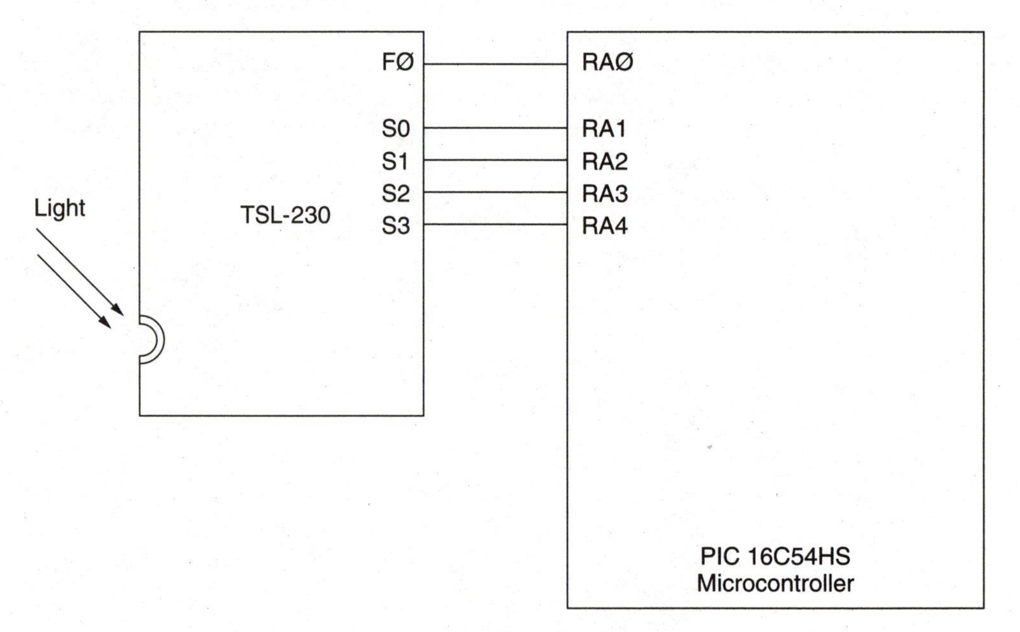

5-17 TSL-230 microcomputer interface. Reprinted by permission from Texas Instruments.

Infrared light-to-frequency/frequency-to-voltage data link

There is often a need to send or relay analog data from one place to another through the air or free space. You can use the analog data link shown in Figures 5-18 and 5-19 to send light-intensity data values from a light-probe IR transmitter to a corresponding remote IR receiver, which can remotely display the light-probe readings. One of the most efficient methods of sending analog light data is to convert light to frequency. Once light has been converted to frequency information, it can be easily sent over long distances through air, wire, or radio.

The analog data link consists of an IR transmitter and an IR receiver unit. The light probe shown in Figure 5-18 can be of either type. The top light probe at RA is used to produce a current when light falls on the photoresistive cell. Note that RA and RB form a voltage divider that is coupled to the input of the voltage-to-frequency (V-to-F) converter. Alternately, the second light probe at RA can be used to produce a voltage at junction (Y) when darkness is over the photocell. Either light probe can be coupled to the V-to-F converter depending upon your particular application. Also note that a resistive temperature detector could be substituted for the photocells, for temperature measurements. The V-to-F converter shown is a Teledyne 9400 chip, which can be used either as a V-to-F converter or conversely as a frequency-to-voltage (F-to-V) converter.

Resistor RS couples the light probe to the actual V-to-F converter, shown in Figure 5-18. This resistor transforms current into a proportional frequency. The voltage reference at pin 5 is coupled to the input pin via a 47-pF capacitor. The comparator input at pin 11 is coupled to amplifier output at pin 12. The amplifier is tied to the input via a .001-µF capacitor. The zero adjustment at pin 2 is balanced between 9 volts and ground via two 33-kilohm resistors at R2 and R3. The light probe's current intensity provides a proportional frequency output at pin 10 of the V-to-F converter. It sends light-probe data to an IR LED, which in turn sends light-probe data to the F-to-V remote receiver.

You can greatly increase the distance from the light transmitter to the light receiver by using IR filters and lenses at both the transmitter and receiver. You can also increase the optimum link distance by using a collimator tube ahead of the detector Q1 in the receiver unit.The analog data transmission receiver shown in Figure 5-19 consists of a second 9400 converter chip, which is used as an F-to-V converter. This unique chip can convert the incoming frequency pulses from the transmitter's light probe back to voltage, which can be monitored at M1. The light pulses are received from the IR transmitter via Q1, a Radio Shack 276-142 IR phototransistor. The incoming frequency is converted by inputting Q1 to pin 11 at the comparator input.

Compared to the transmitter unit, pins 3 and 5 of U1 are connected together via a 47-pF capacitor, and the voltage output from the F-to-V converter is presented at amplifier output on pin 12 of the 9400 chip. You can then use a 0- to 1-mA meter from ground to pin 12 to display the voltage values representing the light-probe input levels. Potentiometer R7 zeros the meter. You could also use the output from pin 12 to drive an A/D board in a personal computer.

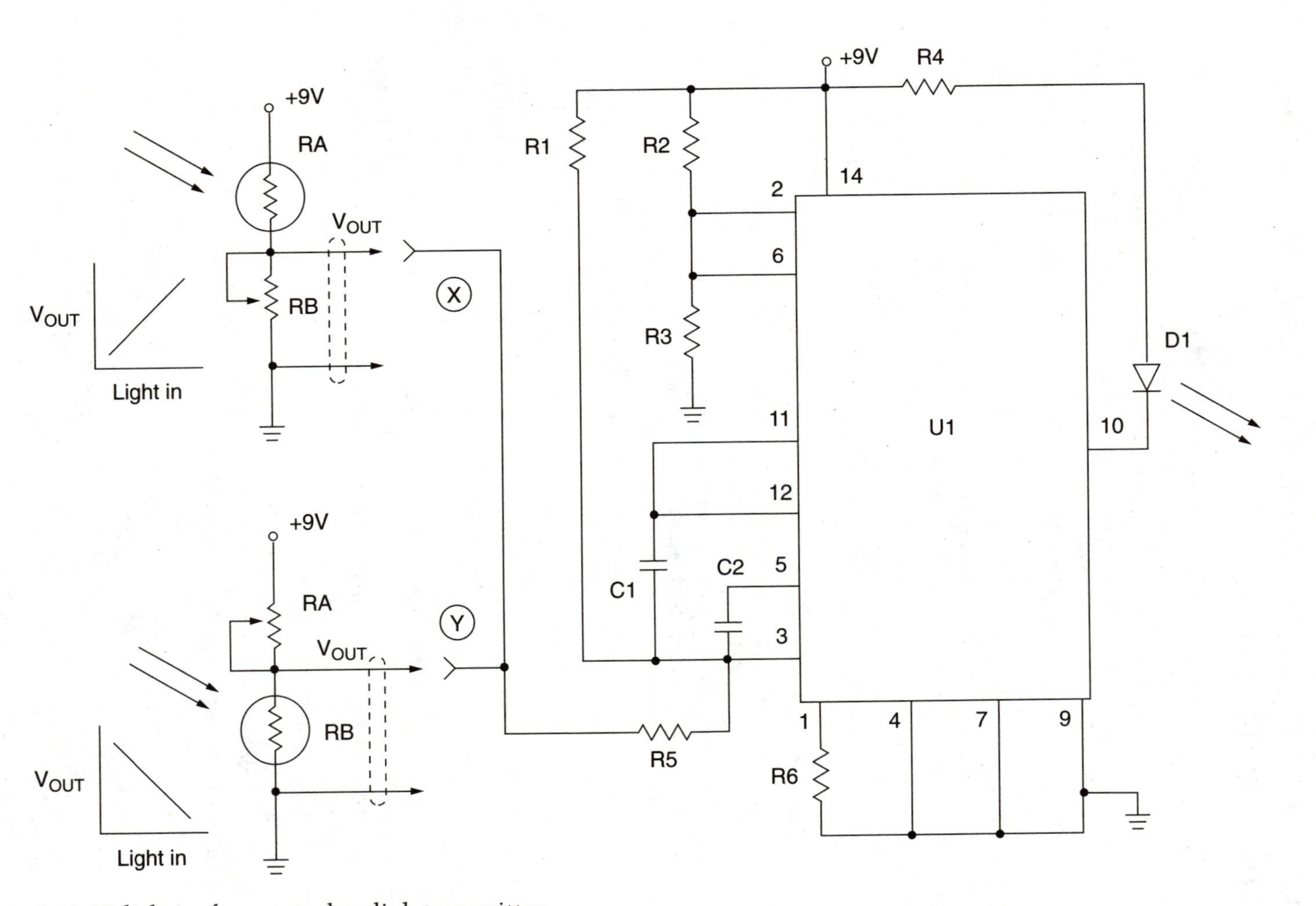

5-18 IR light-to-frequency data-link transmitter.

Analog data transmitter parts list

RA1,RA2	100-kilohm photoresistive cell
RB1,RB2	100-kilohm photoresistive cell
R1	1-megohm, ¼-watt resistor
R2,R3	33-kilohm, ¼-watt resistor
R4	270-ohm, ¼-watt resistor
R5	1-megohm, ¼-watt resistor
R6	100-kilohm, ¼-watt resistor
C1	.001-µF, 25-volt ceramic disc capacitor
C2	47-pF, 25-volt disc capacitor
D1	IR LED
U1	9400 V-F/F-V converter (Teledyne)

Analog data receiver parts list

R1,R5,R6	10-kilohm, ¼-watt resistor
R2	33-kilohm, ¼-watt resistor
R3,R4	100-kilohm, ¼-watt resistor
R7	5-kilohm potentiometer
C1	47-pF, 25-volt disc capacitor
C2	.001-µF, 25-volt disc capacitor
Q1	IR phototransistor (RS 276-130 or equivalent)
U1	Teledyne 9400 V-F/F-V converter
M1	0- to 1-mA meter

Light-wave temperature link

You can measure temperature many ways. The simplest means is by using a common glass thermometer. You can also measure temperature by electronic means by using a thermocouple or a thermistor. To remotely monitor temperature, you could connect solid-state sensors to a remote monitoring station via a cable or wire. In some applications, however, it is not practical, safe, or desirable to connect sensors via wire to a remote monitoring station. If, for example, a sensor is mounted near high-voltage circuits or on a movable platform, you would not want to have electrical cables present. Outdoor wires and cables are also susceptible to lightning. Other special applications of temperature measurement include temperatures taken at various altitudes via balloon, rocket, or aircraft.

The remote temperature link uses a voltage-to-frequency (V-to-F) converter in the transmitter unit to convert the sensor's voltage to a frequency signal that modulates an IR LED. The transmitter or sending link can also be coupled to the receiver section either through free space or via a tiny fiber-optic cable. The receiver link unit converts the incoming frequency pulses back to a voltage, which is then displayed on the analog meter. An analog-to-digital converter card in a personal computer could also be used to collect voltages.

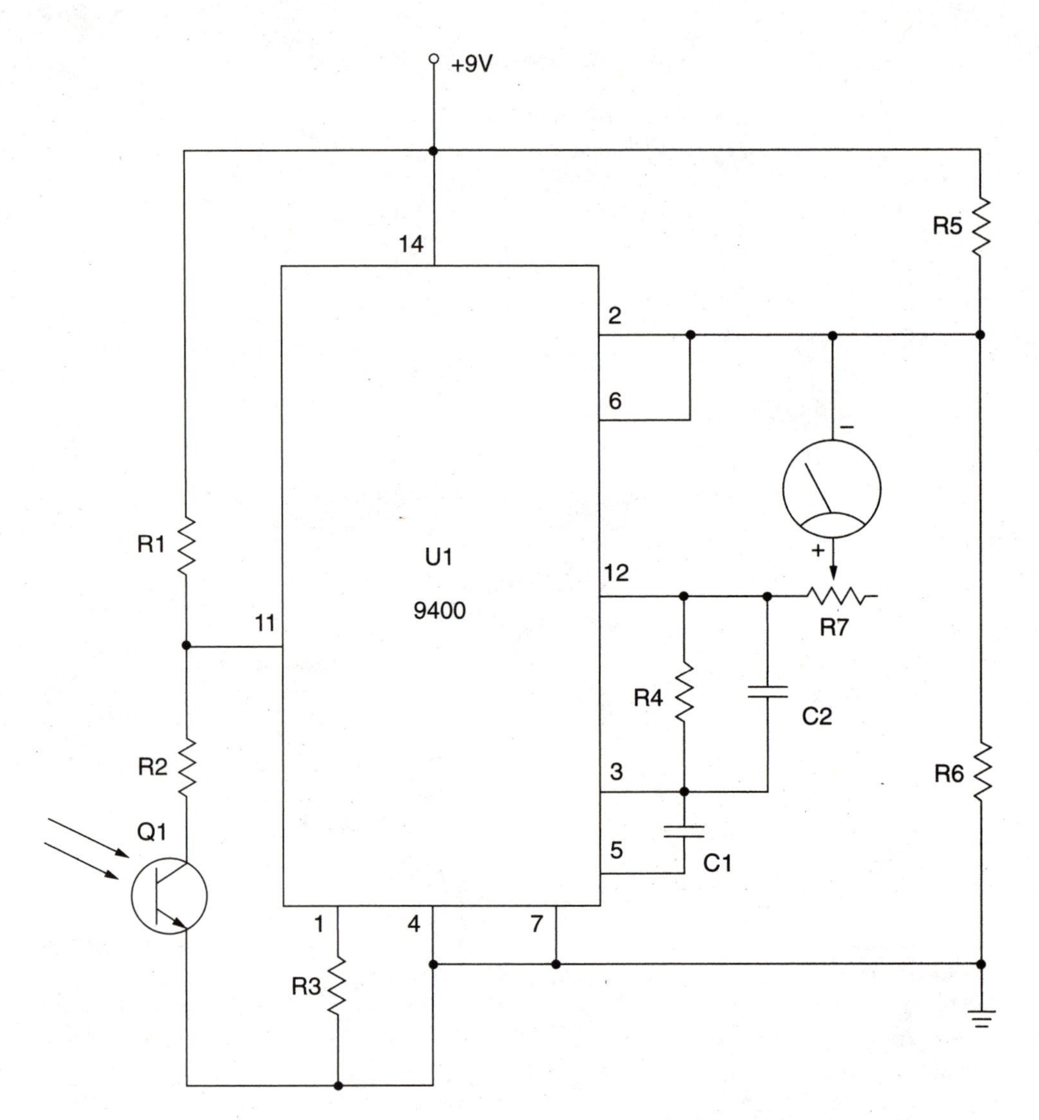

5-19 IR frequency-to-voltage receiver/decoder.

The heart of the infrared temperature link depicted in Figure 5-20 is the thermistor, which drives the V-to-F converter. A 2500-ohm room-temperature Fenwall glass-bead thermistor is connected in series with resistor R4, which forms a voltage divider. As the thermistor's resistance changes with the temperature, the voltage across the junction of the thermistor and resistor R4 changes. The resistances of thermistors are generally inversely proportional to temperature; i.e., the resistance of a thermistor falls as the temperature rises. Therefore, the output of the thermistor/ resistor voltage divider increases as the temperature increases. The temperature-dependent voltage from the thermistor divider is then applied to the input of the V-to-F converter via R2.

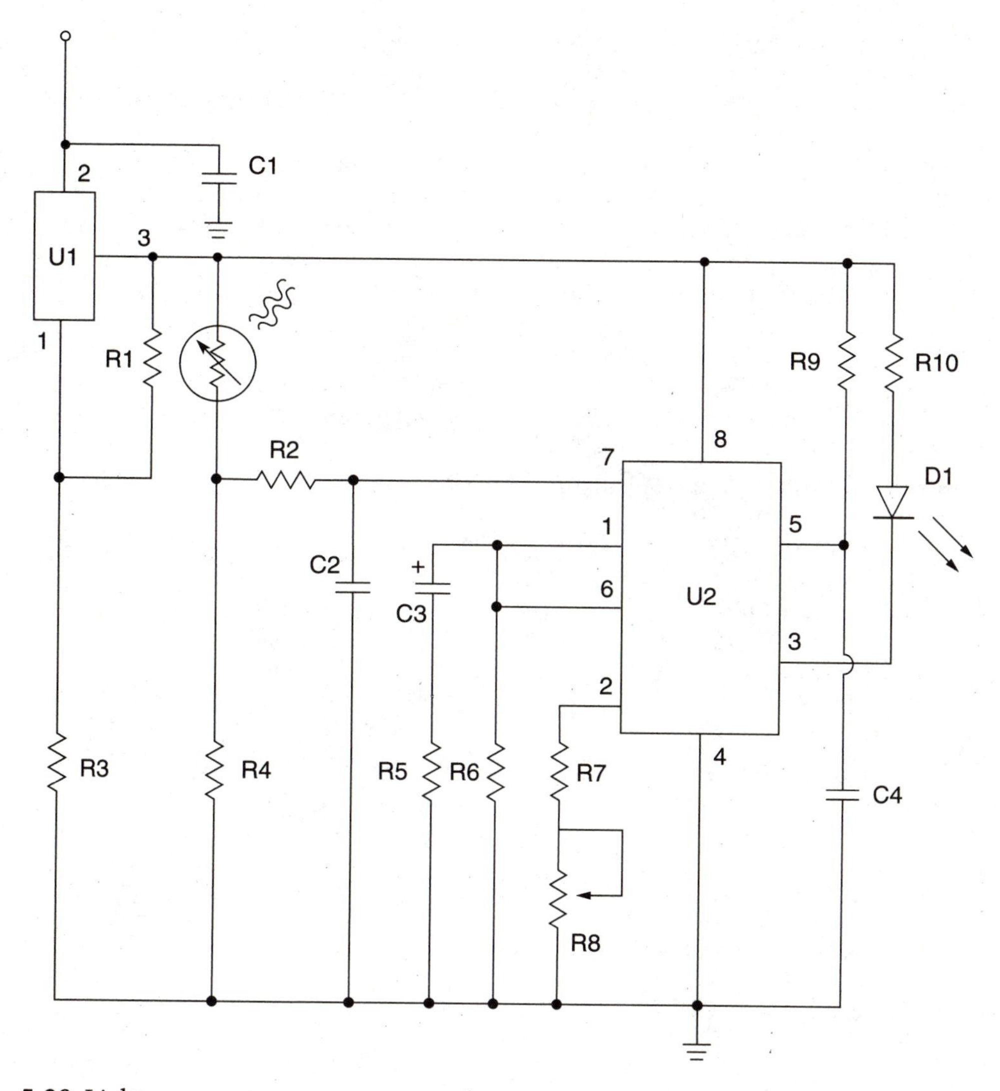

5-20 Light-wave temperature transmitter. From Forrest Mimms' *Circuit Scrapbook* (Howard W. Sams & Co., 1987). Copyright by Forrest M. Mimms III. Used with permission.

The V-to-F converter, a National LM311 chip, also functions as an F-to-V converter. The linear response of the LM311 varies from 1 Hz to 1 kHz, is adjusted via R8, and serves to aid in calibrating the system. The output of the V-to-F converter drives an IR LED resistor R10, a 10-ohm resistor. The IR LED chosen for the IR temperature link was a GE GFOE1A1. Power is applied to the voltage-sensitive V-to-F converter via U1, an LM350T regulator. The regulator provides a stable voltage for both the thermistor and V-to-F converter.

The IR temperature-link receiver depicted in Figure 5-21 receives incoming temperature data from the transmitter via photodiode D1, which is coupled to a standard LM741 op-amp. A small solar cell could be substituted for the photodiode, if desired.

The output of the light-wave receiver at U1 is coupled via C2 to the input of U2. The LM555 in the receiver's circuitry converts the incoming frequency pulses to a dc voltage, which can be read on a meter. Potentiometer R7 calibrates the system and R6 zeros the meter.

In order to calibrate your IR temperature-link system, you will need a half cup of boiling water, a cup of ice water, and a thermometer. First, record both the temperature of the hot water and the frequency produced by the transmitter when the thermistor is immersed in the hot water, keeping the thermistor leads out of the water if possible for these measurements. Now add some ice water to the hot water and repeat the measurements. Continue adding cold water and making measurements until the cup is full. Then start making measurements in the cup of ice water and record the temperature and frequency readings to complete the calibration procedure.

For reliable measurements, stir the water after adding water of a different temperature. Allow a half minute or so before taking actual measurements since the IR temperature transmitter uses a thermistor as a temperature sensor, and the output curve is nonlinear over its range.

You can use the IR temperature link to relay temperature measurements from your pool to your house, from an outdoor animal shelter to your home, or from outdoor storage tanks to your home or shop. If you plan on using the IR transmitter link through free space, then consider using IR filters at both ends of the link, as well as lenses at the IR LED and the photodiode sensor. For best results, use a collimator on the photodiode.

You could also use a fiber-optic cable to couple the IR transmitter to the IR temperature-link receiver between buildings or from a ground-monitoring station to a tethered weather balloon. A temperature link is a useful addition to any remote monitoring station.

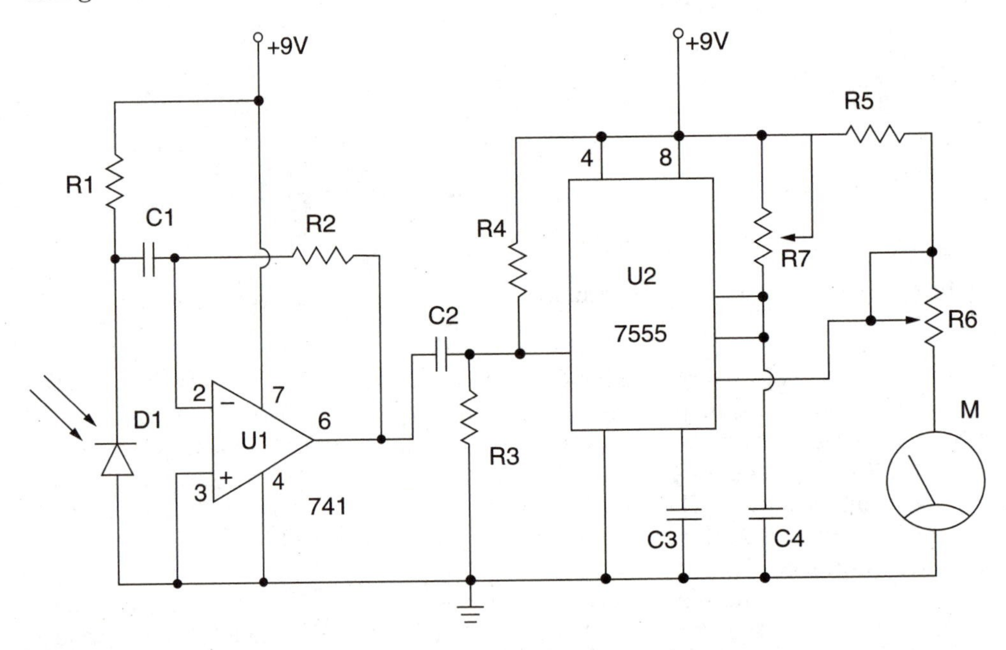

5-21 Light-wave temperature receiver. From Forrest Mimms' *Circuit Scrapbook* (Howard W. Sams & Co., 1987). Copyright by Forrest M. Mimms III. Used with permission.

IR temperature-link transmitter parts list

TH	Fenwal thermistor #107 (Newark Electronics)
R1	1.4-kilohm, ¼-watt resistor
R2	1.2-kilohm, ¼-watt resistor
R3	150-ohm, ¼-watt resistor
R4,R6	100-kilohm, ¼-watt resistor
R5	47-ohm, ¼-watt resistor
R7	12-kilohm, ¼-watt resistor
R8	5-kilohm potentiometer
R9	6.8-kilohm, ¼-watt resistor
R10	10-ohm, ¼-watt resistor
C1	.1-µF, 25-volt disc capacitor
C2,C4	.01-µF, 25-volt disc capacitor
C3	1-µF, 25-volt electrolytic capacitor
D1	IR LED GEOE1A1 (GE)
U1	LM350T regulator
U2	LM311 V-to-F converter (National)

IR temperature-link receiver parts list

R1,R2	1-megohm, ¼-watt resistor
R3,R4,R5	4.7-kilohm, ¼-watt resistor
R6	10-kilohm, potentiometer (zero)
R7	250-kilohm, potentiometer
C1,C2	.01-µF, 25-volt disc capacitor
C3	.1-µF, 25-volt disc capacitor
C4	.22-µF, 25-volt disc capacitor
D1	Photodiode or photovoltaic cell
U1	LM 741 op-amp
U2	LM7555 timer IC
M	0- to 1-mA meter

Hunter's Companion

Hunter's Companion is a unique project designed to assist game hunters in locating large downed animals such as deer, moose, or bear after they have been shot. This tracking device helps pinpoint wounded animals in thickly wooded areas. Hunter's Companion is a heat-sensitive or pyroelectric detector, sensitive to the body heat of humans and large animals. Shown in Figure 5-22, it consists of an Eltec 406-3 single-element pyroelectric detector ahead of an LM324 amplifier/filter coupled to a 20-segment LED bargraph display. The Eltec 460-3 detector is housed in a TO-5 transistor package, which houses a lithium tantalate crystal detector element, a high-megohm resistor, and a FET transistor. The sensor package has a square window that is covered with a special optical filter ahead of the crystal detector. You can specify the filter transmission characteristics when ordering the detector, since they are designed for particular wavelengths of the spectrum.

A bias is applied via R2 to the drain of the FET at pin 1 of the detector. A source resistor is placed between pin 2 of the sensor and ground. The coupling capacitor at C1 couples the signal output from the pyroelectric detector to the first stage of the LM324 op-amp at U1. The first stage of amplification is set by resistors R5 and C2, and the network acts as a high-frequency filter. The -3-dB points of the filter are computed by the following formula:

$$HF = 1 / 2 (Pi) R5 / C2$$

Capacitor C3 and resistor R4 are placed into the circuit and act as a low-frequency filter, computed by this formula:

$$LF = 1 / 2 (Pi) R4 / C3$$

The output of the first stage of the LM324 is coupled to U1:B via R6. The gain of the second stage of the amplifier is configured by R7. Pin 5 of U1:B is used to control the output offset adjustment in order to zero the output voltage. Bypass capacitors are used at both pins 4 and 11 of the LM324. Note that a plus and minus 9-volt supply voltage is used to power the circuit. The output of the op-amp at pin 7 is fed to the bargraph drivers through potentiometer R11. This potentiometer is used to set up the bargraph to display the absence of segments when no signal is present. Resistors R12 and R14 set the input scaling for each display driver. The low end of the display begins to light up at pin 1 U2 and travels to pin 10 or D20 at full signal. You use the DPDT toggle switch at SW1 to select the bar or dot mode of operation. You could eliminate this switch if you prefer a particular mode by selecting a jumper.

In order to provide long-range detection, the Eltec 406-3 sensor requires a Fresnel filter array ahead of it. A low-cost Fresnel Technology Inc. LR1.2GZI12VI flat filter array was used in the prototype of the Eltec pyroelectric detector. The Fresnel filter was placed 1.2 inches in front of the detector to allow long-range detection. This filter was located as a sample from the manufacturer, but other filters could be used. Pay particular attention to the focal length of the filter.

In operation, turn on the power to Hunter's Companion and adjust potentiometer R11 until the lowest sensitivity lamp at D1 is just extinguished. Now pass your hand quickly in front of the detector and the whole range of LEDs should come to life. Adjust the potentiometers for the full range of display from no light to full bargraph when your hand is moved in front of the detector. Hunter's Companion is now ready for operation.

In the field, simply sweep the detector across the field of view, low along the ground. Continually sweep back and forth in front and around you as you slowly walk to where the big game was last seen. Happy hunting!

Hunter's Companion parts list

S1	406-3 pyroelectric detector (Eltec)
R1	6.2-kilohm, ¼-watt resistor
R2	50-kilohm, ¼-watt resistor
R3	1-megohm, ¼-watt resistor
R4	100-kilohm resistor (see text)
R5	10- kilohm, ¼-watt resistor

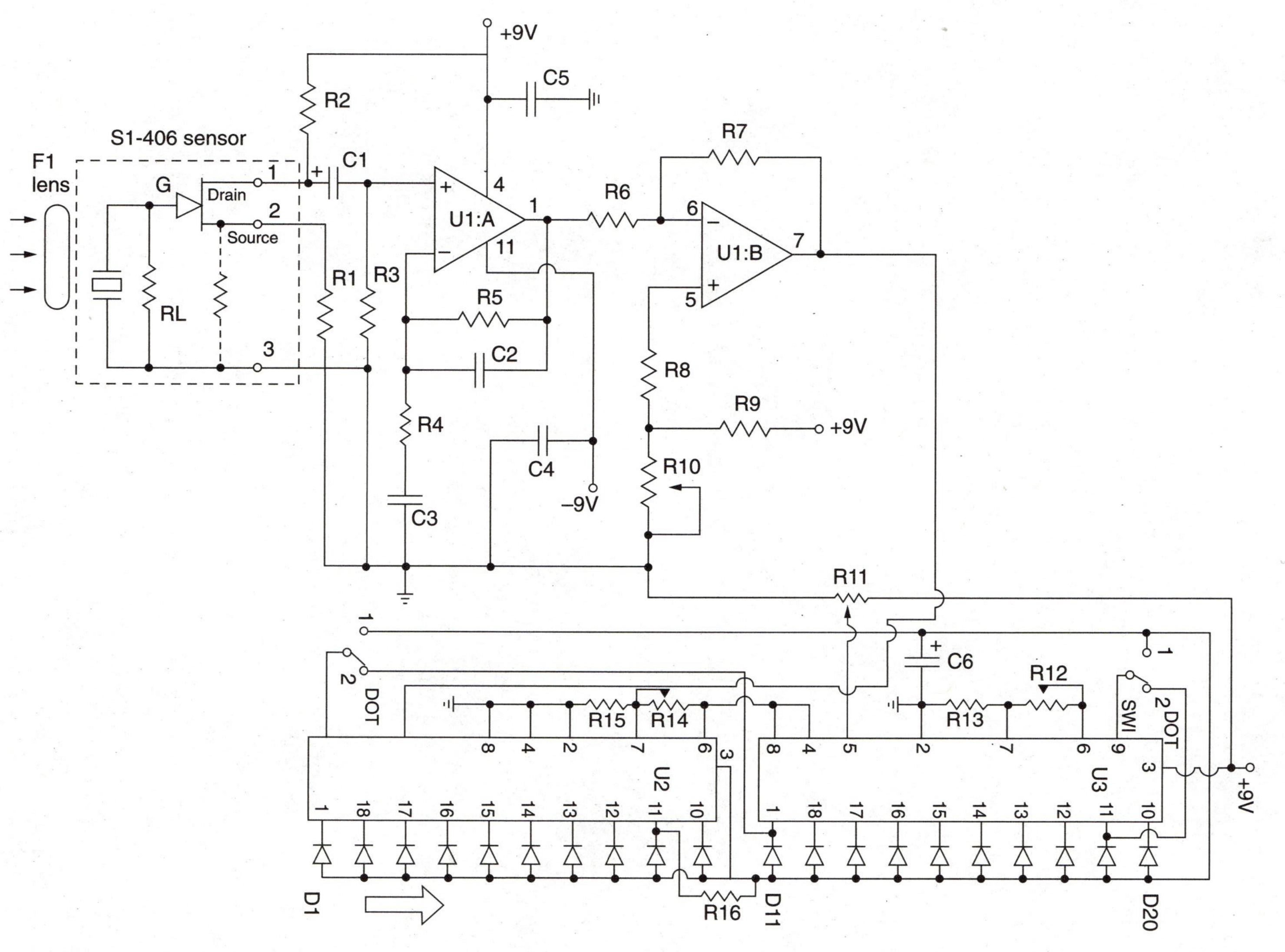

5-22 Hunter's Companion.

Hunter's Companion parts list continued

R6	5-kilohm, ¼-watt resistor
R7,R9	200-kilohm, ¼-watt resistor
R8	33-kilohm, ¼-watt resistor
R10,R12,R14	5-kilohm trimmer
R11	100-kilohm potentiometer
R13,R15	1-kilohm, ¼-watt resistor
R16	22-kilohm, ¼-watt resistor
C1	1-µF, 25-volt tantalum capacitor
C2	10-nF, 25-volt Mylar capacitor
C3	.01-µF, 25-volt disc capacitor
C4,C5	.1-µF, 25-volt disc capacitor
C6	2.2-µF, 25-volt electrolytic capacitor
U1	LM324CN op-amp
U2,U3	LM3914 LED display driver
SW1	DPDT toggle switch
D1-D20	20 LEDs or two ten-segment bargraphs
F1	Long-range Fresnel lens LR1.2GI12VI (Fresnel Technology)

Long-range heat/cold leak detector

The unique long-range window/door leak detector is a sensitive infrared sensing circuit designed to "sniff" hot and cold leaks. You can use the IR leak detector to detect cold drafts along window sills or door jambs, or to detect cold leaks along outside walls. You could also use the IR leak detector to find cold leaks in freezer or refrigerator seals. The versatile IR leak detector can also be pressed into service to detect heat, such as "hot" walls during the early stages of fire detection.

The sensing portion of the long-range IR window/door leak detector, shown in Figure 5-23, centers around the Eltec 406-3 single-element pyroelectric detector and chopper assembly. Pyroelectric or IR heat detectors are most generally used to detect movements of warm bodies passing in front of the detector for burglar alarms. The pyroelectric detector can also be used to detect heat/cold leaks, but the operation of the detector is more active since the detector is now moved instead of moving a body in front of the detector. While you can use a lensless pyroelectric sensor to detect leaks from strong sources of hot or cold by quickly scanning back and forth over a suspected leak, the sensitivity and usefulness of the detector is limited. In order to greatly enhance the sensitivity of the detector for long range, place a Fresnel lens in front of the detector to chop or interrupt the input signal. The purpose of chopping the input signal is to distinguish small differences of heat and cold by referencing the input signal against the ambient or background signal. The chopping allows the detector to become more sensitive to small changes over time, and greatly increases the overall system gain and discrimination by interrupting the input signal to the detector.

The Eltec 406-3 single-element lithium tantalate crystal detector element is housed in a TO-5 transistor-type can along with a high-megohm resistor and an FET transistor. The pyroelectric crystal is placed behind an optical filter, which is specifically selected for the wavelength of interest. For long-range applications, a Fresnel lens or array is placed in front of the pyroelectric detector.

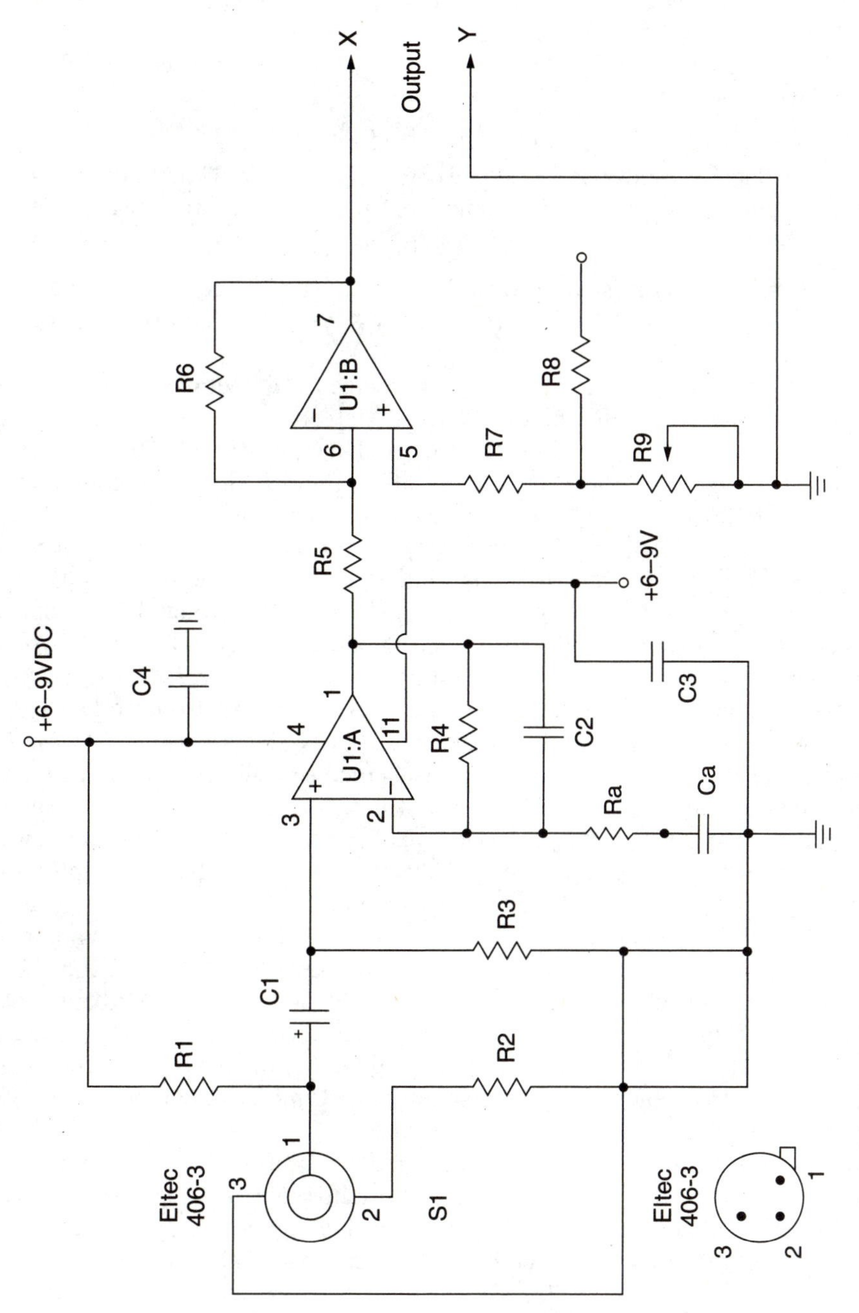

5-23 Infrared hot/cold leak-detector sensor electronics.

The pyroelectric detector requires a bias resistor at R1, which supplies a current to the drain lead of the FET at pin 1. A source resistor of 6.2 kilohms is placed at R2 between pin 2 of the sensor and ground. Capacitor C1 is then used to couple the detector to the first stage of the LM324 op-amp.

At U1:A, a high-frequency filter is formed by C2 and R4, which can be calculated by the following formula:

$$HF = 1 / 2 (Pi) C2R4$$

An optional low-pass filter is formed by components RaCa. The values chosen are typical. You can calculate the specific component requirements with this formula:

$$LF = 1 / 2 (Pi) RaCa$$

The first stage of U1:A is coupled to the second stage via R5. A gain path is set up between pins 6 and 7 of U1:B with resistor R6. A zero offset control is configured with R7, R8, and R9, which is fed to pin 5 of U1:B.

The output of the second stage of the pyroelectric detector is coupled to the LCD display section of the IR window/door leak detector, as shown in Figure 5-24. The LCD display is a 4½-digit voltmeter with an 11-megohm input. The heart of the digital voltmeter section is the Maxim ICL7129A A/D chip. The A/D chip features a resolution of 20,000 counts. The Maxim voltmeter is a dual slope conversion A/D converter that uses a crystal oscillator tuned to 120 kHz to provide 60 Hz of power line rejection. The input voltage to 1C1 is coupled via pins 32 and 33.

The input is first fed to a voltage divider consisting of R2 through R5, providing a 10:1 reduction in voltage presented to 1C1. Components R10 and C5 form a low-pass filter to attenuate any noise or ac component from appearing across the input. A 4½-digit triple multiplexing liquid-crystal display module allows you to control all 37 segments, including continuity and low-voltage indication, with just 15 connections. This feat is accomplished by separating the various elements of the display into three sections. Three backplane terminals are used for the triple multiplexing. The segments of the display are in three groups, each controlled by its own backplane square wave voltage. The LCD driver in 1C2 generates the backplane signals that cause the appropriate elements of the display to be activated in sequence. This process takes place so fast that the display appears to be constant. The low-voltage indicator is automatically energized when the supply voltage on pins 23 and 24 falls below 7.2 volts.

To calibrate the display section of the IR window/door leak detector, connect the x and y inputs of the LCD meter to a known 1.5-volt test battery while adjusting R8 for the correct display reading. You can use a digital multimeter to compare display readings in order to calibrate the new meter. The LCD meter circuit consumes little power, and its 9-volt battery should last for some time.

In order to reference or chop the input signal for greatest sensitivity and resolution, as previously discussed, you have a few options (shown in Figures 5-25, 5-26, and 5-27). The diagram in Figure 5-25 illustrates a chopper assembly using a small brushless dc motor to drive a slotted disk placed in front of the pyroelectric detector. The slotted disk rotates to allow the detector to "look" out into space only part of the time when the slot is open to the detector's view. When the detector is blocked by the disk, it is referenced to the ambient background.

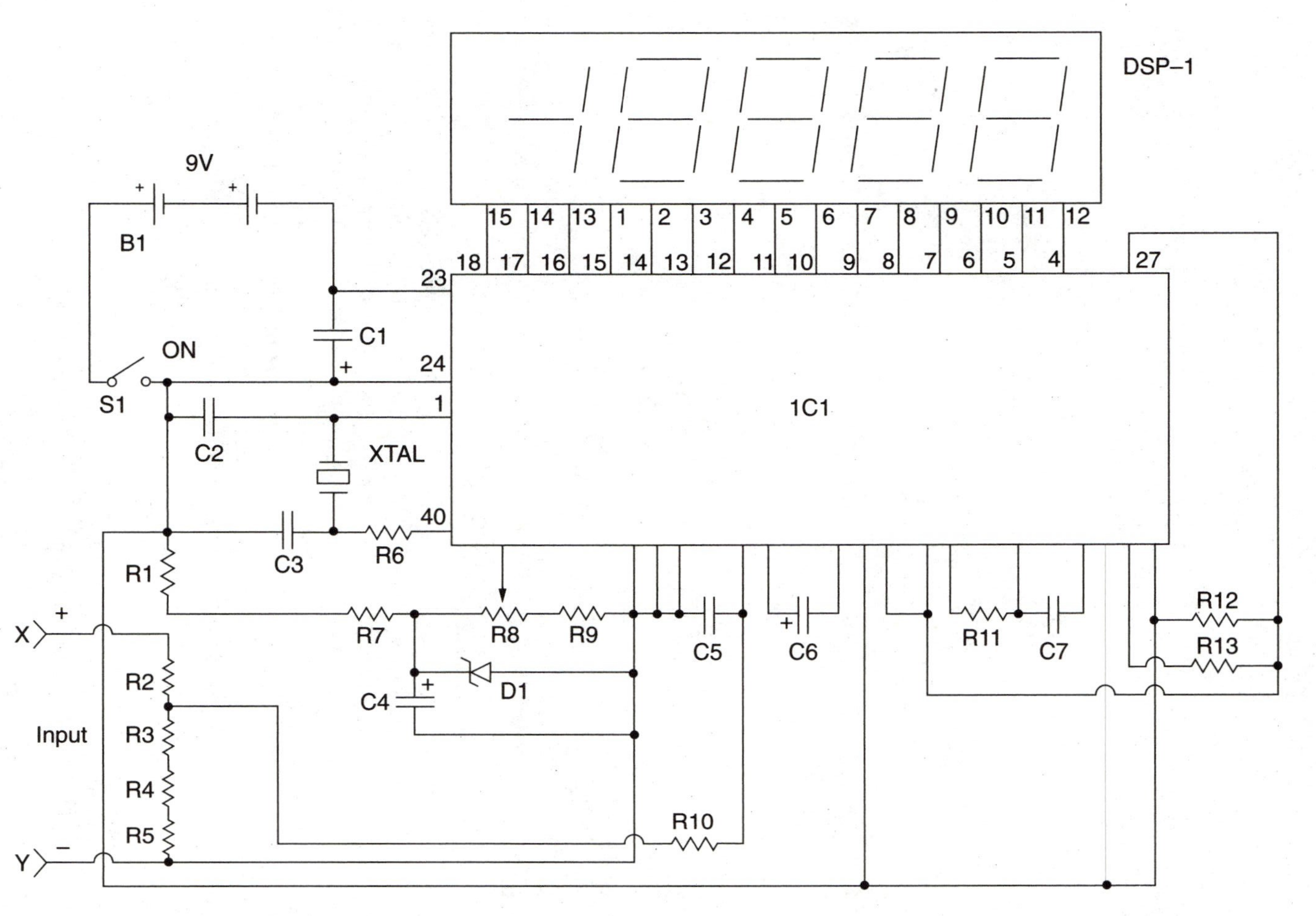

5-24 IR leak-detector digital display.

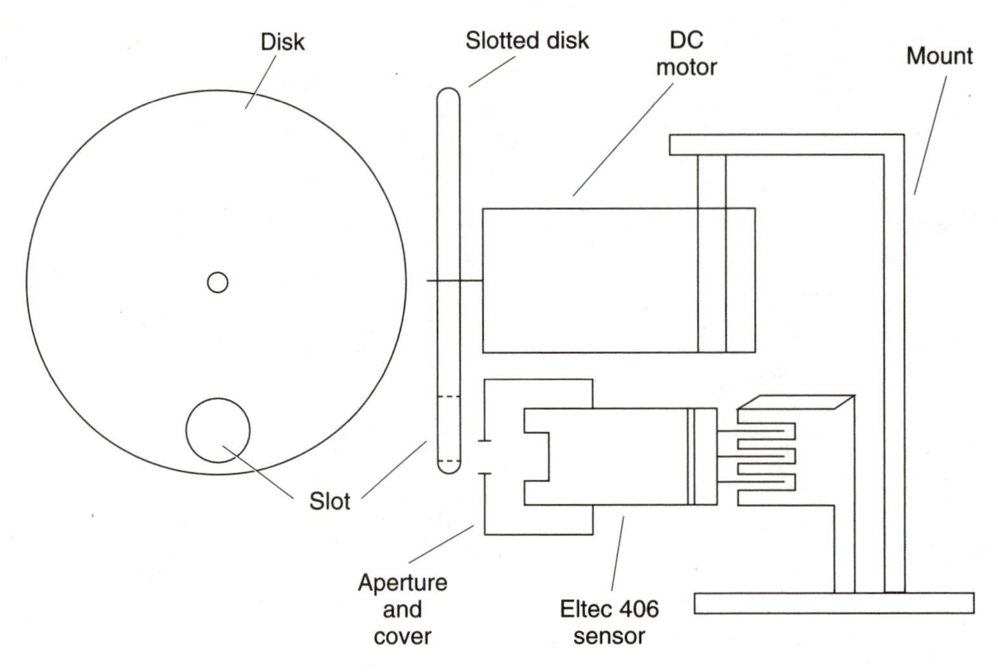

5-25 Chopper motor assembly.

Make sure to use the dc motor with a regulated motor-speed controller so you can control the speed of the motor and thus accurately control the speed of the rotating disk (see Figure 5-26).

The Eltec 406-3 pyroelectric detector is mounted via a transistor socket to the main bracket, as shown. A small aperture disk is placed in front of the detector to limit the sensor's view angle and time. An aperture of about $\frac{1}{4}$ inch is recommended. An aperture sleeve was created and placed over the detector to allow the detector only a small view through the aperture disk.

A second method of chopping or interrupting the input signal is shown in Figure 5-27. This method uses a dc piezoelectric fan or piezo-resonant blade (see the Appendix). This solid-state method of chopping features long life and no rotating parts. The piezo blade vibrates back and forth, as shown by the dotted lines. A chopper blade is epoxied to the vibrating element and allowed to vibrate in front of the detector aperture periodically. The piezo-resonant blade assembly is specially constructed with multiple layers to provide a rather large movement as opposed to small piezo buzzers, which move only a minuscule amount. In normal operation, an ac voltage of about 30 volts is applied to the resonant blade, which allows the blade to rapidly vibrate back and forth.

A simpler method is to try to locate a low-cost dc piezoelectric fan; these were quite ubiquitous a few years ago. Using the dc fan eliminates an ac source and simplifies the overall design of the system.

Once you have chosen and assembled the chopper system, you can wire the detector and electronic circuits together. Locate the detector and chopper assembly as close as possible to the detector's electronic circuitry. The entire assembly of the de-

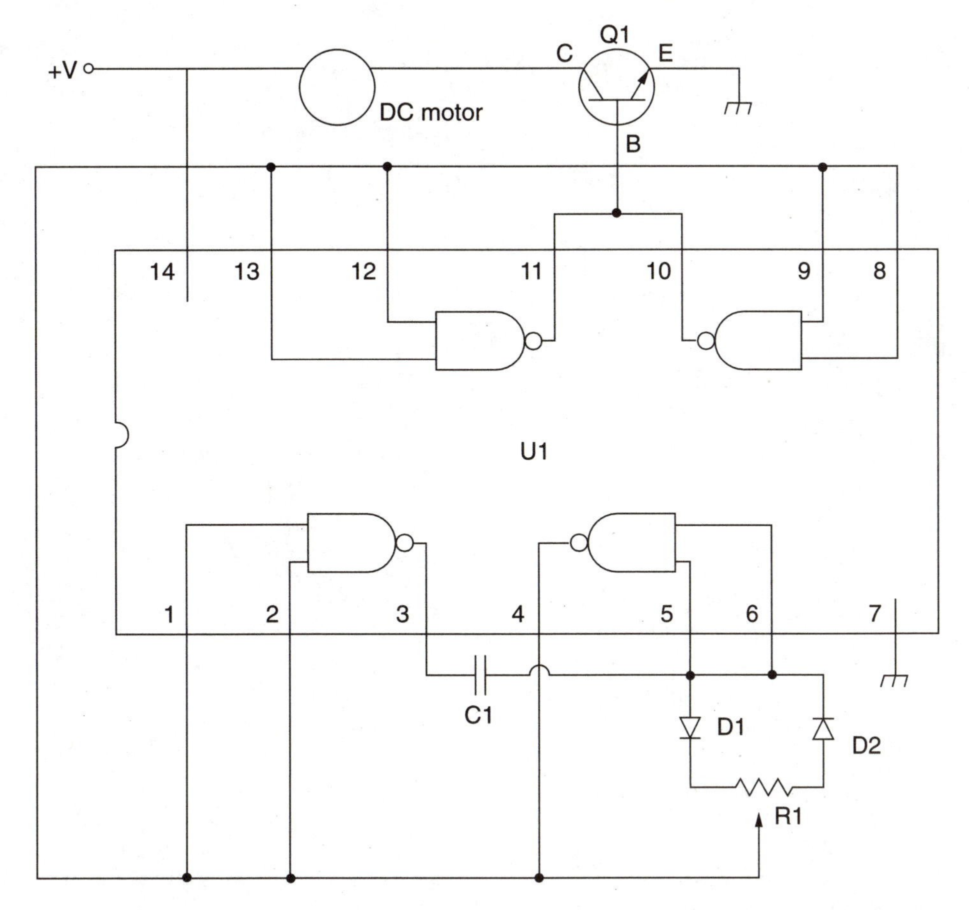

5-26 Motor-speed control circuit.

tector, chopper, electronics, and display can be mounted in a metal chassis box. A metal box is better than plastic because it reduces and possibly eliminates RF interference, which can affect the pyroelectric detector. Remember that the detector must be allowed to "see" the outside world, so drill a small-diameter hole into the chassis for this purpose.

Power to the IR window/door leak detector is supplied via two dc sources. A 6- to 9-volt dc power supply is used to power the pyroelectric detector, chopper, and display circuit. A second 6- to 9-volt power supply provides a minus voltage to pin 11 of U1. The plus supply provides a larger current than the secondary or minus supply. You can measure the current of the entire circuit assembly once it is finished to determine the necessary current, which of course depends on the chopper method you have chosen. The prototype used C cells to power the entire system.

You can test each assembly separately and treat each subsystem as a block. First, apply power to the IR detector and connect an external voltmeter or the Maxim IC voltmeter to the output of the pyroelectric detector's electronic circuitry. Then adjust the zero potentiometer, R9, to ensure a zero baseline output from the

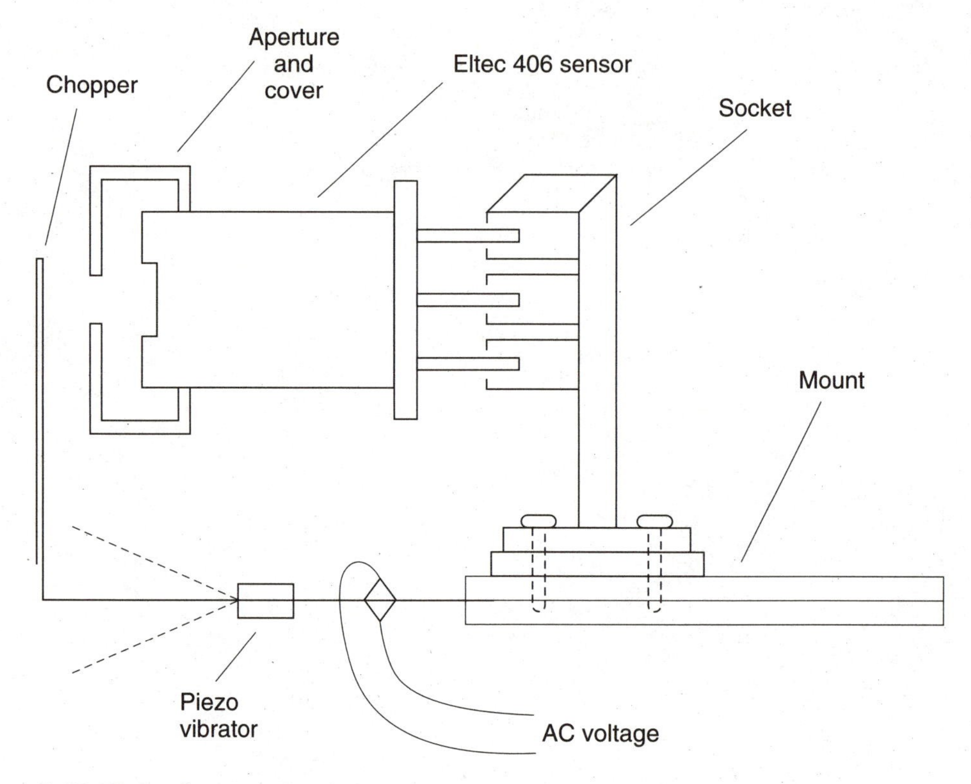

5-27 Piezo chopper circuit.

detector. Quickly move your hand in front of the detector and you should see an output voltage on your voltmeter. If you are using the Maxim IC voltmeter, you might have to adjust R13 to reference the input.

Once the preliminary testing is complete, you can begin testing the chopper assembly. Place the aperture on the detector and then bring the chopper assembly in front of the detector. Once the entire system is operational, check to ensure that the detector looks at the ambient background first. Then bring the IR window/door leak detector in front of a hot/cold source and check your voltmeter. You might have to make a few minor adjustments of R9 for proper operation.

If you want to use the IR window/door leak detector for long-range detection as a hot/cold telescope, then consider a long-range Fresnel lens array such as the Fresnel Technology LR1.2GI12VI. This low-cost lens is placed 1.2 inches in front of the pyroelectric detector. A commercial version of the long-range heat telescope is shown in Photo 5-6. You now have a sensitive, long-range, noncontact, thermal detector for many applications around your home, office, or laboratory.

Long-range heat/cold leak detector parts list

S1	406-3 pyroelectric sensor (Eltec)
R1	50-kilohm, ¼-watt resistor
R2	6.2-kilohm, ¼-watt resistor
R3	1-megohm, ¼-watt resistor

Long-range heat/cold leak detector parts list continued

R4	10-kilohm, ¼-watt resistor
R5	5-kilohm, ¼-watt resistor
R6,R8	200-kilohm, ¼-watt resistor
R7	33-kilohm, ¼-watt resistor
R9	5-kilohm trimmer potentiometer
Ra	100-kilohm resistor (see text)
C1	1-µF, 25-volt tantalum capacitor
C2	2-µF, 25-volt polyester capacitor
C3,C4	.1-µF, 25-volt disc capacitor
Ca	.01-µF, 25-volt disc capacitor (see text)
U1	LM324CN op-amp

DC motor-speed control parts list

R1	1-megohm potentiometer
C1	.01- to .05-µF, 25-volt disc capacitor
D1,D2	1N4148 silicon diode
Q1	2N3055 transistor
U1	CD40113 CMOS Nand gate
M1	6-volt dc brushless motor

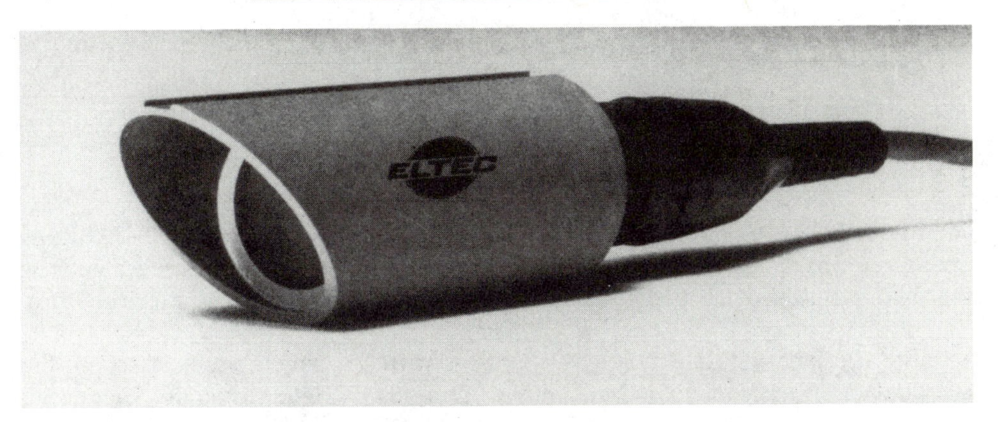

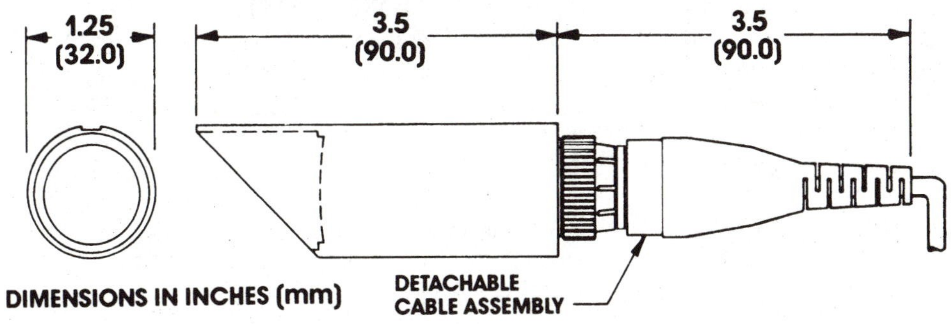

Photo 5-6 Pyroelectric sensor with Fresnel lens.

Leak detector display section parts list

R1,R7	5.11-kilohm, 1-percent, ¼-watt resistor
R2	10-megohm, 1-percent, ¼-watt resistor
R3	1-megohm, 1-percent, ¼-watt resistor
R4	110-kilohm, 1-percent, ¼-watt resistor
R5	1.1-kilohm, 1-percent, ¼-watt resistor
R6	270-ohm, ¼-watt resistor
R8	10-kilohm trimmer
R9	10-kilohm, 1-percent, ¼-watt resistor
R10	100-kilohm, ¼-watt resistor
R11	150-kilohm, ¼-watt resistor
R12,R13	47-kilohm, ¼-watt resistor
C1	10-µF, 25-volt electrolytic capacitor
C2	5-pF, 50-volt ceramic capacitor
C3	10-pF, 50-volt ceramic capacitor
C4	4.7-µF, 25-volt electrolytic capacitor
C5	.01-µF, 25-volt disc capacitor
C6	1-µF, 25-volt electrolytic capacitor
C7	.1-µF, 25-volt disc capacitor
D1	1CL8069CCZR 1.2-volt band gap reference diode (Harris)
1C1	1CL7129ACPL A/D converter (Maxim)
DSP-1	353R3R036HZ1 4½-digit LCD module (LXD)
X1	120-kHz crystal
S1	SPST switch
B	9-volt battery

Lightning monitor circuit

The lightning monitor featured in Figure 5-28 is an experimental UV lightning detection system that can be used both to alert you of an oncoming electrical storm and to turn off sensitive electronic equipment like computers for the duration of the storm.

The UV lightning detection system begins with a collimator tube, fitted with a UV bandpass filter ahead of a Centronics OSD series UV detector diode. The output of the UV silicon photodiode is fed to the input of an OPA11 Burr-Brown op-amp, which is set up for high-gain amplification. The gain resistor/filter is formed by the R1/C1 combination across pins 2 and 6 of U1. The output of the op-amp, U1, is directed into U2 via R2, as shown. The IC at U2 is an LM211 comparator. The comparator is then fed to a CD4528 one-shot, which acts as a time-window discriminator. The "time window" is formed by C3 and R4. The output from the detector head assembly consists of D1, U1, and U2. The signal is combined with the output of the one-shot, gated by U4:A, and then coupled to U4:B, which briefly sounds the piezo buzzer. The gated output from U4:A is also sent to timer U5, an LM555 timer chip. Once the buzzer sounds, you have about two to four minutes in which to close computer files before the off-timers at U6 begin their cycle. The two- to four-minute

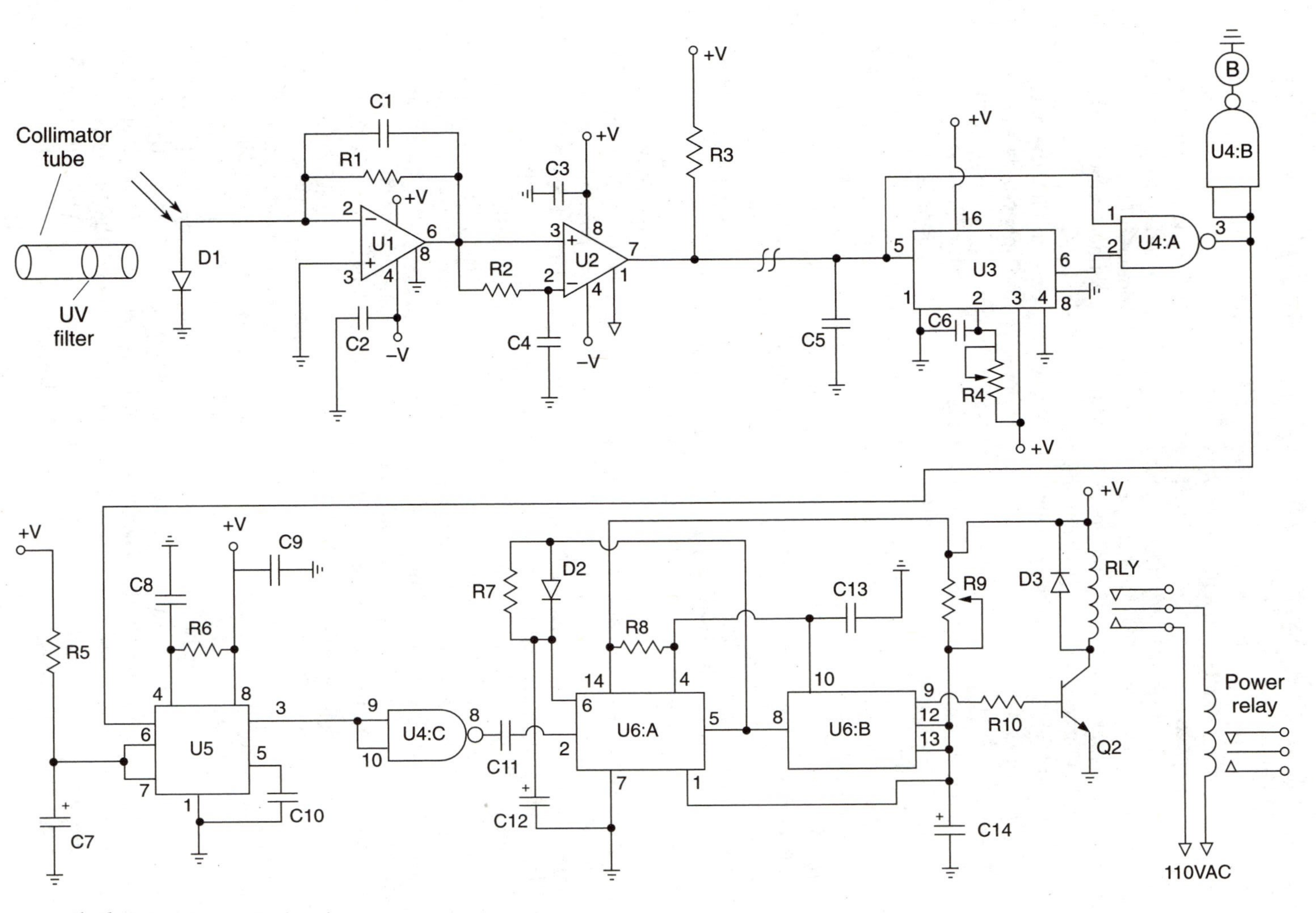

5-28 Lightning monitor circuit.

timer is set up by U5; you can adjust it by changing the value of R5/C7 or substituting a potentiometer for R7.

Once the two-minute time period has ended, U6 is triggered. The relay pulls in for 15 to 20 minutes and your equipment or computer is shut off for the duration. If more lightning occurs during the off time, then the additional triggers accumulate and the off-timer is retriggered until no more lightning is detected. You can adjust the off-timer at U6:B via R9 to increase or decrease the equipment off time. You can also use the experimental lightning detection system to disconnect a radio transceiver from an outdoor antenna system using the proper RF relay.

The experimental UV lightning monitor is composed of two circuits: the detector head assembly and the timer relay board. The detector assembly consists of the collimator tube, UV filter, and detector diode, as well as U1 and U2. There are a number of UV silicon diodes that can be used for the lightning detector. The Centronics model OSD5.8-7Q or OSD35-7Q UV detectors are suitable for this project and are available in various packages. Note the Q in the part designation, which denotes a quartz window in front of the detector diode, which is recommended. I chose the quartz window version, which peaks at between 254 and 340 nm. The diodes mentioned are also sensitive to other light wavelengths, so a UV bandpass filter is required ahead of the detector diode (see the Appendix for filter sources). The Centronics OSD5-7-254C is another UV detector diode, sensitive only to 254 nm, that contains its own UV filter in a housing assembly. You will obtain the best results by using large, open-frame diodes that can "see" a large portion of the sky at once. The detector head assembly should be mounted in a small metal chassis box and sealed against rain and severe weather. The collimator should be allowed to exit the chassis box so the detector can "see" the outside world. The detector head assembly should be mounted outdoors, with the UV silicon detector diode pointed in the direction from which most electrical storms originate. In the United States, most storms travel from west to east. The detector should be mounted at an upwards angle, pointing to the unobscured sky. The relatively short collimator tube, measuring one to two inches in length, should be used ahead of the UV detector. The output signal from the detector head assembly is directed to the main relay board via a length of RG-174 coaxial cable. The detector head assembly requires both a plus and a minus power supply. Power can be supplied from the main timer board using a four-wire cable. For best results, the circuit should be powered from batteries, with the batteries trickle-charged from a plus/minus power supply. In this way, the UV lightning detector is more isolated and free from power-line noise and interference. You must be patient when setting up and aligning the experimental lightning monitor. Once the lightning detector and power supply are built and tested and placed in their respective enclosures, you can begin the setting up and adjustment. Locate a UV lamp or light source that contains some UV light. With power applied to the lightning detector, place the UV lamp in front of the detector head and turn the UV lamp on and off a few times in succession. Experiment with the setting of R4 at U3 until it takes a few flashes in a row to activate the buzzer at U4:B. Once the time-window discriminator is working, you can move on to testing the rest of the circuit. The timer at U5 has a fixed time period, but you can vary this time by replacing R5 with a potentiometer. Once the buzzer is triggered and the buzzer sounds, U5 should trigger about two to four minutes later, with an output seen at pin 3 of U5.

Once U5 sends its output to the timers at U6, the equipment off-timers should begin. The off-timer is set for about 15 to 20 minutes and can be adjusted with R9. The output at U6:B at pin 9 is used to activate Q2 via R10, and you can use the relay at RLY to turn off the equipment. Make sure to note the current ratings of the relay before trying to switch heavy power loads. You might have to use the relay to drive an external power relay. The contacts on RLY can be connected in series with 110 volts of ac and a second power relay to control large appliances. You could also use the lightning detector to disconnect a radio transceiver from an outdoor antenna by using an appropriate RF relay.

The experimental lightning monitor is a highly useful and preventative form of insurance against the fury of electrical storms. Why not protect your home electronic equipment with this lightning detector circuit?

Experimental lightning monitor parts list

R1	200-megohm, ¼-watt resistor
R2	20-kilohm, ¼-watt resistor
R3	10-kilohm, ¼-watt resistor
R4	500-kilohm potentiometer
R5,R6,R8	24-kilohm, ¼-watt resistor
R7	47-kilohm, ¼-watt resistor
R10	1-kilohm, ¼-watt resistor
C1	1-pF, 25-volt mica capacitor
C2,C3,C4,C8,C9,C13	.1-µF, 25-volt disc capacitor
C5	.01-µF, 25-volt disc capacitor
C6	100-µF, 25-volt electrolytic capacitor
C7	47-µF, 25-volt electrolytic capacitor
C10,C11	.047-µF, 25-volt disc capacitor
C12	.47-µF, 25-volt disc capacitor
C14	200-µF, 25-volt electrolytic capacitor
D1	Centronics OSD5.8-7Q UV diode, OSD35-7Q, or OSD5-7-254C
D2,D3	1N914 silicon diode
U1	OPA11 op-amp (Burr-Brown)
U2	LM211 comparator (National)
U3	CD4528 monostable multivibrator
U4	SN7400 quad Nand gate
U5	LM555 timer
U6	LM556 dual timer
Q1	2N3904 transistor
RLY	6- to 9-volt relay
Misc	UV filter, collimator, and chassis boxes

6
CHAPTER

Optical control circuits

This chapter runs the gamut of light-controlled circuits, from simple photoresistive and phototransistor light/dark-activated relays to a multichannel IR remote-control system. Covered in this chapter are numerous control circuits that can be triggered by visible and IR light, as well as UV or laser light, depending upon your particular application. An optoisolator relay, a 120-volt line monitor, and SCR and triac control circuits are illustrated, which can be used to control ac and dc power circuits or appliances. Computer control circuits, an auto night light, a photo flash circuit, and a fiber-optic control circuit are also presented in this chapter. Finally, an infrared control link and a five-channel IR remote-control system close the chapter.

The control circuits in this chapter and the sensing circuits in the previous chapter can be used in a stand-alone configuration or they could be combined with one another for more complex systems—whatever is required for particular applications. You can sense and control objects via visible, IR, UV, or laser light and then send the signals over fiber-optic or free-space links for ultimate long-distance control or signaling.

Photoresistive light-activated relay

The first light-activated relay circuit is depicted in Figure 6-1. A photoresistive cell is placed in series with a 1-kilohm potentiometer, which is fed to a 9-volt battery or power source. Resistor R1 is used to change the sensitivity of the light input, while resistor R2 is connected from the base of Q1 to ground. Transistor Q1, a 2N2222, is used to drive a 500-ohm general-purpose relay. Since the response of photoresistive cells is relatively slow, the relay will remain energized briefly after the light source is removed. You can use this simple circuit to turn on appliances or to activate displays or toys.

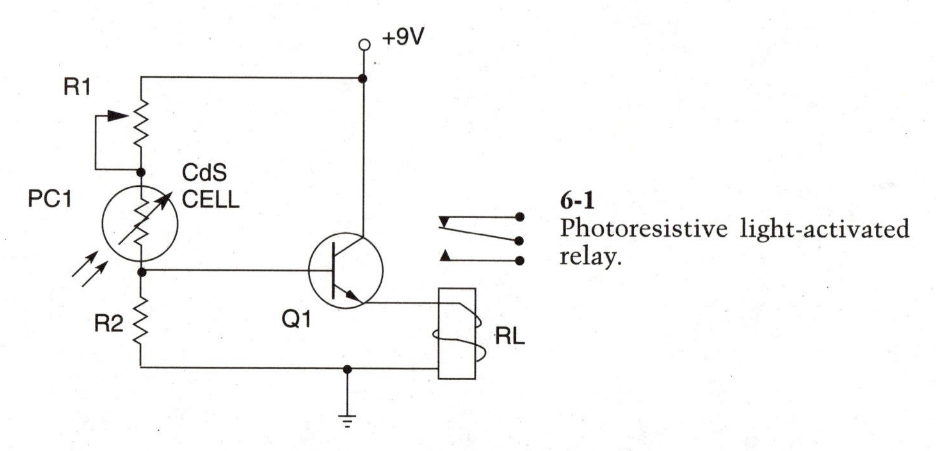

6-1
Photoresistive light-activated relay.

Photoresistive light-activated relay parts list

R1	1-kilohm potentiometer
R2	4.7-kilohm, ¼-watt resistor
Q1	2N2222 transistor
PC1	CdS photoresistive cell
RLY	500-ohm relay (RS 275-005)
B	9-volt transistor radio battery

Phototransistor light-activated relay

This second light-activated relay circuit substitutes a phototransistor for the CdS photoresistive cell. The phototransistor-actuated relay circuit shown in Figure 6-2 responds much faster than the previous circuit. Potentiometer R1 adjusts the sensitivity to incoming light. In this circuit, the phototransistor and potentiometer form a balanced input or voltage divider connected to the input of Q2's base. The emitter of Q2 drives a 500-ohm low-current relay. The phototransistor light-activated relay is powered from a 9-volt battery or power source. For optimum results, use a light shield or collimator tube ahead of the CdS or phototransistor to prevent stray light sources from falsely triggering both this and the previous circuit.

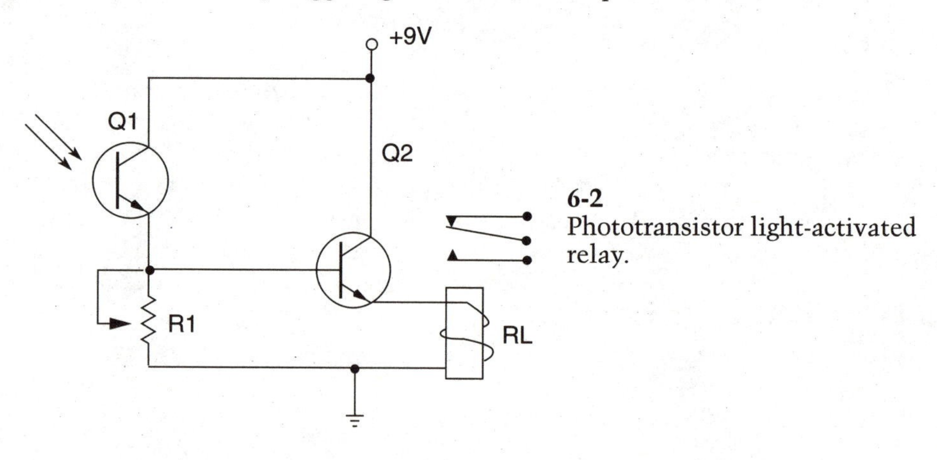

6-2
Phototransistor light-activated relay.

Phototransistor light-activated relay parts list

R1	100-kilohm potentiometer
Q1	Phototransistor (RS 276-145)
Q2	2N2222 PNP transistor
RLY	500-ohm relay (RS 275-005)
B	9-volt battery

Photoresistive dark-activated relay

The next two relay circuits are dark-activated. It is often necessary to activate a circuit with the absence of light, and the circuit shown in Figure 6-3 does just that. Notice the similarity between this circuit and the one in Figure 6-1. In the dark-activated relay circuit, the light-detecting sensor is connected between the base of the switching transistor and ground, rather than between the base and plus voltage supply. Resistor R1 is a 100-kilohm potentiometer, used to adjust the overall sensitivity of the circuit. Transistor Q1 is designed to drive a low-current 500-ohm relay. This circuit is powered from a 9-volt source.

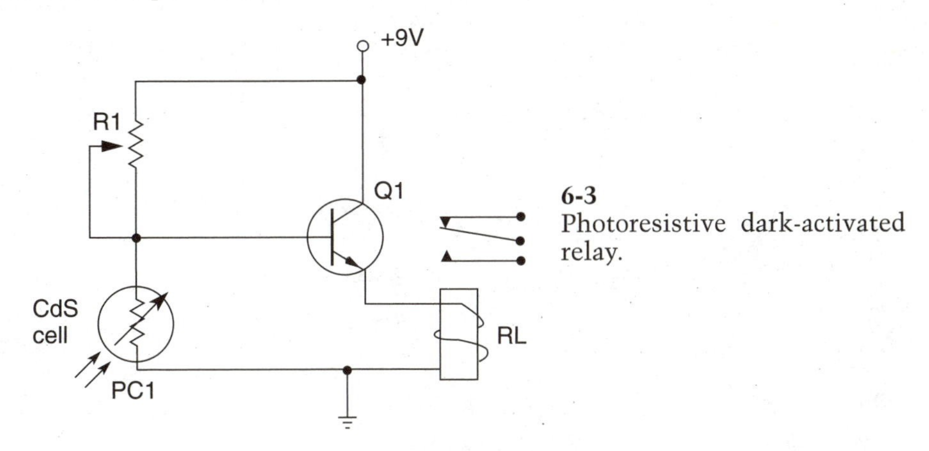

6-3 Photoresistive dark-activated relay.

Photoresistive dark-activated relay parts list

R1	100-kilohm potentiometer
PC1	CdS cadmium-resistive photocell
Q1	2N2222 transistor
RLY	500-ohm relay (RS 275-005)
B	9-volt battery

Phototransistor dark-activated relay

The second dark-activated relay circuit, shown in Figure 6-4, uses a phototransistor in a circuit similar to the one in Figure 6-2. Note that the phototransistor is connected between the base of Q1 and ground, opposite that of Figure 6-2. Potentiometer R1 is used to adjust the input sensitivity. The relay responds to the absence of light, so it is energized when light is taken away from the detector. This circuit responds quite a bit faster than the photoresistive dark-activated relay circuit. A 9-volt battery or power supply can operate this circuit.

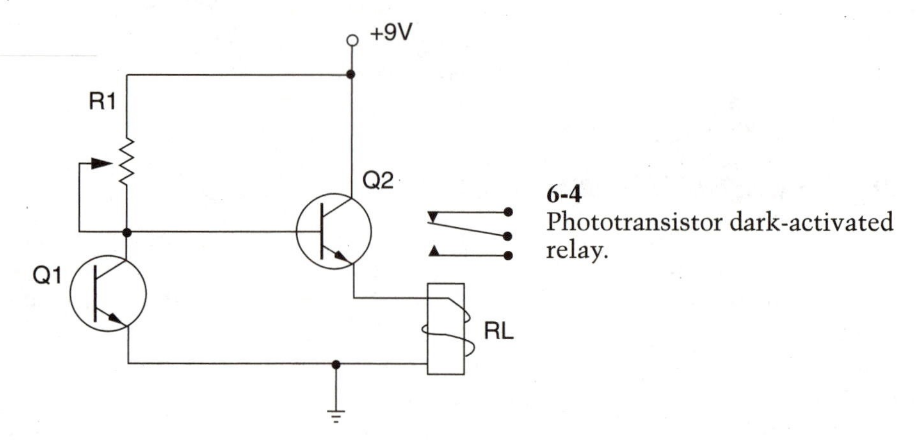

6-4 Phototransistor dark-activated relay.

Phototransistor dark-activated circuit parts list

R1	100-kilohm potentiometer
Q1	Phototransistor (RS 276-145)
Q2	2N2222 PNP transistor
RLY	9-volt power source

Linear bargraph relay driver

You can use a linear bargraph LED driver like the one shown in Figure 6-5 to control a relay. The relay circuit can be inserted at any LED output to control the relay with a specific voltage or incoming light level if the bargraph is used as a light-meter. Refer to the section *Bargraph light-meter* in Chapter 4. You could use this simple relay circuit with the bargraph light-meter described in that chapter to control an external circuit.

This bargraph is useful for dc or varying-input ac signals. The LM3914 shown is a linear bargraph, while LM3915 is useful for logarithmic applications such as power meters since each step between LEDs is 3 dB. The LM3916 bargraph display driver is pin-for-pin replaceable with the previous drivers, but it is a semi-log driver, which means it can be used for audio applications.

The relay drive circuit shown with the linear bargraph is comprised of a capacitor, a diode, and a relay. In effect, this circuit can perform a number of ac-

Linear bargraph relay parts list

C1	47-µF, 25-volt electrolytic capacitor
D1	1N914 silicon diode
U1	LM3914, LM3915, and LM3916
RLY	Low-current relay (RS 275-005)

tions, including detecting an overvoltage condition. It has a number of control possibilities.

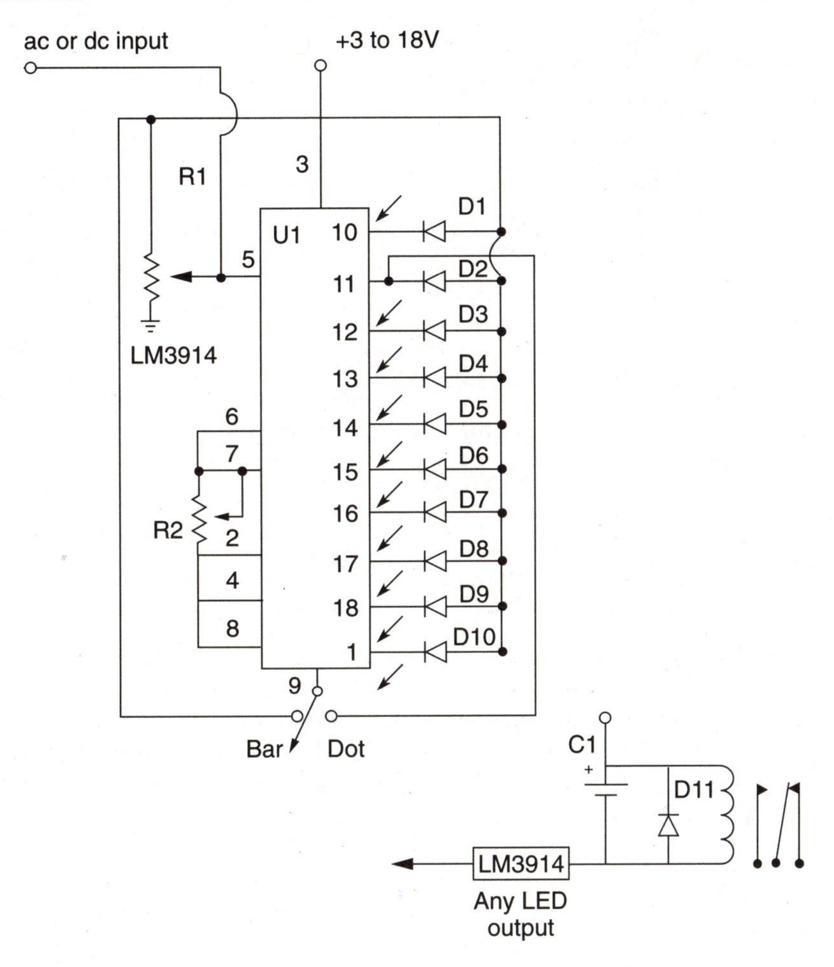

6-5 Linear bargraph LED relay driver.

Optical Schmitt trigger relay

The optical Schmitt trigger circuit shown in Figure 6-6 provides a clean switched output using an LM555 timer IC. A Schmitt trigger, as you might recall, will accept an input signal within a range of input voltages and pulse shapes, and provide a clean switched output pulse. This optical Schmitt circuit can be used in many applications where a high input impedance and a low output impedance are required with a minimum component count. The LM555 chip has both its trigger and threshold inputs tied together and connected to the voltage divider formed by R1 and PC1. A 12-volt relay is energized when light falling on the CdS cell falls below a preset input. The optical Schmitt trigger relay circuit can be powered by any 12-volt power source.

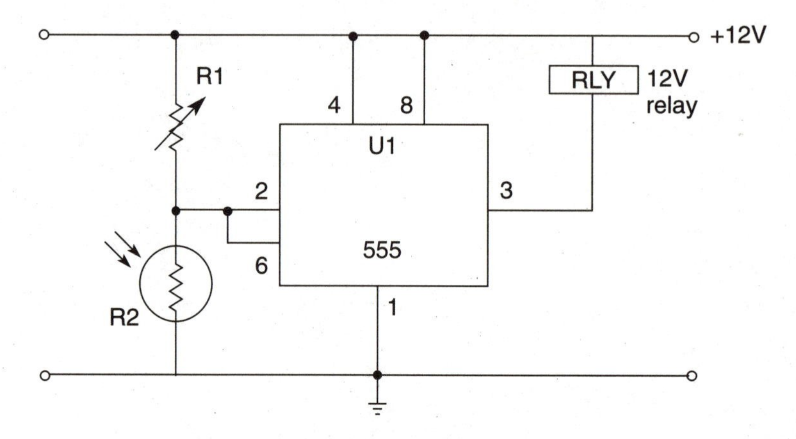

6-6 Optical Schmitt trigger relay.

Optical Schmitt trigger parts list

PC1	CdS photoresistive cell
R1	10-kilohm potentiometer
U1	LM555 IC timer
RLY	12-volt relay

120-volt ac line monitor

You can use an optocoupler relay circuit to monitor a 120-volt ac line voltage, or you can use the circuit in Figure 6-7 to determine when the ac line voltage is interrupted or has failed. This circuit is somewhat different than the previous optoisolator circuit since the input resistor at R1 uses a higher value and diode D1 is connected across the opto-

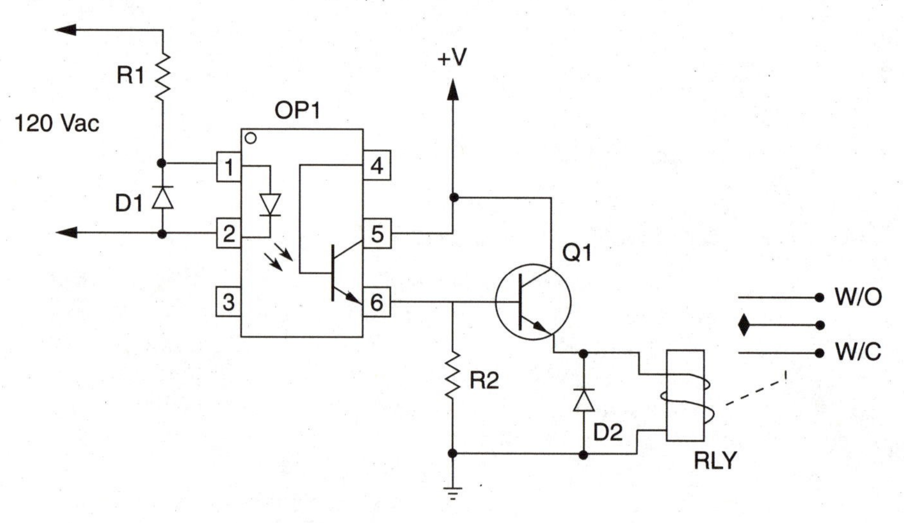

6-7 120-volt optoisolator line monitor.

coupler's LED, as shown. The transistor in the optoisolator drives Q1, which in turn drives the relay. You could use the output relay to sound an alarm or activate a phone dialer. The 120-volt line monitor can be powered from a 9-volt power source or battery.

120-volt ac line monitor parts list

R1	220-kilohm, ½-watt resistor
R2	4.7-kilohm, ¼-watt resistor
D1,D2	1N914 silicon diode
OP1	Texas Instruments TIL-111 optoisolator
Q1	2N2222 transistor
RLY	500-ohm relay (RS 275-005)

Optically isolated dc control circuit

The circuit shown in Figure 6-8 features an isolated dc control circuit. You can use this circuit for a variety of dc control applications where pulsating dc voltage is present at the load. Table 6-1 illustrates different dc input voltages and the corresponding values for input resistor R1. Table 6-2 shows the relationship between dc load voltage and values selected for the SCR and the value of R2.

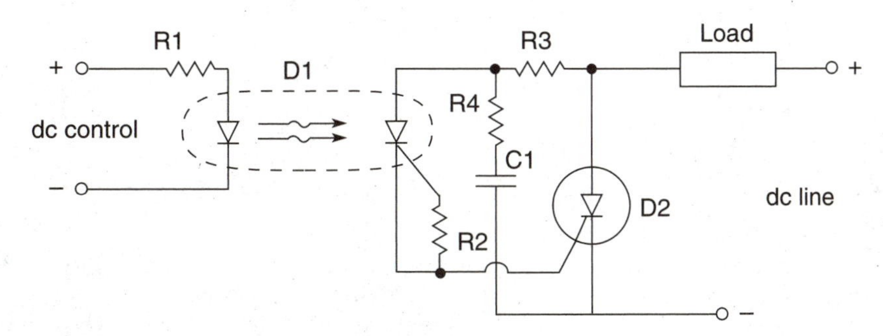

6-8 Optically isolated dc control circuit.

Table 6-1 Optically isolated dc control circuit.

dc input voltage (volts)	R1 value
6	470 ohms
12	1.1
24	2.4
48	4.7
120	12

Table 6-2 Optically isolated dc control circuit.

Line voltage (volts dc)	C122 SCR port #	R2 value
12	-U	200 ohms
24	-F	470 ohms
48	-A	1
120	-B	2.2

This SCR latch can be used for dc latch functions and provides reverse polarity blocking for currents to 300 mA. The gate cathode resistor is supplemented by a capacitor at C1, which minimizes the sensitivity to transients for pulsating dc applications. The capacitor value must be designed to either retrigger the SCR at the application of the next pulse or prevent retriggering at the next power pulse. For higher-current operation, you could use the optoisolator to trigger a higher-current SCR or a relay at the load to control a higher-current dc load.

Multivoltage dc SCR latch parts list

R1	See Table 6-1
R2	27-kilohm, ¼-watt resistor
R3	See Table 6-2
R4	100-ohm, ½-watt resistor
D1	H11C1 optocoupler
D2	C122 SCR (see Table 6-2)
C1	.01-µF, 50-volt disc capacitor
Load	dc load (see Table 6-2)

Triac optoisolator ac controller

Why not use your computer to control ac devices such as lamps, motors, or appliances by using the circuit shown in Figure 6-9? You could use the output of most logic or computer I/O ports to drive the LED in the triac optoisolator via resistor R1. A com-

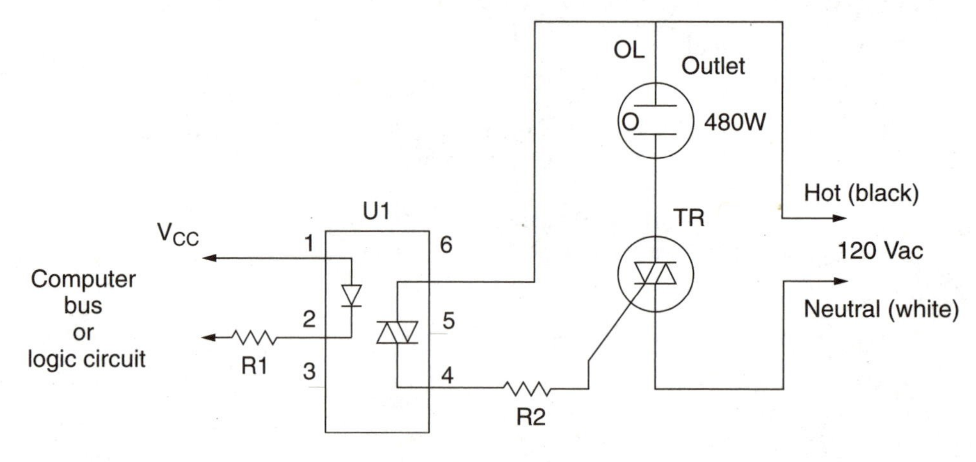

6-9 Triac optoisolator ac controller.

puter I/O bus can readily supply a 5-volt signal to control an ac load, as you will see. The optoisolator consists of an LED light source and a triac inside a six-pin chip. The LED under control of a PC or logic circuit will activate the triac via light. You can then use the low-current triac to control a higher-current ac load by using the triac at TR. You can then use the higher-current triac to control up to 4 amps of current or up to 480 watts at the outlet. If you anticipate running at the rated 480 watts, plan on using a large heatsink to dissipate heat from the triac. This useful ac control can be used in a host of applications where you want to use a small dc signal to control a large ac load device.

Triac optoisolator ac controller parts list

R1	560-ohm, ¼-watt resistor
R2	10-kilohm, 2-watt resistor
U1	MOC304215 triac optoisolator (Motorola)
TR	4-amp Q4004 (triac)
Misc	Outlet and heatsink

High-current inductive triac controller

The optoisolator circuit in Figure 6-10 can be used to control high-current inductive ac devices from any TTL logic device or computer port. Any 5-volt logic gate can activate the LED inside the MOC 3010 triac optoisolator via R1. The low-current triac in U1 is used to control or gate a second higher-current triac at D1. The triac at D1 is selected to pass the required current for your application. The high-current triac could then control an inductive load of your choice, such as motors, fans, and pumps.

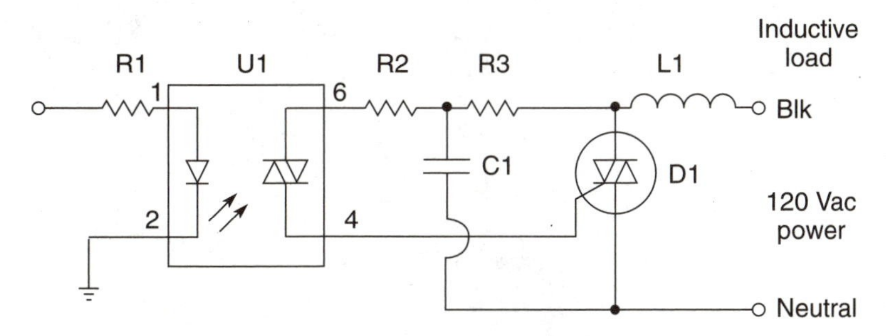

6-10 High-current inductive triac controller.

Inductive triac controller parts list

R1	680-ohm, ¼-watt resistor
R2	22-ohm, ¼-watt resistor
R3	2.4-kilohm, ¼-watt resistor
C1	0.1-µF, 100-volt disc capacitor
U1	MOC 3010 triac optoisolator
D1	Q4004-NO high-current triac
L1	Inductive load

Bistable ac control switch

The bistable ac control switch is shown in Figure 6-11. It uses an LM555 timer IC and a pair of CdS photoresistive cells to control appliances such as coffee pots or heaters with the sun rising or falling on the photoresistors. You could also use the circuit to turn on driveway lamps or walkway lights when your car headlights enter the driveway. The circuit can be used in a variety of remote-control applications, so let your imagination run wild.

In operation, light striking one of the photocells will switch the LM555 timer from a high to a low state, or vice versa. If initially in the high state, light striking PC1 will cause a positive input pulse that switches the timer, which in turn activates the lamp in the optocoupler. If the LM555 timer is initially in the low state, light striking PC2 will do the reverse and the circuit will switch states. The lamp in the optocoupler gates the triac via photoresistor R3. Once triggered, the triac can sustain power to an ac load.

The optocoupler can be constructed from an LED. If you use an LED for the light source, remember to include a series resistor. A third CdS photocell is used with the LED to form the optocoupler, as shown. You can use heat-shrink tubing to couple the optocoupler components together and seal the optocoupler from external light sources. To control a larger load, be sure to size the triac accordingly.

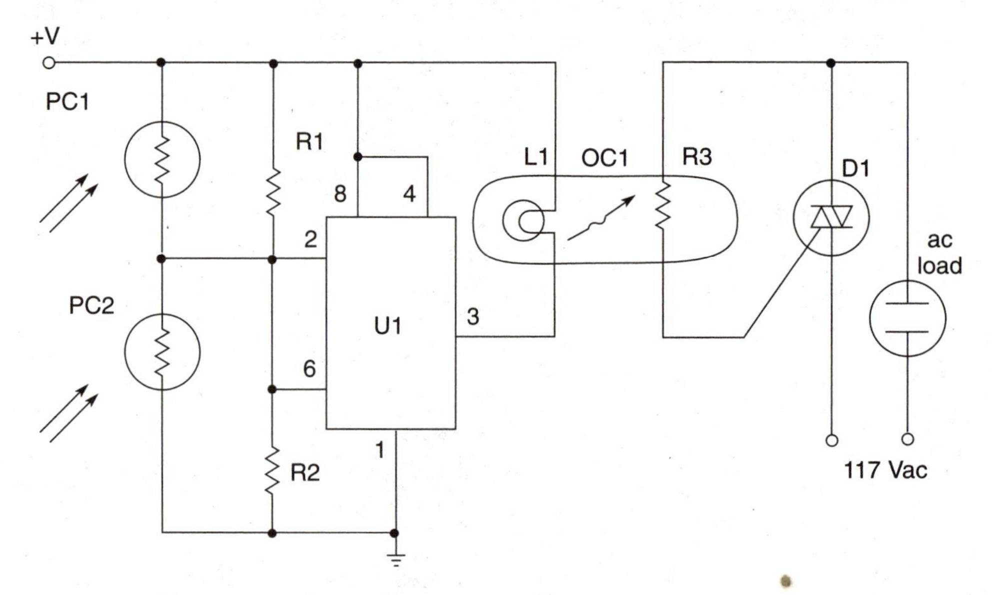

6-11 Bistable ac control switch.

Bistable ac control switch parts list

R1,R2	100-kilohm, ¼-watt resistor
PC1,PC2	CdS photoresistive cell
U1	LM555 timer IC
OC1	Optocoupler (see text)
D1	C106B SCR (or equivalent)

Line-powered automatic night light

Walking around your home or apartment in the dark of night can be dangerous, unless you have an automatic night light to light your way. The line-powered automatic night-light circuit is illustrated in Figure 6-12. The heart of the automatic night light is the L14C1 optotransistor. This circuit permits stable threshold operation based on the dependence of the current through the optotransistor, which generates a base emitter voltage drop across the threshold potentiometer. The double-phase shift network supplies voltage to the ST-4 trigger, which ensures that the triac triggers at line voltage phase angles that are small enough to minimize RFI/EMI problems. This eliminates the need for a large, expensive inductive filter.

The diode bridge supplies dc to the optotransistor and to Q2's circuitry. The varying light input signal then activates the ST-4 trigger, which in turn triggers the triac—thus lighting the lamp. You can select the triac from Table 6-3 based on the current requirement of the load lamp. Select capacitor C4 by the operating voltage (see the parts list). You can also use the automatic night light to light walkways or dark areas outside your home, provided you enclose the circuit in a watertight box. The line-powered automatic night light is a useful circuit that can easily eliminate broken bones and stubbed toes for all you sleepwalkers out there.

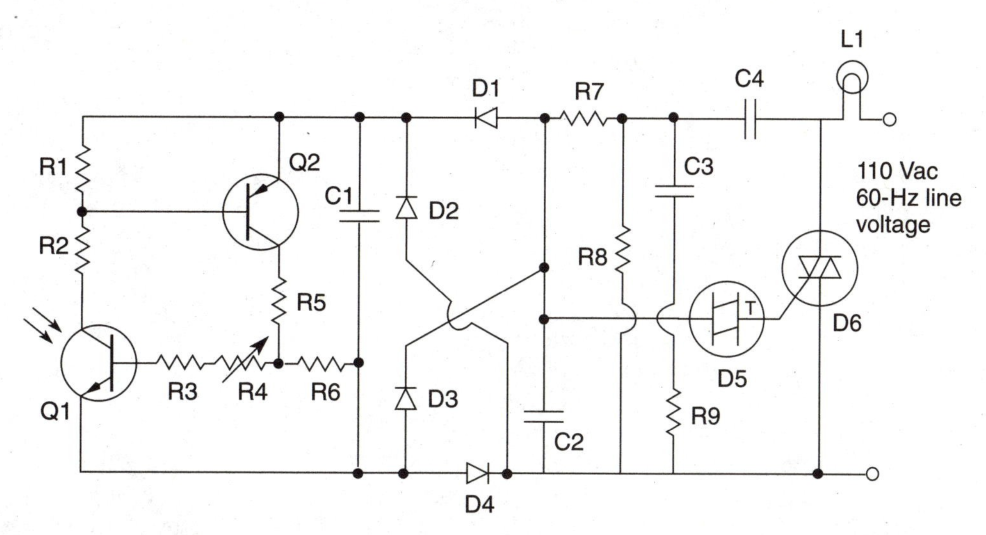

6-12 Line-powered automatic night light.

Table 6-3 Line-powered automatic night light.

Triac	120-volt (ac) lamp	220-volt (ac) lamp
SC141D	400 watts	800 watts
SC146D	550 watts	1100 watts
SC151D	750 watts	1500 watts
SC260D	1200 watts	2500 watts
SC265D	2000 watts	4000 watts

Automatic night light parts list

R1,R2	18-kilohm, ¼-watt resistor
R3	330-kilohm, ¼-watt resistor
R4	5-megohm potentiometer
R5	1-kilohm, ¼-watt resistor
R6	100-ohm, ¼-watt resistor
R7	33-kilohm, ½-watt resistor
R8	62-kilohm, ½-watt resistor
R9	22-ohm, 1-watt resistor
C1	1-µF, 25-volt electrolytic capacitor
C2	.1-µF, 25-volt disc capacitor
C3	.1-µF, 100-volt disc capacitor
C4	.1-µF, 100-volt capacitor for 120 volts ac and .068-µF, 100-volt capacitor for 220 volts ac
D1-D4	DHD806 or 1N4001 silicon diode
D5	ST-4 trigger
D6	Triac (see Table 6-3)
Q1	L14C1 optotransistor
Q2	2N6076 transistor
L1	Incandescent lamp (see Table 6-3)

Photo slave flash trigger

The photo slave flash trigger circuit shown in Figure 6-13 is a classic circuit used to trigger a remote photo flash unit at the same time as the the flash unit on a camera. The remote slave flash is commonly used for additional "fill" or "back" lighting.

The remote slave flash is remotely triggered via phototransistor Q1. The remote slave flash trigger unit can be used with any commercial photo flash unit. This circuit is designed to connect to the trigger cord or "hot-shoe" connector, which is found on almost all commercial flash units, and to be triggered by the light of your camera's flash unit. This provides remote operation without the need for messy trailing wires between the camera and the remote photo flash.

The heart of the remote slave flash unit is the L14C1 phototransistor. This phototransistor has a wide viewing angle, so alignment is not crucial. If you want a more sensitive or long-range directional flash, you could replace the phototransistor with an L14G2 lensed phototransistor. The L14G2 provides a 10-degree angle of view and gives over a 10-to-1 improvement in light sensitivity, and therefore a 3-to-1 range improvement. Note that the phototransistor is connected in a self-biasing circuit arrangement, which is relatively insensitive to slow-changing ambient light yet discharges the .01-µF capacitor at C2 into the SCR when illuminated by a remote flash. The diode bridge arrangement at the input of the circuit provides power to the circuit, as well as a means to activate or trigger the remote slave flash. The C106D SCR could be replaced by a C205D SCR if size is a concern. You could construct a few of these slave flash units to add yet more "fill" lighting. The completed slave flash unit can be built into a small plastic box for portability.

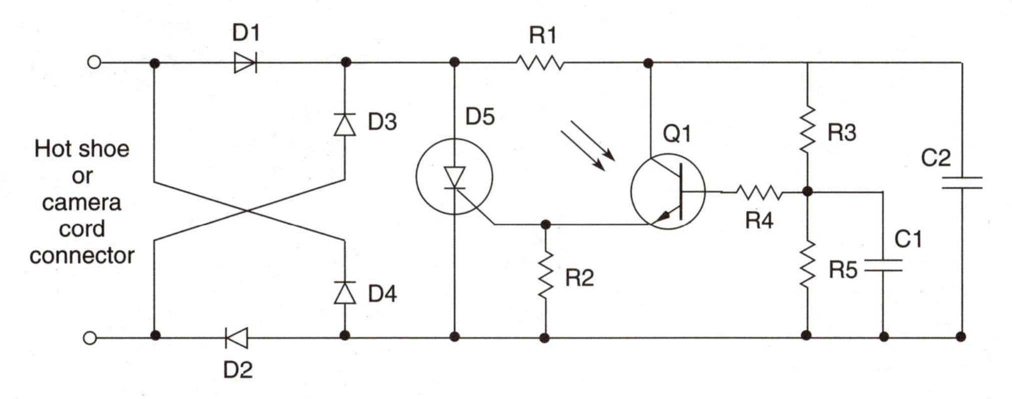

6-13 Photo slave flash trigger.

Photo slave flash trigger unit parts list

R1	2.2-megohm, ¼-watt resistor
R2	3.3-kilohm, ¼-watt resistor
R3	22-megohm, ¼-watt resistor
R4	1-megohm, ¼-watt resistor
R5	750-kilohm, ¼-watt resistor
C1	.1- µF, 200-volt disc capacitor
C2	.01-µF, 200-volt disc capacitor
Q1	L14C1 or L14G2 phototransistor
D1-D4	1N5060 silicon diode or equivalent
D5	C106D or C205D SCR

Fiber-optic solid-state ac controller

Fiber optics offers many advantages in power control systems. Since electrical signals cannot flow along nonconducting plastic fibers, shock hazards are prevented to both operators and equipment. EMI/RFI pickup on fibers is nonexistent, so you can implement long control lines and virtually eliminate noise pickup. Both ac and dc power circuits can be controlled by fiber-optic techniques similar to the circuit shown in Figure 6-14. A 5-volt control signal is applied to a GF0E1A1 LED or equivalent. A fiber-optic coupler then couples the light source to a reasonable length of fiber-optic cable. At the opposite end of the fiber-optic cable, phototransistor Q1 is also placed in a fiber-optic coupler housing. The incoming light modulates the GF0D1A1 phototransistor, which in turn activates a 2N4256 NPN transistor. Transistor Q2 drives an SCR at the junction of the voltage divider formed by R4 and R5. The ac triac is triggered from a low-gate voltage trigger current SCR. Switching line voltage is derived from a full-wave bridge formed by diodes D3 through D6. A varistor is placed across the triac for absorbing transients upon switching. The ac load is connected in series with the circuit, and power for the fiber-optic control circuit is supplied via the line's current input. This zero-voltage switch controller circuit is ideal for resistive loads such as lamps and heaters.

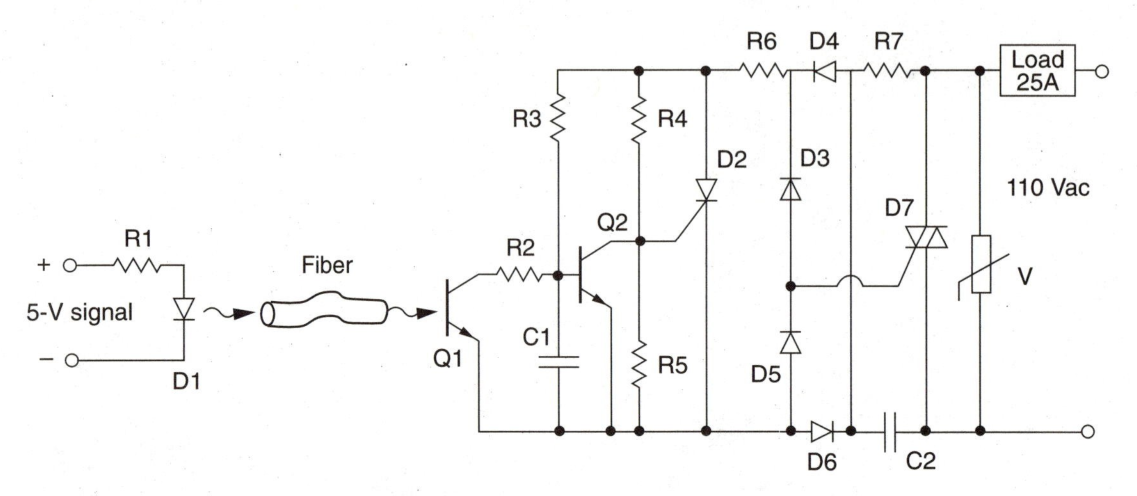

6-14 Fiber-optic solid-state ac controller.

Fiber-optic ac controller parts list

R1	33-ohm, ¼-watt resistor
R2	22-kilohm, ¼-watt resistor
R3	180-kilohm, ¼-watt resistor
R4	100-kilohm, ¼-watt resistor
R5	47-kilohm, ¼-watt resistor
R6,R7	100-ohm, 2-watt resistor
C1	330-pF, 50-volt disc capacitor
C2	.01-µF, 200-volt disc capacitor
D1	LED GF0E1A1 or equivalent
D2	C106 BX301 SCR
D3,D4,D5,D6	DT230B silicon diode or equivalent
D7	SC160B triac
Q1	2N4256 NPN transistor
V	V130LA20A 130-volt 20-amp surge varistor

IR remote-control system

There are numerous applications for an IC remote-control system. The infrared (IR) remote-control system shown in Figures 6-15 and 6-16 illustrates a simple yet effective IR control link that could be used to control appliances, toys, or audio systems.

The infrared transmitter portion of the IR link, shown in Figure 6-15, consists of an LM555 timer IC configured as a stable or free-running oscillator, which drives an IR LED at D1. The oscillator's frequency is determined by R1, R2, and C1. In operation, you need apply only a 9-volt battery through a momentary push-button switch to start the oscillator. Simply push the button to start the oscillator and adjust R1 until the receiver relay is activated. For optimum results and increased range, use IR filters and lenses for the transmitter and receiver sections, as shown.

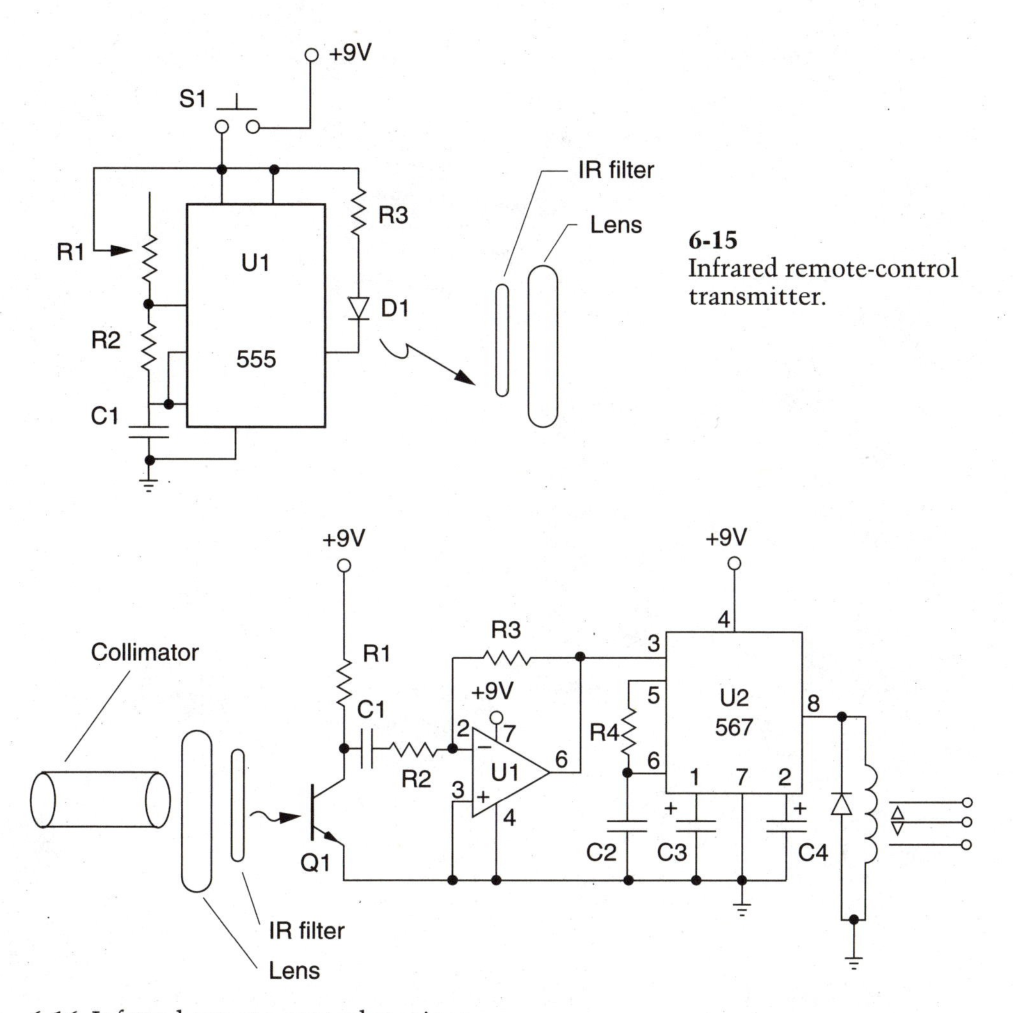

6-15
Infrared remote-control transmitter.

6-16 Infrared remote-control receiver.

The IR remote-control receiver shown in Figure 6-16 uses a phototransistor at Q1 to detect the IR transmitter's tone signal. A bias is first applied to Q1 via R1. Capacitor C1 and resistor R2 couple the modulated light from the phototransistor to the input of an LM741 op-amp at U1. The LM741 acts as an amplifier. Resistor R3 is used to set the gain of the op-amp. The output of the op-amp is then passed on to the tone decoder at U2. The LM567 tone decoder is configured to decode a specific tone frequency used by the transmitter. The decoder's frequency control is accomplished by resistor R4 and capacitor C3. The output of the LM567, when activated by the IR transmitter, can be used to control an LED optocoupler or a low-current relay, as shown. A standard 6-volt low-current relay could be used to control a higher-current relay in order to control higher-current loads such as appliances. The IR receiver can be powered from an ordinary 9-volt transistor radio battery or a 9-volt dc wall cube. As mentioned earlier, for increased range, use IR filters on both the transmitter and receiver, as well as lenses at both ends of the link. You should also place a collimator light tube ahead of the phototransistor Q1.

IR remote-control transmitter parts list

R1	10-kilohm trim potentiometer
R2	1.2-kilohm, ¼-watt resistor
R3	100-ohm, ¼-watt resistor
C1	.33-µF, 25-volt disc capacitor
D1	IR LED (high-power LED for longer range)
U1	LM555 timer IC
S1	Push-button switch
B	9-volt battery
Misc	Lens, IR filter, and collimator tube

IR remote-control receiver parts list

R1,R3	100-kilohm, ¼-watt resistor
R2,R5	1-kilohm, ¼-watt resistor
R4	10-kilohm, ¼-watt resistor
C1,C2	.1-µF, 25-volt disc capacitor
C3	2.2-µF, 25-volt electrolytic capacitor
C4	1-µF, 25-volt electrolytic capacitor
Q1	Phototransistor (RS 276-130) or equivalent
U1	LM741CN op-amp
U2	LM567 tone decoder
RLY	6-volt relay (RS 266 266-004)
B	9- volt battery
Misc	IR filter, lens, and collimator

Five-channel IR remote-control system

The five-channel infrared remote-control system in Figures 6-17 through 6-20 illustrates a reliable five-function remote control that you can use to control either multiple devices or multiple functions on a single complex device. The 40-kHz IR system could adapt itself to numerous applications around your home or office.

The 40-kHz IR remote-control transmitter is based on an LM556 dual-timer chip, shown in Figure 6-17. The first timer section is set up to oscillate between 100 and 1000 Hz. The momentary push buttons S2 through S6 and potentiometers R7 through R11 determine the tone selected for control purposes. The second oscillator in the LM556 is used to modulate the IR LED at the 40-kHz carrier-wave frequency. The five-channel IR remote transmitter can be placed in a small chassis box and can run from a 9-volt transistor radio battery. You can increase the range of the IR remote-control transmitter from the normal 8- to 10-foot range by using an IR filter, lens, and collimator arrangement ahead of the IR LED.

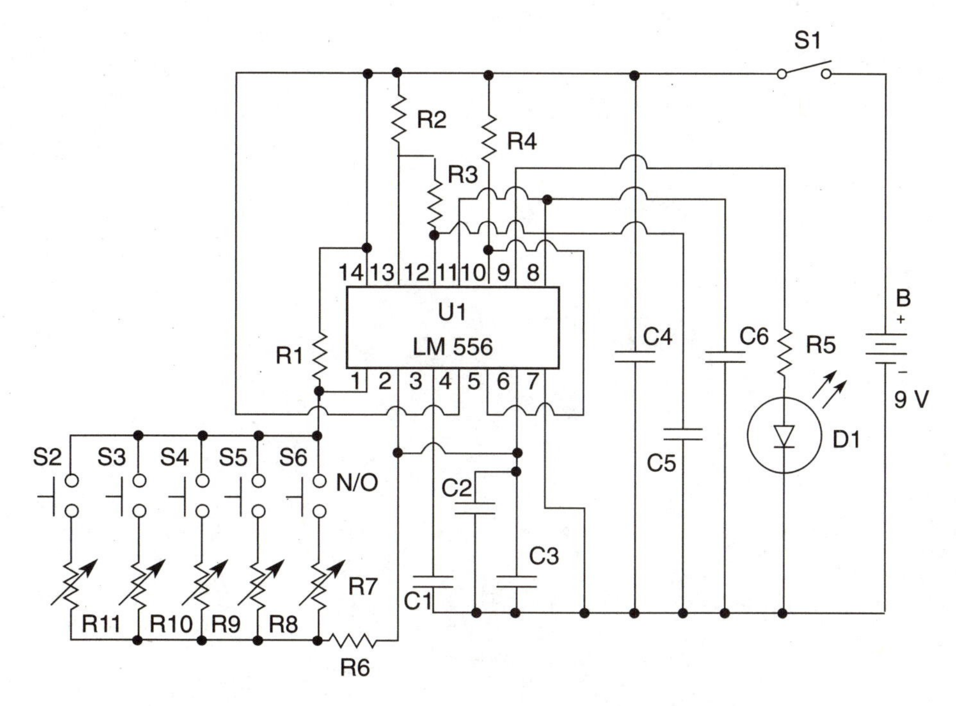

6-17 40-kHz five-channel IR remote-control transmitter.

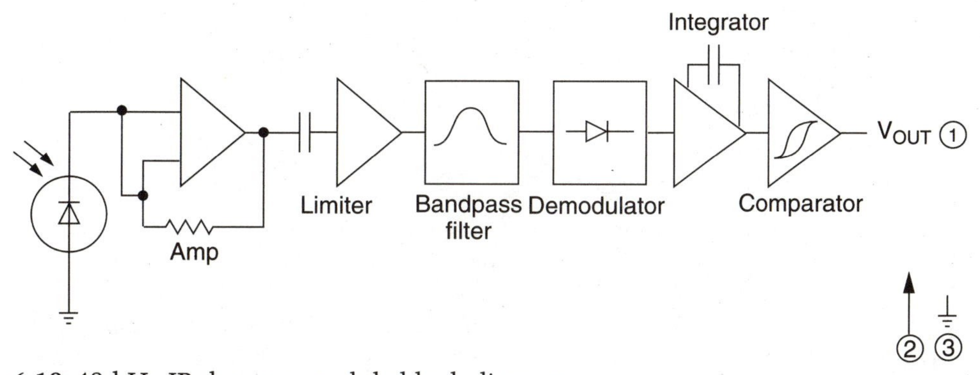

6-18 40-kHz IR detector module block diagram.

The heart of the five-channel IR remote-control receiver is the GPIU52X IR receiver/demodulator. The tiny IR receiver module shown in Figure 6-18 features an IR PIN photodiode detector followed by a preamplifier and limiter. The output from the limiter is fed directly into a bandpass filter tuned to 40 kHz. The bandpass filter rejects all signals outside ± 4 kHz of the 40-kHz carrier frequency. The resultant output signal from the bandpass filter is coupled to a demodulator, then to an integrator, and finally to a comparator at the output of the IR module. The final output of the IR receiver module is a clean, well-filtered signal. The 40-kHz receiver module is

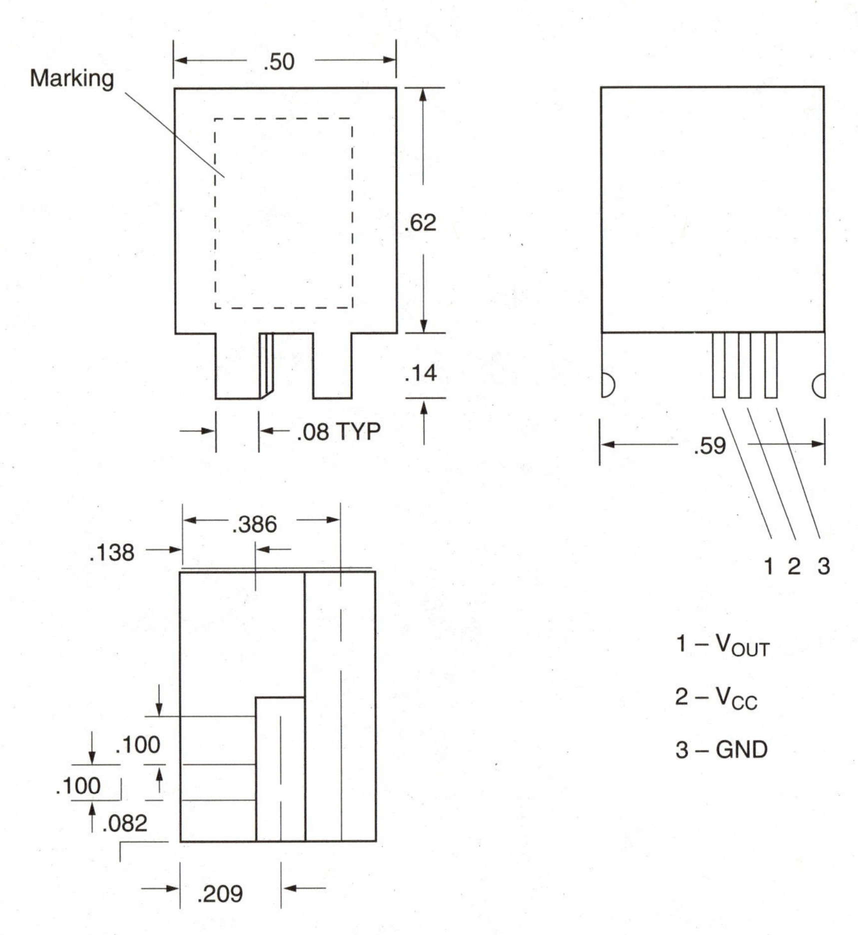

6-19 40-kHz IR module pinout.

packaged in a small metal case measuring about ½ inch square, as depicted in Figure 6-19. The IR receiver module consumes only 5 mA, at 5 volts dc.

The output of the IR receiver/demodulator module at D1 drives a one-transistor amplifier that is capacitively coupled via C1 to five LM567 tone decoders, as shown in Figure 6-20. Each decoder can be adjusted to decode the five tones produced by the IR transmitter. Each LM567 tone detector features a potentiometer at pin 5, which matches a particular tone from the IR transmitter. Each tone decoder activates a small, low-current, 5-volt relay to control a particular function. The output contacts of each relay can directly power a low-current device, or the relay contacts can activate a large-current relay in order to power an even larger-current device such as a home appliance. The five-channel IR receiver system can be housed in a metal chassis box and powered from a 5-volt power source of your choice.

Remember to keep the leads from each tone decoder input as short as possible and use good printed circuit techniques. The five-channel, 40-kHz IR remote-control system, when completed, must be calibrated for proper operation. Each of the

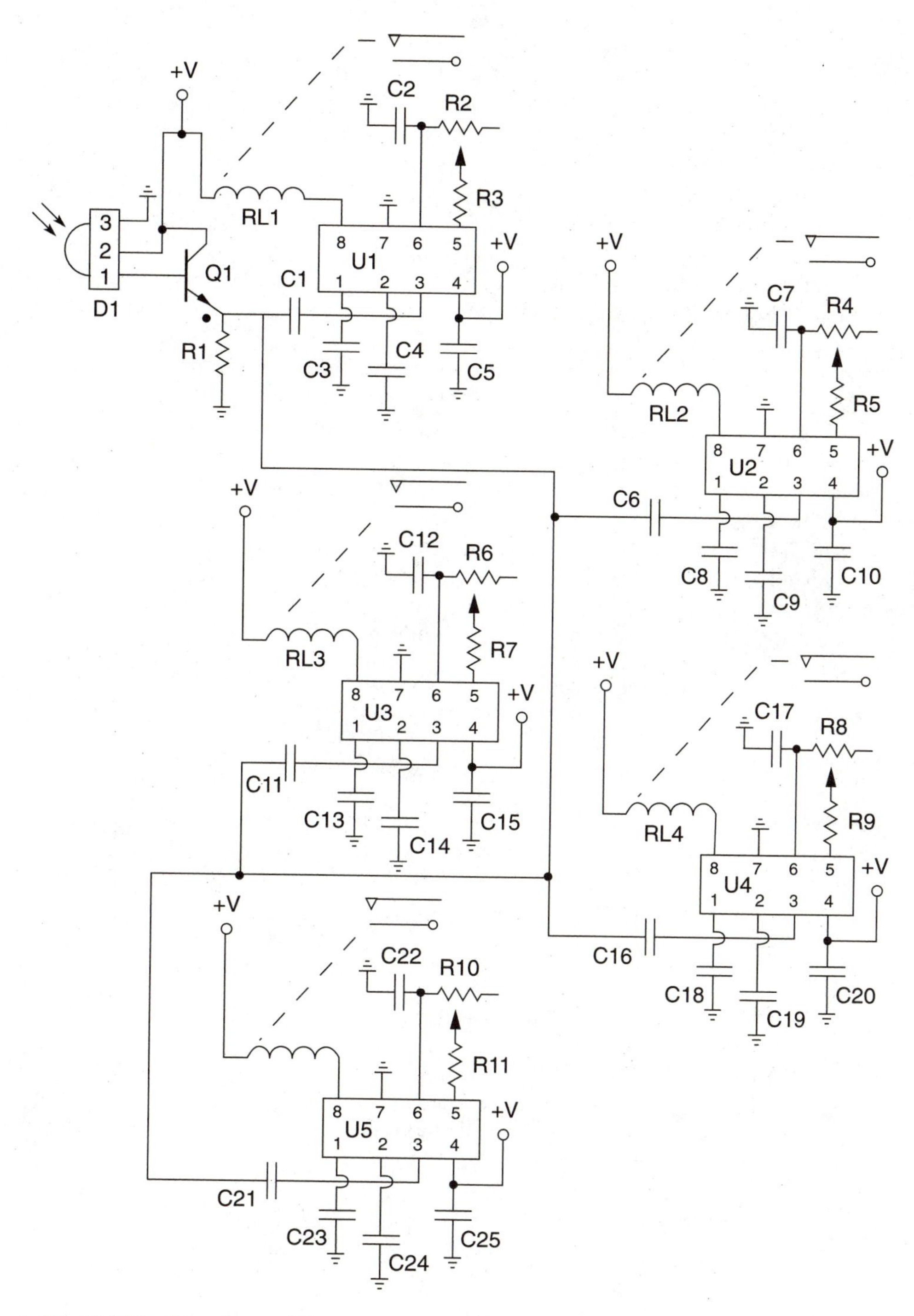

6-20 40-kHz five-channel IR remote-control receiver.

transmitter tones must be between 100 and 1000 Hz for correct operation, and each tone should be spaced at least 150 Hz apart for good detection and reliable results. Make sure to view the output of the first oscillator at pin 5 of the LM556 with the oscilloscope before making tone adjustments. Then you can adjust each of the five tones separately.

Now adjust the second oscillator stage of the transmitter for 40-kHz operation. Ground pin 6 and adjust for 40 kHz, ± 2 kHz, from pin 9 with a high-impedance frequency counter. The transmitter section is now ready for operation.

The IR receiver subsystem must also be calibrated for proper operation, which can be done in two different ways. The first method is to use a sine-wave generator coupled to the emitter of Q1. If the five tones you selected for the transmitter were recorded during calibration, you can simply dial up your signal generator for each tone used. The second method of receiver calibration dispenses with the audio generator and uses the transmitter itself to generate the tone. The transmitter and receiver are both placed on a flat, smooth table at a close distance to each other. The transmitter is turned on S1 and each tone is selected on an ascending scale, from lower to higher frequencies.

As each tone is selected from the transmitter, you adjust a selected potentiometer at one of the LM567 tone decoders to the corresponding tones until all five channels are adjusted. For optimum results, use an IR filter, lens, and collimator with both the IR LED in the transmitter and the input of the IR receiver module.

Five-channel IR remote-control transmitter parts list

R1,R2	1-kilohm, ¼-watt resistor
R3	33-kilohm, ¼-watt resistor
R4,R6	5.6-kilohm, ¼-watt resistor
R5	120-ohm, ¼-watt resistor
R7,R8,R9,R10,R11	.01-µF, 25-volt disc capacitor
C2	.047-µF, 25-volt disc capacitor
C3	.022-µF, 25-volt disc capacitor
C4	.1-µF, 25-volt disc capacitor
C5	470-pF, 25-volt Mylar capacitor
C6	.01-µF, 25-volt disc capacitor
S1	SPST toggle or slide switch
S2,S3,S4,S5,S6	Normally open push-button switch
D1	IR LED
U1	LM556 dual-tone chip
B	9-volt battery

Five-channel IR remote-control receiver parts list

D1	GPIU52X IR receiver module (RS 276-137)
R1	1-kilohm, ¼-watt resistor
R2,R4,R6,R8,R10	100-kilohm trim potentiometer
R3,R5,R7,R9,R11	5.6-kilohm, ¼-watt resistor
C1,C6,C11,C16,C21	.01-µF, 25-volt disc capacitor
C3,C8,C13,C18,C23	4.7-µF, 25-volt electrolytic capacitor
C4,C9,C14,C19,C24	4.7-µF, 25-volt electrolytic capacitor
C5,C10,C15,C20,C25	.047, 25-volt disc capacitor
U1,U2,U3,U4,U5	LM567 tone decoder chip
Q1	MPS2222A transistor or equivalent
RL1 to RL5	5-volt miniature relay (RS 275-240)

7
CHAPTER
Optical alarm circuits

Alarm circuits play an ever-increasing role in today's modern world. Crime and pilferage are all around us in almost every neighborhood, which has prompted many law-abiding citizens to take personal action against burglars and criminals.

The collection of circuits in this chapter includes power-outage sensors, personal alarms, object-sensing alarms, smoke alarms, body-heat sensors, and a long-range laser perimeter alarm system. Many of these alarm circuits can be used independently as self-contained alarm systems, or you can combine a number of circuits with an alarm control box to form a complete home or office alarm system.

You can use many of the circuits described in this chapter to serve as warning devices to keep kids or adults away from particular objects, or you can use the sensing and alarm circuits together to announce personal arrivals or departures. Alarm sensing circuits can also be used in point-of-sale displays to activate automated demonstrations when customers approach the display.

There are many uses for alarm sensors and detectors. Let your imagination inspire you to find applications for the circuits presented.

Line-operated power-outage/emergency light

The first circuit in this chapter is a 110-volt, line-operated, power-outage and emergency light, shown in Figure 7-1. The power-outage/emergency light is a flexible circuit that senses 110-volt ac power via a neon lamp. Neon lamps come in many voltages, so this circuit can also detect the presence of higher or lower voltages. A series resistor is connected to the neon lamp to limit current and extend the lamp's life. Phototransistor Q1 should be positioned to maximize coupling of both the neon lamp and ambient light without allowing self-illumination from the 6-volt lamp. The phototransistor light detector is coupled to Q2, which amplifies the light level in order to activate Q3, the semiconductor lamp switch. Transistor Q3 drives a 6-volt lamp or series of lamps around the house. A 6-volt battery can power the entire emergency-light circuit.

The 6-volt lamp serves as an emergency lamp when power fails, but you can easily adapt this circuit to serve other functions by replacing the lamp with a 5- to 6-volt relay. Once a relay is installed, the power outage detector can be coupled to an alarm control box, which can sense power loss to crucial instruments, freezers, or refrigerators. You could also use the power-outage detector to activate a battery-powered automatic telephone dialer, which could be programmed to call you or a relative in a remote location.

Line-operated power-outage/emergency light parts list

R1	51-kilohm, ¼-watt resistor
R2	1-megohm, ¼-watt resistor
R3	1.5-megohm, ¼-watt resistor
R4	1-kilohm, ¼-watt resistor
R5	120-ohm, ¼-watt resistor
C1	1-µF, 50-volt ceramic capacitor
Q1	L14R1 phototransistor
Q2	D16P1 transistor
Q3	D41E1 transistor
L1	NE-2 neon lamp
L2	6-volt lamp or relay
B1	6-volt battery

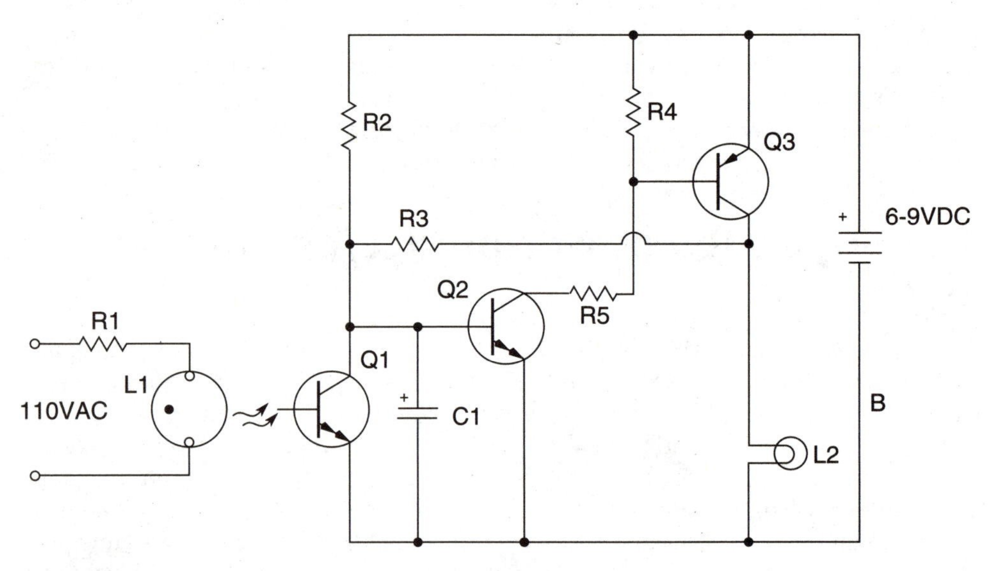

7-1 Line-operated power-outage/emergency light.

Personal alarm and strobe

The personal alarm circuit shown in Figure 7-2 is a very handy safeguard for people who walk or run in secluded areas by day or night. The personal alarm features both a very loud siren, or sonalert sounder, to attract attention (which no mugger or criminal wants) and a flashing strobe lamp to momentarily blind an assailant.

When the power switch is turned on, power is applied directly to the 110-dB sonalert sounder or solid-state siren. At the same instant, power flows into the direct coupled oscillator formed by Q1 and Q2. The NPN and PNP transistor oscillator's frequency is controlled by C1 and R2. The oscillator drives one side of the transformer T1. The secondary winding of T1 produces over 100 volts. This high voltage is rectified and applied to capacitor C2. A neon relaxation oscillator in this secondary circuit in turn activates the SCR. The SCR causes C3 to discharge through T2. As soon as the voltage applied to the neon bulb is high enough to ionize the lamp, the charge on C3 is applied to T2 through the SCR, D2. The trigger coil T2 converts the 100 volts from T1 to the 4-kilovolt pulse needed to ionize the xenon-gas strobe tube. When the gas in the xenon flash tube is ionized, a brilliant flash is produced. The cycle repeats itself over and over to keep the strobe flashing.

The personal alarm and strobe can be powered from any 6- to 9-volt battery source. The components in this circuit are not particularly crucial except for T1, T2,

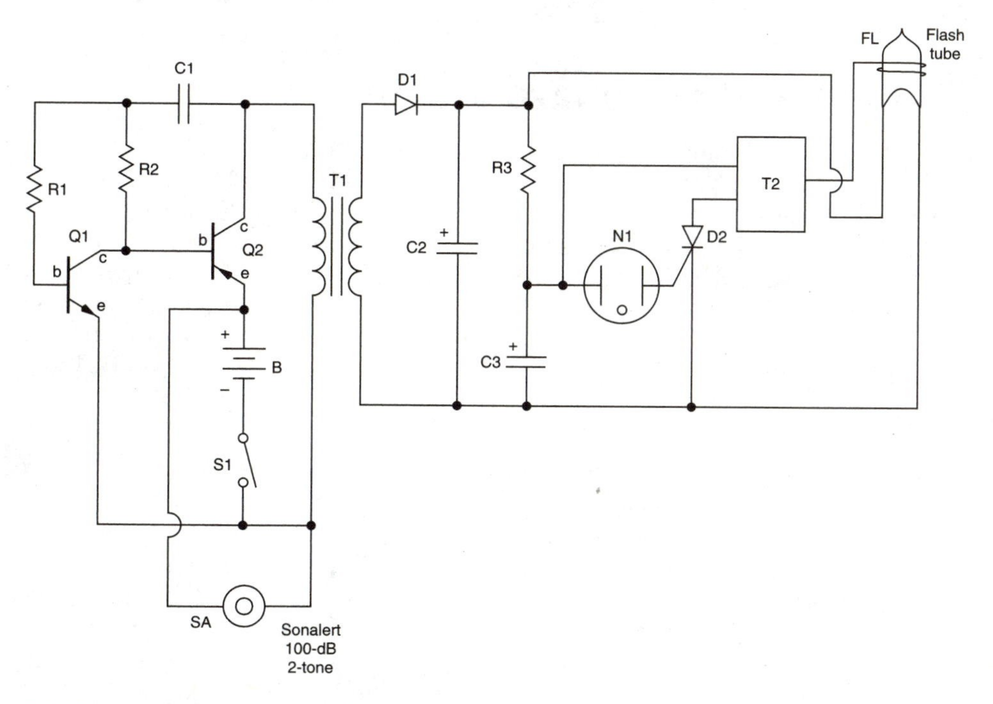

7-2 Personal alarm and strobe circuit.

and the flash tube, which is available from the sources listed in the Appendix. The personal alarm and strobe can also be used for other applications. If you replace S1 with a small relay, you could remotely control the unit from an alarm control panel and use the circuit outside under the roof overhang in front of your home to announce that your alarm system has been activated.

Personal alarm and strobe parts list

R1	9.1-kilohm, ¼-watt resistor
R2	430-kilohm, ¼-watt resistor
R3	22-megohm, ¼-watt resistor
C1	.001-µF, 25-volt dc disc capacitor
C2	1.8-µF, 200-volt Mylar capacitor
C3	.1-µF, 200-volt Mylar capacitor
D1	1N4007 silicon diode
D2	NTE5408 SCR
Q1	2N3904 NPN transistor
Q2	2N3906 PNP transistor
T1	100- to 6-volt transformer
T2	4-kilovolt trigger pulse transformer
SA	6- to 9-volt sonalert/siren (RS 273-070 or equivalent)
S1	SPST toggle switch
B	6- to 9-volt battery

Drawer and cupboard alarm

You can use the compact light alarm shown in Figure 7-3 and Photo 7-1 in a variety of applications, protecting your medicine cabinet, cash drawer, cupboards, cookie jars, or closets from unwanted hands. The light alarm will sound as soon as light falls on the Darlington phototransistor, i.e., when a drawer or cabinet is opened to ambient light. The heart of the sensitive light alarm is the MEL12 Darlington phototransistor, which consists of two super-alpha phototransistors in a single package. This particular Darlington transistor is extremely sensitive to light; the MEL12 is 10 times more sensitive than a conventional phototransistor and 100 times more sensitive than a photodiode.

In dark conditions, the MEL12 photo-Darlington is virtually an open circuit. No current flows through the 500-kilohm potentiometer to the base of Q2, and hence the detector is off. In this condition, the logic level at U1 is low. Once light strikes the photo-Darlington, current begins to flow through Q2, which in turn activates the oscillator formed by the CD4011 quad Nand gate array. The oscillating output from the CD4011 turns transistor Q3 on and off. When Q3 is on, the coil L1 and buzzer BL charge up, and the piezo buzzer strangely enough acts as a capacitor. When Q3 is turned off momentarily, the piezo buzzer makes a click as it discharges. Once the piezo buzzer discharges, the magnetic field around the coil at L1 collapses, the piezo sounder again charges up, and the cycle begins again. This action is called *ringing*. Normally there are resistive losses and the amplitude of oscillation is slightly less

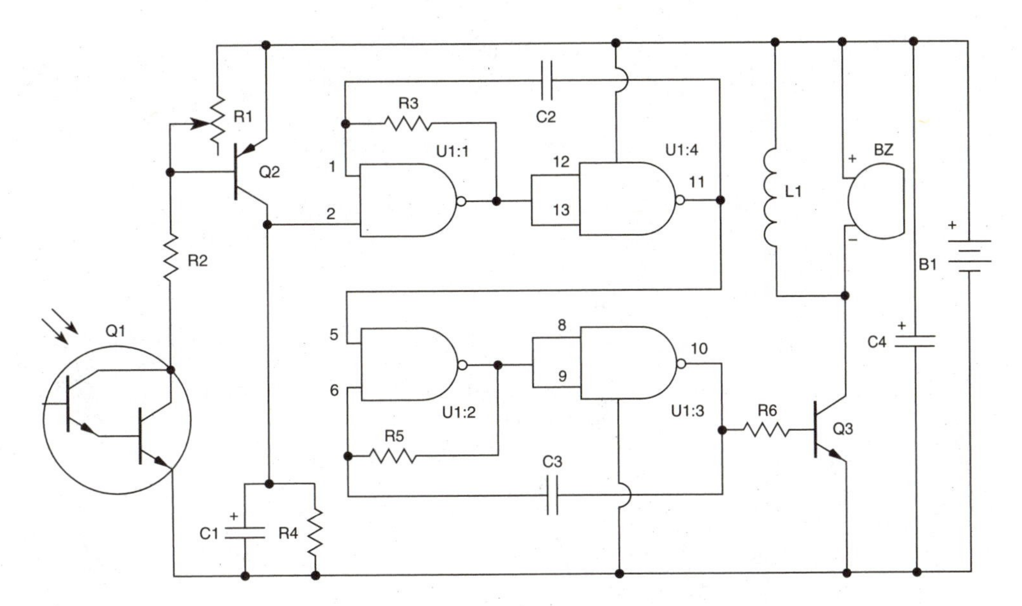

7-3 Drawer/cupboard alarm.

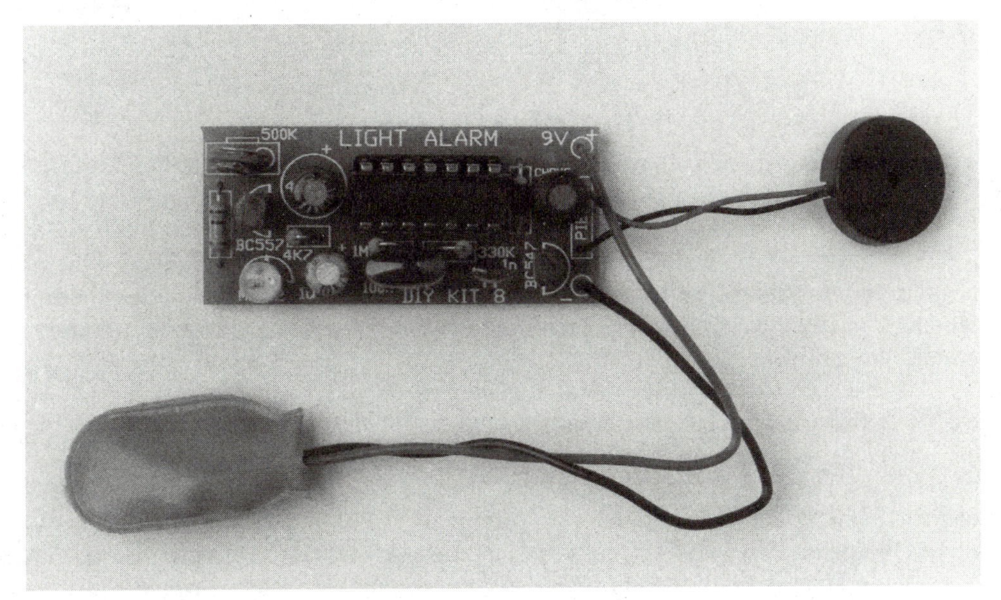

Photo 7-1 Drawer/cupboard alarm.

with every cycle, but the ringing does not decay away completely before Q3 turns on again and recharges the tank circuit formed by B2 and L1.

Once the alarm is put back into the dark condition, the alarm continues for about three seconds before it goes completely quiet due to capacitor C1. The light alarm consumes very little power with the absence of light falling on the photo-Darlington;

therefore no power switch is included. The light alarm circuit is powered by a 9-volt transistor radio battery, which will last for quite a long time.

Drawer and cupboard alarm parts list

R1	500-kilohm potentiometer (trimmer)
R2	100-kilohm, ¼-watt resistor
R3	1-megohm, ¼-watt resistor
R4	4.7-megohm, ¼-watt resistor
R5	330-kilohm, ¼-watt resistor
R6	4.7-kilohm, ¼-watt resistor
C1	1-µF, 25-volt tantalum capacitor
C2	100-nF, 25-volt polyester capacitor
C3	1-nF, 25-volt polyester capacitor
C4	47-µF, 25-volt electrolytic capacitor
Q1	MEL12 photo-Darlington transistor
Q2	BC557 NPN transistor
Q3	BC547 PNP transistor
U1	CD4011 quad Nand gate
L1	10- to 20-MHz coil
BZ	Piezo buzzer
B1	9-volt battery

Reflective IR optical alarm chip transceiver

The reflective infrared optical transceiver alarm circuit shown in Figures 7-4 and 7-5 illustrates a compact alarm unit that can be used to protect doors or windows in an intrusion alarm configuration. The IR transceiver alarm can also be used for "spot" alarms, dust detectors, particulate or smoke alarms, or industrial parts-counting applications.

The heart of the IR transceiver alarm is a new chip from Cherry Semiconductor, the CS-258A optical transceiver, shown in Figure 7-4. The CS-258A transceiver chip is a complete alarm system that incorporates a three-pole filter network with a gated pulse detector scheme that prevents false alarm conditions. When the IR source is blocked from the detector, no output is produced, but as soon as the IR beam is detected by the phototransistor (i.e., an object moved from the light path), an output is presented on pin 11 of the chip. The output on pin 11 drives Q2, which activates the SPDT relay, which in turn could activate an alarm bell or siren. You could create a sensor module on a small circuit board, incorporating the IR source D2, the photodiode D1, and the CS-258A chip.

The IR source and IR detector should both face outward from the circuit board, but angled from 40 to 50 degrees to receive the reflected signal, as shown in Figure 7-5. The CS-258A reflected-light optical transceiver can be run from 5 to 6 volts. The chip has a 7-volt maximum rating, so be careful and use a regulator or zener diode to

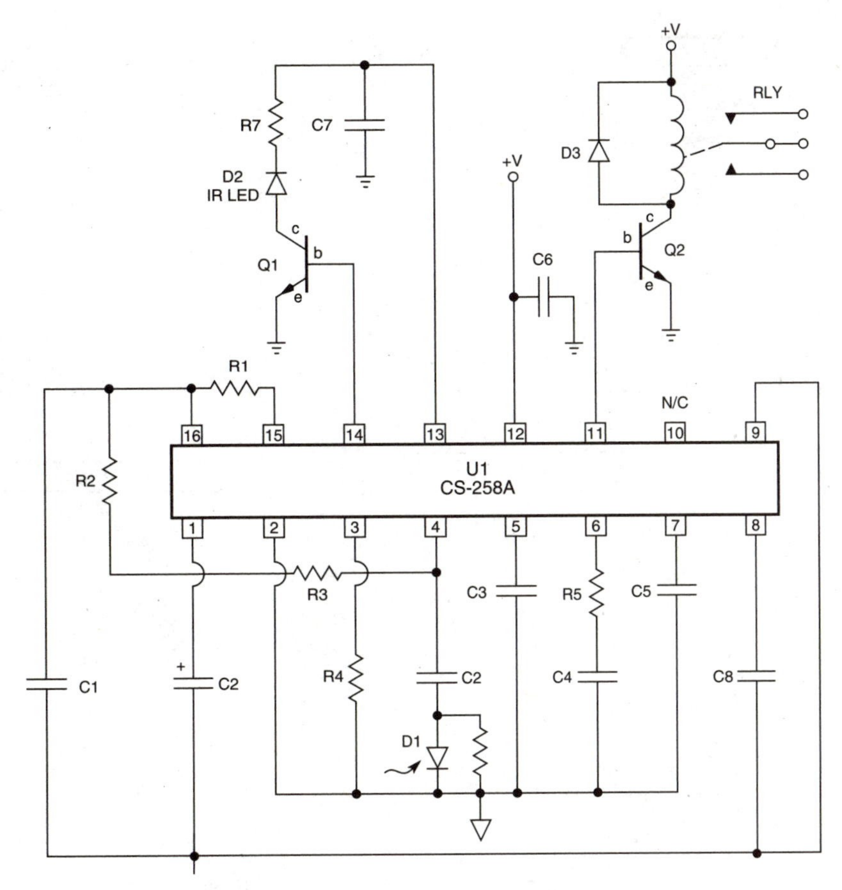

7-4 Reflective IR optical alarm transceiver chip. Cherry Semiconductors. Reprinted by permission.

regulate the supply voltage. The circuit is quite straightforward, but pay particular attention to the dual ground system. The high-level ground is on pin 9, while the low-level ground is on pin 2.

Reflective IR optical alarm chip transceiver parts list

R1	220-ohm, ¼-watt resistor
R2	820-ohm, ¼-watt resistor
R3	680-kilohm, ¼-watt resistor
R4	22-kilohm, ¼-watt resistor
R5	1-kilohm, ¼-watt resistor
R6	1-ohm, ¼-watt resistor
R7	1.8-kilohm, ¼-watt resistor
C1	.1-µF, 25-volt ceramic capacitor
C2,C6	.002-µF, 25-volt disc capacitor
C3,C4,C5,C8	.22-µF, 25-volt capacitor
C7	1000-µF, 25-volt electrolytic capacitor

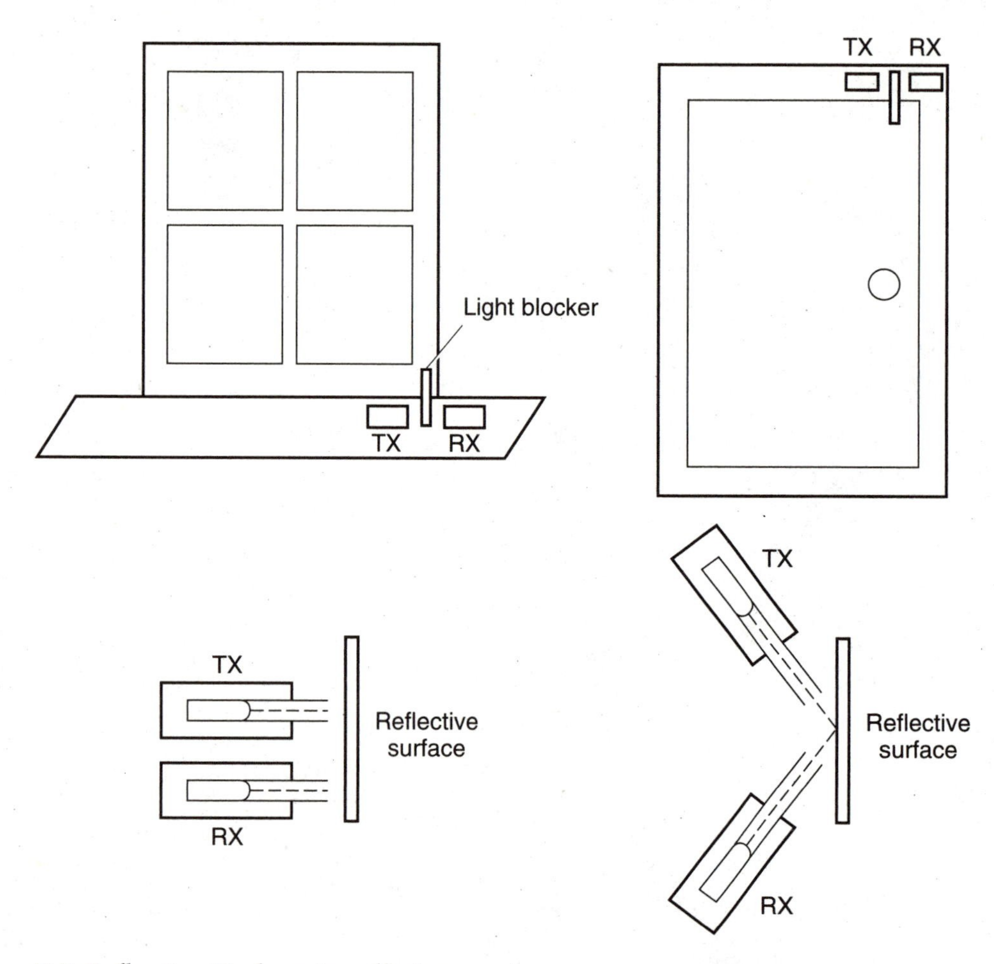

7-5 Reflective IR alarm installation.

Reflective IR optical alarm chip transceiver parts list continued

D1	Photodiode or optocoupler
D2	IR LED or optocoupler
C3	1N4001 silicon diode
Q1	2N2222 transistor
U1	CS-258A optical transceiver chip
RL	5- to 6-volt relay

Long-range object detector

The long-range object detector shown in Figure 7-6 can be used for many applications, including object counting, alarm detection, and control systems. The system shown can be configured with about 50 inches of distance between the IR LED and phototransistor. When long-range detection of objects is your problem and you need a reliable system, the long-range object detector might just be your answer. You can

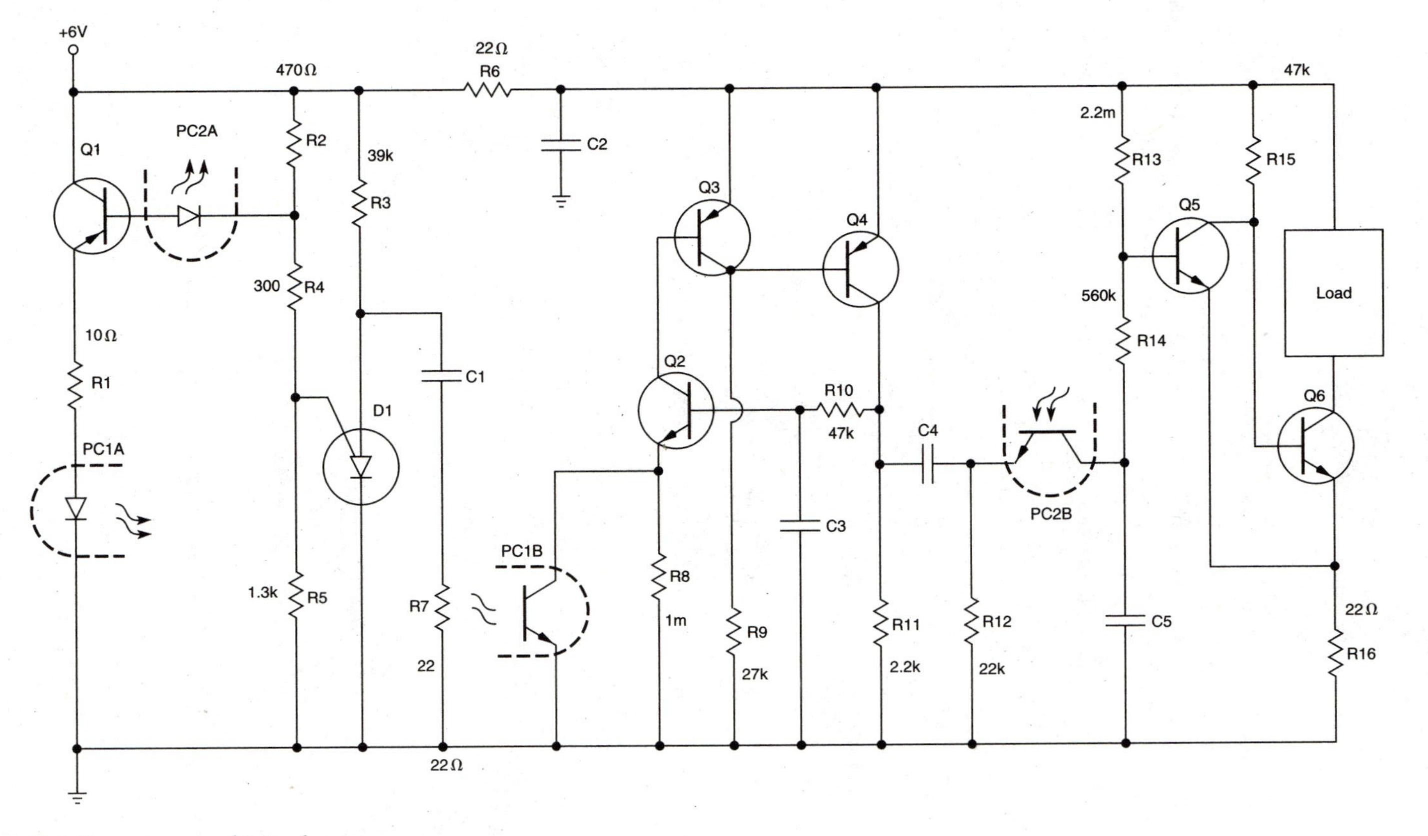

7-6 Long-range object detector.

attain additional reliability by synchronously detecting the photodetector current, as is done in this long-range object detector.

Photodetector PC-1 is a discrete IR light source and phototransistor pair, chosen for range; the specifics are listed in Table 7-1. You can achieve relatively long distances by using 10-microsecond pulses at a 2-millisecond period. The phototransistor current is amplified by transistors Q1 and Q2. PC-2 and the SCR help to increase the range of the synchronous detector, providing a fail-safe noise-immune signal to the Schmitt trigger pair of transistors at Q5 and Q6. This circuit was designed to operate from 4.5 to 6.5 volts dc. The range of the system can be increased in two ways: another stage of amplification driving the IR LED can boost the range by 5 to 10 times, and a higher supply voltage for the IR LED in PC-1 can double the range. You can probably think of a host of applications for this reliable long-range object detector around your home or office.

Table 7-1 Long-range object detector/alarm.

PC-1	Transmission range	Reflective range
H23A1	5 inches	1 inch
LED56 & L14Q1	12 inches	3 inches
LED56 & L14G1	18 inches	4 ½ inches
LED55C & L14G1	32 inches	8 inches
1N6266 & L14G3	48 inches	12 inches
F5D1 & L14G3	80 inches	20 inches
F5D1 & L14P2	200 inches	50 inches

Long-range object detector parts list

R1	10 ohm ¼w resistor
R2	470 ohms ¼w resistor
R3	39K ohm ¼w resistor
R4	300 ohms ¼w resistor
R5	1.3K ohm ¼w resistor
R6, R7, R16	22 ohms ¼w resistor
R8	1 megohm ¼w resistor
R9	27K ohm ¼w resistor
R10,R15	47K ohm ¼w resistor
R11	2.2K ohm ¼w resistor
R12	22K ohm ¼w resistor
R13	2.2 megohm ¼w resistor
R14	560K ohm ¼w resistor
C1, C4	.05µf 25v disc capacitor
C2	10µf 25v electrolytic capacitor
C3	.002µf 25v disc capacitor
Q1	D29E2 transistor or equivalent
Q2, Q5, Q6	2N5249 transistor
Q3, Q4	2N5356 transistor
D1	2N6027 SCR
PC-1	See Table 7-1
PC-2	H11A5 optocoupler

Long-range object detector parts list continued

Q2,Q5,Q6	2N5249 transistor
Q3,Q4	2N5356 transistor
D1	2N6027 SCR
PC-1	See Table 7-1
PC-2	H11A5 optocoupler

Balanced-light motion detector

A balanced-light or optical motion detector is illustrated in Figure 7-7. This simple yet effective alarm, if properly set up, can readily detect humans passing tens of feet in front of its field of view. Although not as sophisticated as the alarm industry's pyroelectric detector (the passive, infrared, body-heat detector described later in this chapter), this simple alarm could be used as the heart of a low-cost burglar alarm system or point-of-sale detector. After experimenting with one of these optical motion sensors, you could use a number of these detectors to form a complete alarm system.

The key to a low-cost optical motion detector is a surplus six-inch Fresnel lens, placed at one end of a light-tight box or enclosure. Two CdS photoresistive cells are mounted about one inch apart, opposite the lens, as shown. The inside of the box is painted flat black to avoid any light reflections inside the enclosure that could interfere with the sensors' operation. Mount the two CdS cells at the focal distance of the Fresnel lens. Experiment to find the correct distance, if necessary.

The two CdS photoresistive cells represent a balanced voltage to pin 2 of the LM741 op-amp. A 50-kilohm threshold potentiometer is used as a balance adjustment control. The op-amp is powered by a single 9-volt battery, or it can be powered

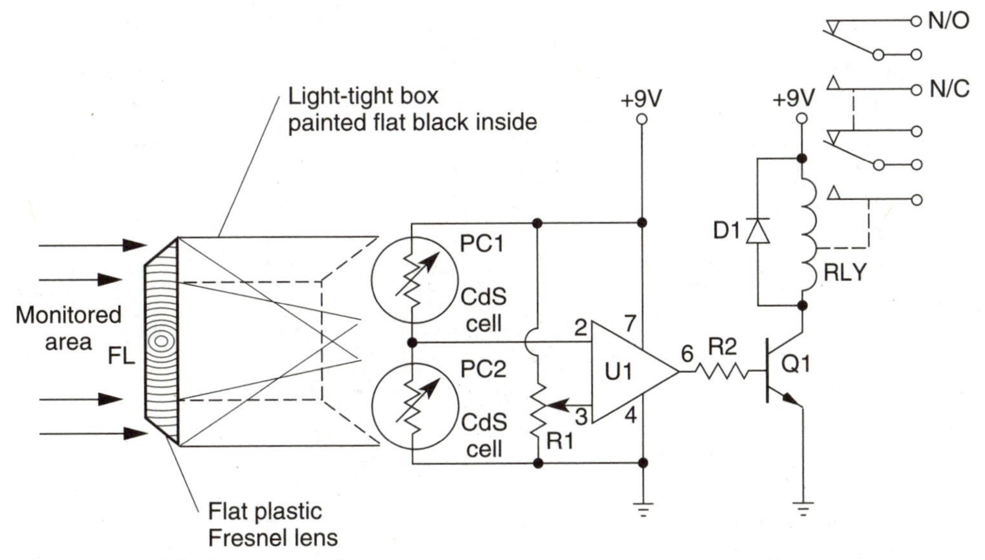

7-7 Balanced-light motion detector.

remotely by an alarm control panel. The LM741 op-amp drives Q1, a 2N3904 transistor, which supplies power to a low-current, 6- to 9-volt relay. The dual set of relay contacts can be used for two functions. One set of contacts can activate a local buzzer or LED to help set up the alarm, and a second set of contacts could activate an alarm control box.

To operate the balanced-light optical alarm, point the lens at the area to be protected and adjust the potentiometer until the relay triggers. Then turn the control back down until the relay drops out. Any large object or person moving in front of the lens will imbalance the op-amp input, thus causing Q1 to activate the relay. The optical motion detector requires ambient background lighting for the detector to function correctly. This balanced optical motion sensor can also be used for other alarm applications, as you will see in the next section.

Balanced-light motion detector parts list

R1	50-kilohm potentiometer
R2	20-kilohm, ¼-watt resistor
R3	2-kilohm, ¼-watt resistor
PC1,PC2	100-kilohm, cadmium-sulphide photocell
D1	1N4002 silicon diode
Q1	2N3904 transistor
U1	LM741CN op-amp
RLY	6- to 9-volt relay
FL	Surplus Fresnel lens

IR TV camera light source

The usefulness of this next project might not be obvious at first glance, but it can be quite beneficial. The TV camera in Figure 7-8 is only part of the story. Television cameras are used in many corporate alarm systems for monitoring employees, specific doors, or particular areas of interest. Television cameras are very useful and can be a good deterrent to pilferage and indiscretions. TV cameras were once the domain of only TV stations and the rich, but advances in electronics have brought the price down to a new low. Small, solid-state TV cameras the size of a pack of cigarettes are low in cost and easy to conceal, and they consume a very small amount of power, thus making them available to many people for a great range of applications (see Photo 7-2).

One little-known benefit of these small CCD TV cameras is that they are sensitive to infrared light and can operate in low light. Did you know that a TV infrared remote can illuminate an object in a dark room, allowing the camera to see quite easily?The IR TV camera light source in Figure 7-9 allows you to more-than-adequately illuminate an area with infrared light so you can observe it in the absence of visible light. Four high-output IR LEDs are wired to a 6- to 9-volt battery bank through series resistors to provide a "flood" of IR light, allowing you to view objects in a dark room. Cut a large hole from the front of a chassis box, and optionally place an IR filter in front of the LEDs. Then you can attach the completed IR light source

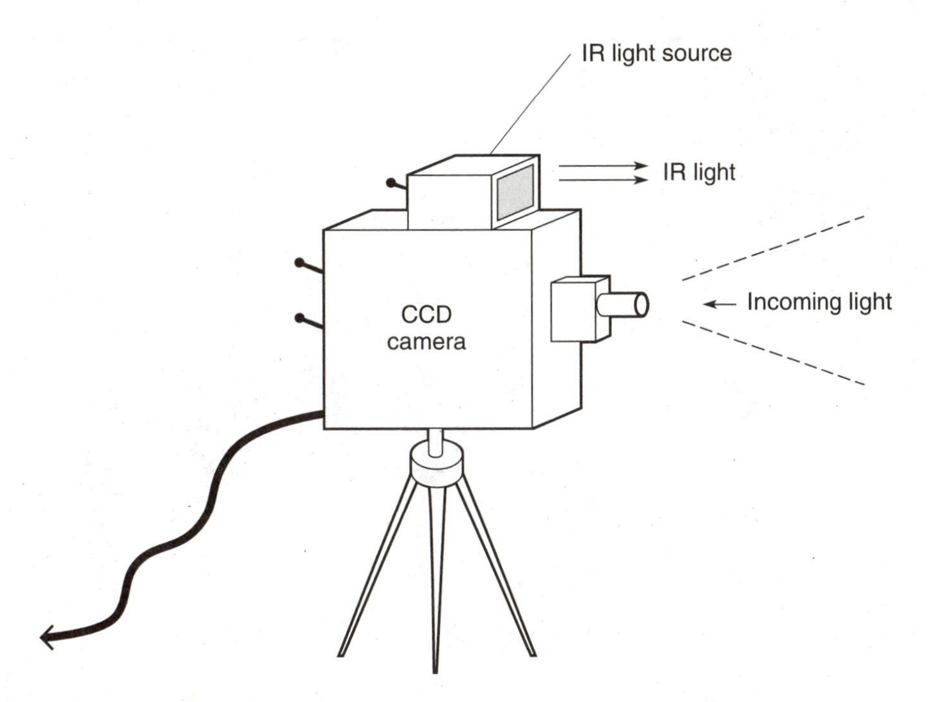

7-8 IR light source and TV camera.

Photo 7-2 Miniature TV camera.

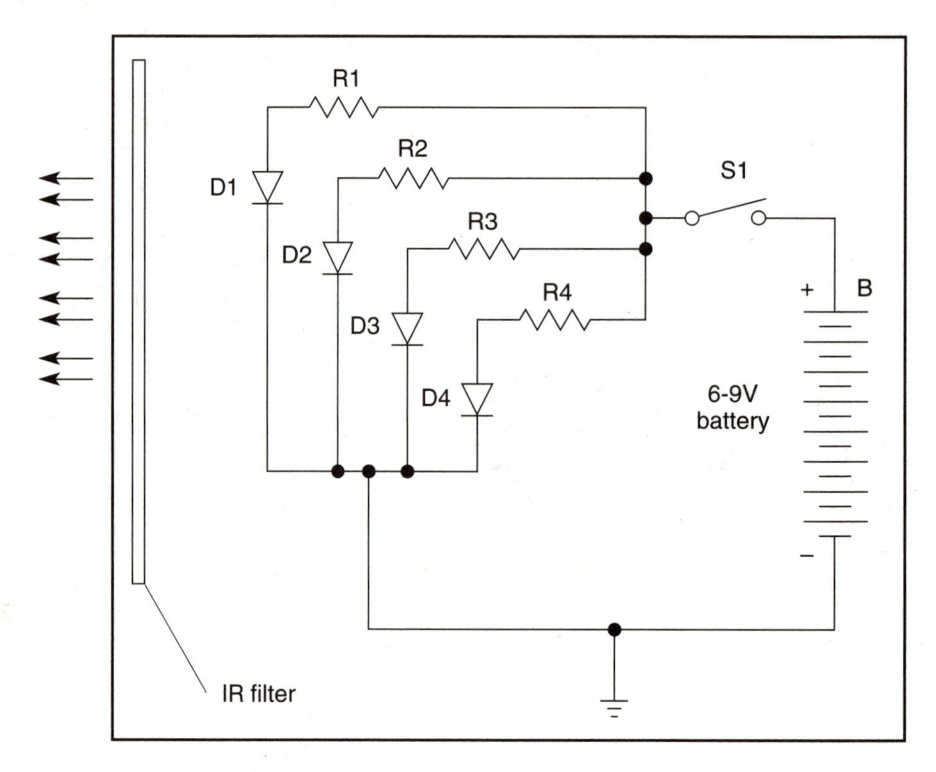

7-9 TV camera IR light source.

to the top of a miniature TV camera. Miniature TV cameras are also available in disguised form, from wall outlets to wall clocks, which help make the operation of the camera a complete secret.

So far you have a low-cost camera for monitoring activity in darkness with an infrared light source. Now you will turn the TV camera into an alarm system. Refer back to Figure 7-7, the balanced-light detector, for the next part of this project. You can modify the balanced-light detector to serve a new purpose.

Connect your miniature TV camera to a monitor at least 9 to 15 inches in size, so you can clearly see what the camera sees. Then take a look at the balanced-light motion detector. You can use it with your TV monitor to watch for movement within a monitored space. Here is how it works. The two CdS photoresistive cells must be placed on the front of your TV monitor, which isn't as difficult as it might seem. One method of attaching a CdS cell to a monitor is to drill out the center of a suction cup and epoxy the CdS cell around the edge carefully, so the CdS cell can be attached to the monitor's screen. Connect four wire miniature shielded cables from the CdS cell detectors to the electronic circuit. Place the electronic circuit board with the relay into a small chassis box near the monitor.

Now for the difficult part. You must affix the two CdS cells to the TV monitor screen so they are lit with a similar amount of brightness, but from different areas of the screen. For example, you could place the two CdS cells over two areas of a hallway, so when the person moves through the hallway scene, the changing brightness sounds an alarm. The balanced-light motion detector can be used both for normal daytime viewing or with IR light. The balanced-light alarm can help you solve your video surveillance needs inexpensively. Commercial digital video motion alarms are available in the $500 range, but for a few dollars and some free time you can experiment with your own low-cost video motion alarm.

IR TV camera light source parts list

R1,R2,R3,R4	330-kilohm, ¼-watt resistors
D1,D2,D3,D4	Super-bright IR LED
S1	SPST toggle switch
B	6-volt battery or C cell pack
Misc	Chassis box and IR filter

Pyroelectric detector

The pyroelectric or body-heat detector has proven itself to the alarm industry as a low-cost, sensitive, and reliable alarm detector. The pyroelectric detector would be an excellent starting point for your own home alarm system. The sensitive pyroelectric detector can sense humans and animals up to 50 feet away. The low-cost Eltec 5192 pyroelectric sensor is shown in Figure 7-10 and Photo 7-3. This detector costs less than $5 and consists of two lithium tantalate crystal detectors connected in a parallel-opposed fashion to cancel out thermal differences between the sensing elements. The thin crystal wafers are coated on both sides, forming two electrodes and thus forming a type of capacitor, with a capacitance of about 30 pF. The pyroelectric material exhibits an internal electric field, which is collected between the electrodes. Monitoring the resultant charge between the electrodes requires a high-impedance amplifier. Therefore, most pyroelectric sensors incorporate a high-value load resistor, which provides an impedance converter as well as a field-effect transistor amplifier, all packaged in a TO-5 transistor case. The front of the pyroelectric detector's case has a quartz window with a quartz IR filter placed front of the sensing elements.

Photo 7-4 illustrates the 5192 pyroelectric sensor, with a wide-angle plastic Fresnel lens.

Pyroelectric sensors respond more rapidly than any other IR heat sensor. They respond to changes in IR intensity rather than just the presence of an IR source. The complete pyroelectric detector/alarm circuit is shown in Figure 7-11. The circuit description begins with a 47-kilohm resistor just after the detector, which sets up the drain current; it is placed across the signal output and ground. The detected IR signal is first fed to a voltage follower, which minimizes loading the sensor. The voltage follower is formed by U1:A. The output of U1:A is coupled via R2 to the next stage at U1:B of the LM324 op-amp. A 500-kilohm resistor at R3 sets the overall system gain. A voltage divider of R5, R6, and R7 is fed to R4 at pin 5 of U1:B. A set-point or zero control is adjusted via R5. The output of U1:B is then sent to pin 2 of an LM555 timer, which acts as a pulse shaper. The "on time" timer is controlled by the R8/C1 combi-

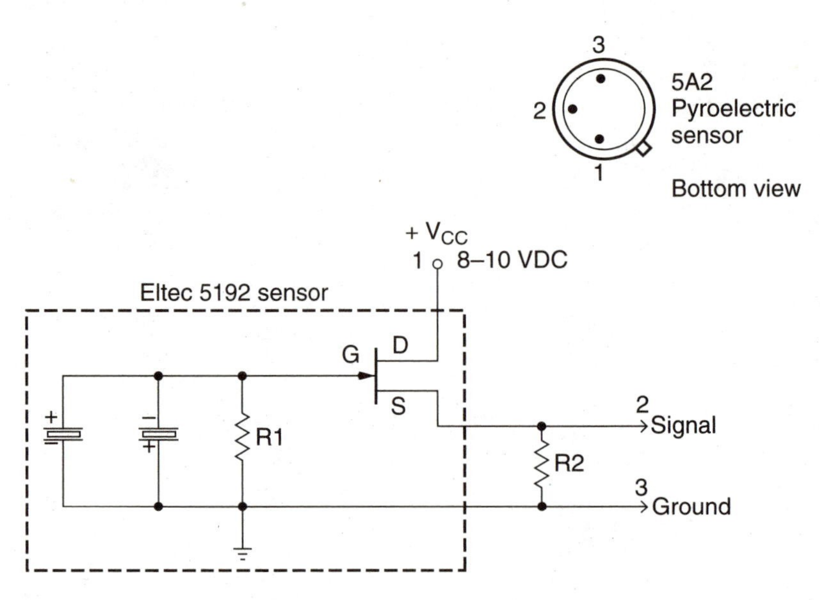

7-10 Eltec 5192 dual element pyroelectric sensor.

nation. A "power on reset" function is performed by C5/R12 on pin 4 of U2. The output of U2 is coupled to pin 5 of a CD4528 dual retriggerable monostable multivibrator. The CD4528 provides a time window discriminator. The output from the LM555 is also fed to a 7400 Nand gate along with the output of the CD4528 chip.

By adjusting the time window via C3/R10, only a certain number of triggers or pulses from the LM555 are allowed through the window detector in a specified time period. The LED at D1 helps while you are setting up and aligning the detector. The output of U4 is fed to R11, which drives a 2N3904 NPN transistor. Transistor Q1 is designed to activate a low-current relay. You could also use the output to activate an SCR in place of Q1.

Constructing the pyroelectric detector is quite straightforward. The electronics and detector are placed on a small PC circuit board, as shown in Photo 7-5. The Eltec 5192 detector should be mounted as close to U1:A as possible. Mount the completed circuit board in a small metal chassis box to eliminate any stray RF energy, thermal noise, and air currents from reaching the detector. Pyroelectric detectors are generally used with low-cost Fresnel lenses ahead of the detector element. Without the Fresnel lens, the detector has a very short range, but can be used to protect specific devices; this is called a "spot" detector. Thin, low-cost, Fresnel lens arrays come in a variety of types to control the coverage of the detector. A wide-angle lens provides wide-angle coverage to protect a 10 × 12-foot room, while a long-range Fresnel array can provide long-range narrow beamwidth coverage up to 50 feet, as shown by the diagram in Figure 7-12. Special Fresnel arrays are also available that avoid detecting pets such as dogs or cats by truncating the low portion of the beamwidth coverage pattern.

One approach to sensor mounting is shown in Figure 7-13. A Fresnel lens is glued to the front of a slotted tuna can, with the pyroelectric sensor mounted

Photo 7-3 Eltec 5192 pyroelectric sensor.

1.2 inches behind the Fresnel lens. A shielded wire cable would then connect the sensor to the electronics circuit board placed in a chassis box mounted below the tuna can.

The pyroelectric sensor circuit can be powered from an 8- to 12-volt dc power source, such as a dc wall-cube supply or a remote alarm panel. The pyroelectric detector could also be used as a self-contained portable alarm system, powered by a 9-volt transistor battery. You could wire together a number of these pyroelectric building blocks in parallel to form a complete alarm system. The pyroelectric detector could also be used for lighting controls, flame sensors, door sensors, animal alarms, and robotics.

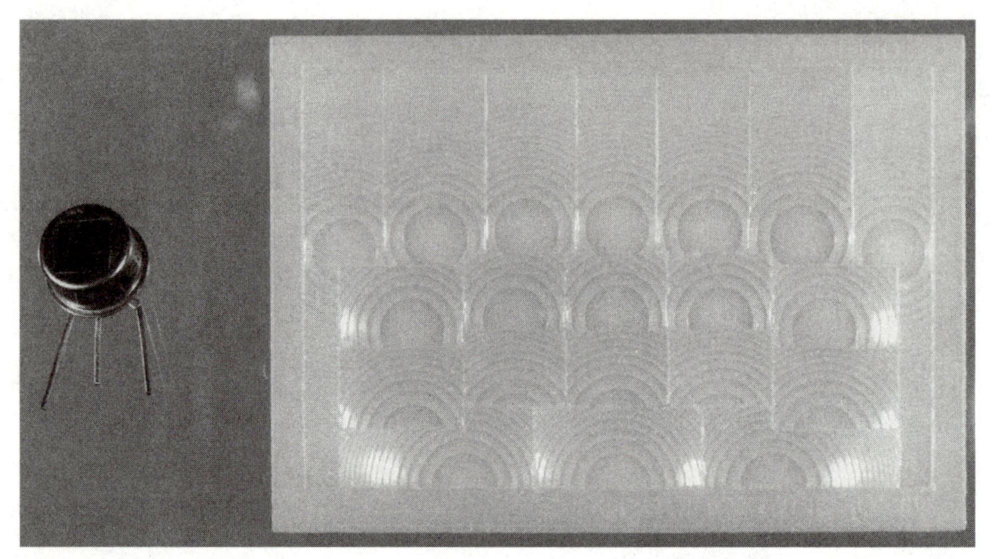

Photo 7-4 5192 integrated wide-angle pyroelectric sensor with Fresnel lens.

Operation of the pyroelectric detector is quite simple. First apply power to the circuit. After a few seconds, adjust R5 to light the LED, and then back off the control until the light just goes out. Now wave your hand in front of the the detector and the LED should light up. When movement stops, the circuit should stabilize and the LED should go out. If the LED does not extinguish, then readjust R5. Set the discriminator to allow two or three pulses through the window detector in a 10-second time window by adjusting potentiometer R10. The pyroelectric detector is now ready to stand guard to protect your home or office.

Pyroelectric motion detector parts list

R1,R7	4-kilohm, ¼-watt resistor
R2	10-kilohm, ¼-watt resistor
R3,R10	100-kilohm trim pot
R4	33-kilohm, ¼-watt resistor
R5	2-kilohm potentiometer (trim)
R6	2.5- kilohm, ¼-watt resistor
R8	100-kilohm, ¼-watt resistor
R9	1-kilohm, ¼-watt resistor
R11	2-kilohm, ¼-watt resistor
C1	1-µF, 25-volt electrolytic capacitor
C2	.01-µF, 25-volt disc capacitor
C3	100-µF, 25- volt electrolytic capacitor
C4,C6	.1-µF, 25-volt disc capacitor
C5	1-µF, 25-volt electrolytic capacitor
D1	LED
D2	1N4002 silicon diode

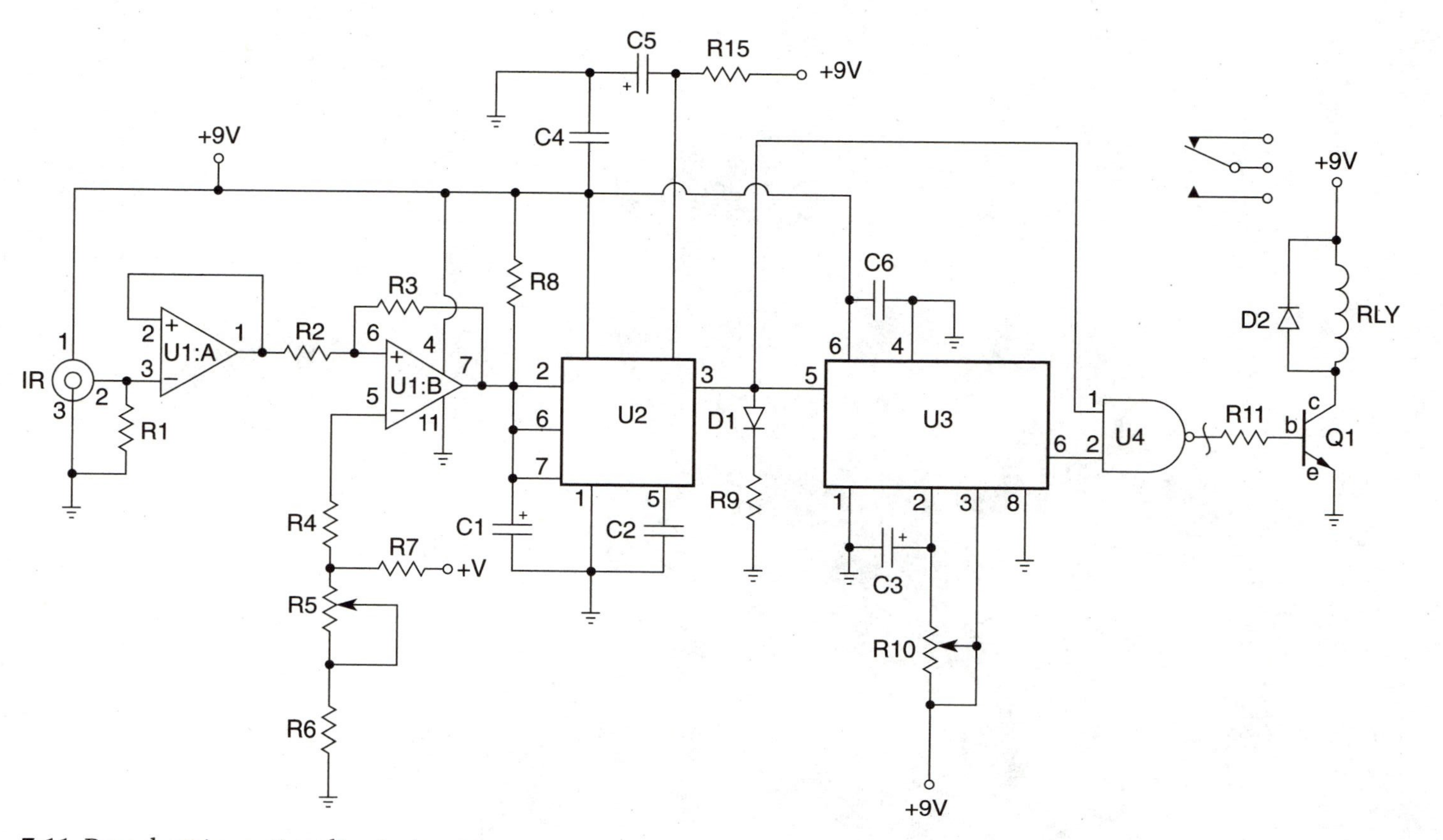

7-11 Pyroelectric motion discriminator.

Photo 7-5 Integrated pyroelectric motion alarm sensor.

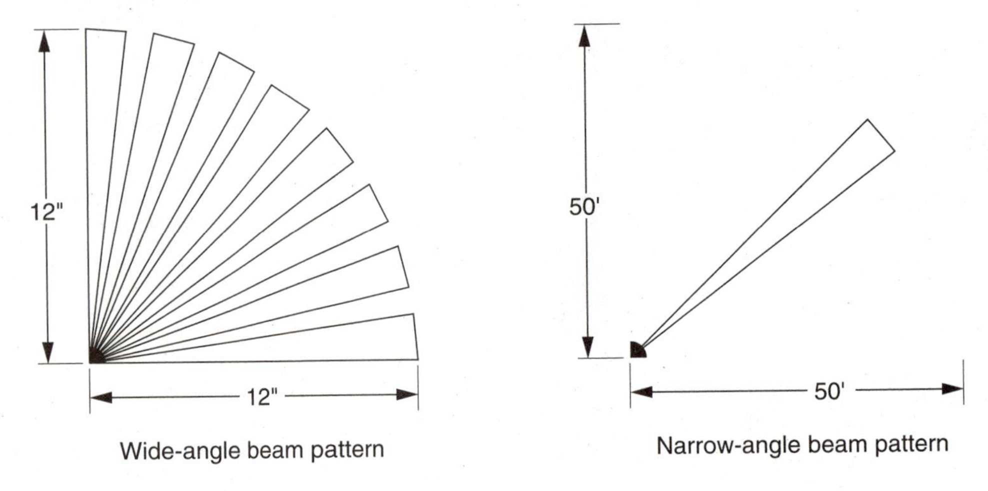

7-12 Pyroelectric sensor coverage.

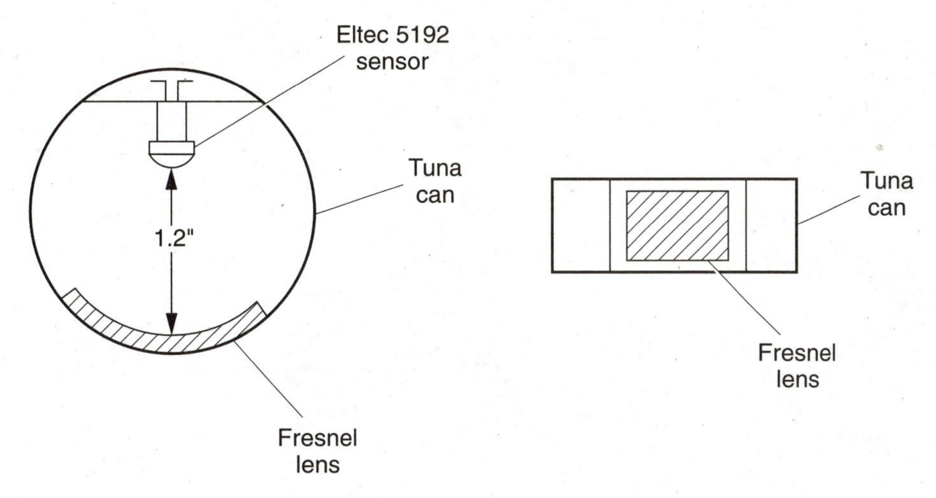

7-13 Pyroelectric sensor mounting.

Pyroelectric motion discriminator parts list continued

U1	LM324 op-amp
U2	LM555 timer IC
U3	CD 4528 one-shot
U4	SN7400 quad Nand gate
Q1	2N3904 transistor
RLY	6- to 9-volt miniature relay
IR	ELTEC 5192 IR sensor
FL	Fresnel lens (see the Appendix)

Infrared beam-break alarm system

The infrared beam-break alarm system shown in Figures 7-14 and 7-15 is a very useful short- to medium-range optical alarm system. The IR beam-break alarm system consists of an optical IR transmitter and IR receiver section, which can be separated by 20 to 30 feet. The simple IR transmitter shown in Figure 7-14 generates a stream of powerful near-infrared pulses. Transistors Q1 and Q2 form part of the oscillator circuit designed to generate pulses of 400 milliseconds in duration, 240 times per second. The transistor oscillator in turn drives a high-output IR LED, which sends IR pulses to the receiver section. The range of the system is several feet, but when used with a lens, it increases to 30 feet. The entire transmitter section can be powered from a 6-volt battery or dc wall-cube supply.

The beam-break receiver depicted in Figure 7-15 receives pulses from the transmitter section via phototransistor Q1. The photocurrent from Q1 is first amplified by U1:A and then passed on to U1:B, which acts as a threshold comparator. The output signal from pin 7 of U1:B is coupled to an LM555 timer IC, which forms a missing pulse detector. The output of the missing pulse detector triggers the low-current, 500-ohm relay when the IR beam is interrupted. For optimum results, shield

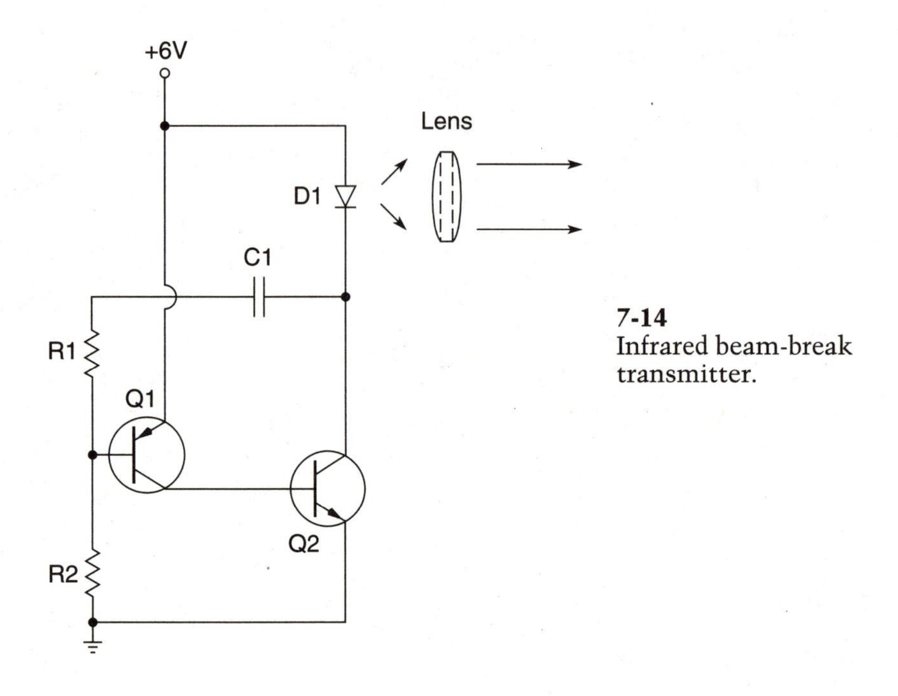

7-14
Infrared beam-break transmitter.

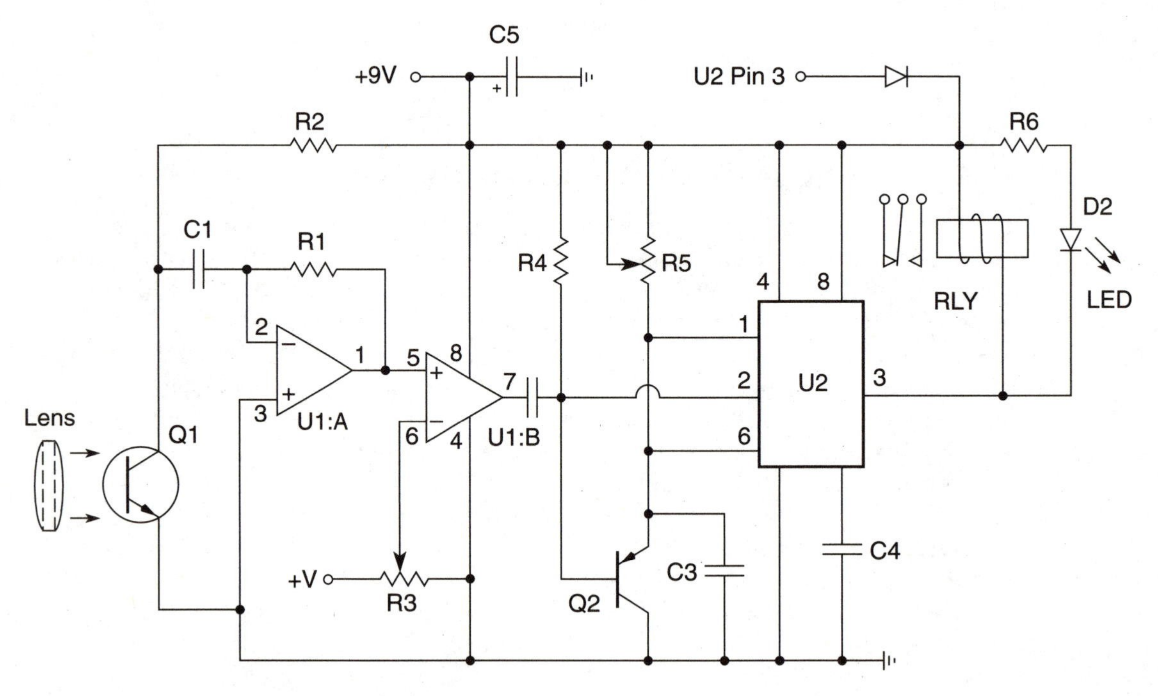

7-15 Infrared beam-break receiver/alarm.

transistor Q1 from ambient light by mounting the receiver in a metal chassis box and using a collimator tube painted flat black inside and out ahead of the phototransistor. For long-range applications, use lenses with both the transmitter and receiver. Mount the lenses in the collimator tube ahead of the phototransistor once the correct focal length is determined.

In operation, the beam-break alarm potentiometer R3 is adjusted for input current threshold and R5 is adjusted to achieve optimum relay operation. The entire receiver section of the beam-break alarm can be powered from batteries for portable use, or from a 6- to 9-volt dc wall-cube power supply. A commercial IR beam-break system is shown in Photo 7-6.

IR beam-break transmitter unit parts list

R1	22-kilohm, ¼-watt resistor
R2	2.2-megohm, ¼-watt resistor
C1	.02-µF, 25-volt disc capacitor
D1	High-output IR LED (RS 276-143)
Q1	2N2907 transistor
Q2	2N2222 transistor
B	6- to 9-volt battery or wall-cube supply
Misc	Chassis box, collimator tube, and lens

IR beam-break receiver unit parts list

R1	1-megohm, ¼-watt resistor
R2	100-kilohm, ¼-watt resistor
R3	10-kilohm potentiometer (trim)
R4	4.7-kilohm, ¼-watt resistor
R5	1-megohm potentiometer (trim)
R6	1-kilohm, ¼-watt resistor
C1,C4	.01-µF, 25-volt disc capacitor
C2,C3	.1-µF, 25-volt disc capacitor
C5	10-µF, 25-volt electrolytic capacitor
D1	1N914 silicon diode
D2	LED
Q1	Phototransistor (276-145 RS)
Q2	2N2907 PNP transistor
U1	LM1458CN op-amp
U2	555 timer IC
RLY	6- to 9-volt power source

Wireless security system

A unique wireless security system is illustrated in Figures 7-16, 7-17, and 7-18. The wireless security system consists of three separate components: a pulsed IR transmitter, an IR receiver/RF transmitter, and an RF receiver/alarm unit. The

Photo 7-6 Photoelectric beam-break system.

pulsed IR transmitter shown in Figure 7-16 forms one half of a beam-break system, which sends a continuous stream of IR pulses to the IR receiver/RF transmitter. The heart of the IR transmitter is U1, a CMOS version of the ubiquitous LM555 timer. The timer is configured as an astable oscillator. Resistors R6 and R7, and C2, set the oscillator's frequency to approximately 1500 Hz. Resistor R7 controls the length of time that pin 3 is low, about 43 microseconds. During that low period, transistors Q1 and Q2 are turned on, thus allowing current to flow through the two IR LEDs. Limiting the "on" time to 43 microseconds reduces the power dissipation in the two LEDs. Resistors R3 through R5 limit the base current of Q1 and Q2. Capacitor C1 provides low-impedance bypassing of the 6-volt power supply. The IR transmitter is quite straightforward and can be constructed on a "perf" board or printed circuit board. The entire IR transmitter can be powered from a 6-volt, 80-mA source. For portable use, the transmitter could be powered from a D-cell battery pack.

The second circuit in the wireless security system is the IR receiver/RF transmitter, shown in Figure 7-17. The IR receiver/RF transmitter receives the pulsed IR beam from the IR transmitter; when the IR beam is broken, the IR receiver/RF transmitter sends an RF signal to the RF receiver/alarm unit. In operation, the IR

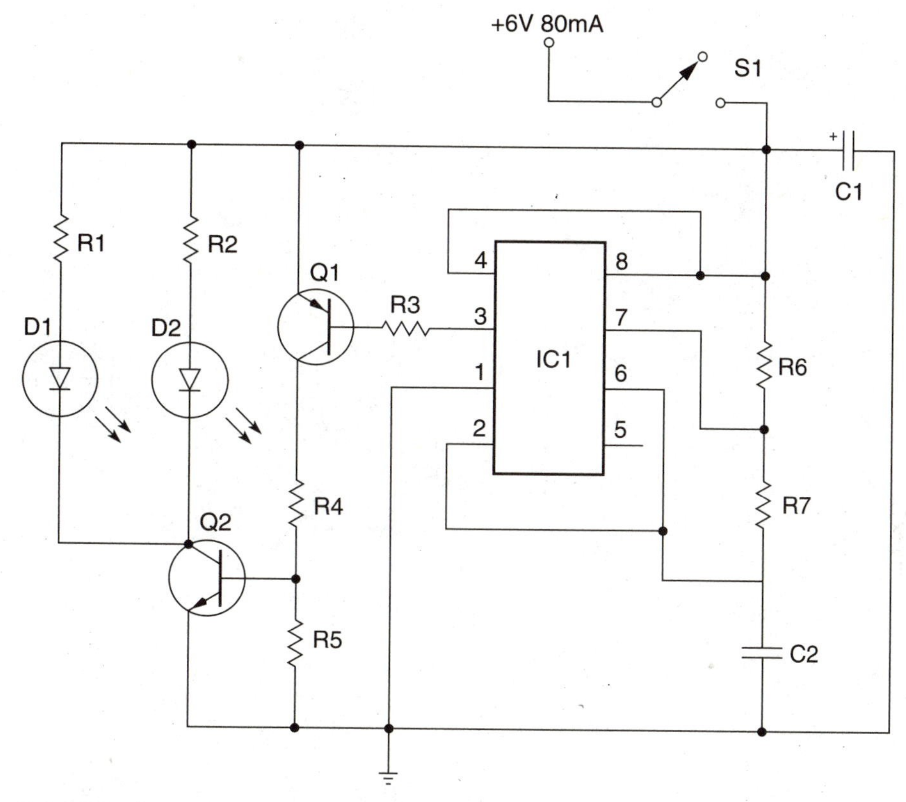

7-16 Wireless IR security system—IR transmitter. Copyright Radio Electronics, 1989. Reprinted with permission.

receiver/RF transmitter's light-dependent resistor LDR1 and transistor Q1 are exposed to the same ambient light. Op-amp U1A together with Q2 and R1 are essentially all in parallel with LDR1, which forms a current source path. Since both Q1 and LDR1 receive the same light, LDR1 automatically adjusts the current to transistor Q1 in order to maximize the sensitivity of the photodetector circuit. Phototransistor Q1 is aimed at the IR transmitter. The phototransistor's emitter-collector voltage fluctuates in step with the received pulses. Capacitor C11 couples the phototransistor to U1B. Resistors R4 and R5 set the gain of U1B to about 51 dB. Resistors R2, R3, R18, and R20 and capacitor C13 form a dc voltage offset that is approximately two-thirds of the power supply voltage, or about 8 volts, which allows the op-amp to operate in a single-ended configuration.

The ac signal and the 8-volt dc offset are fed to pin 2 of U2. As long as the voltage on pin 2 is greater than two-thirds of Vcc, U2 operates as a monostable multivibrator whose time delay is determined by R6 and C2. The 7555 at U2 and transistor Q3 act as a "missing pulse" detector. Provided pin 2 of U2 is held above 8 volts dc, Q3 is biased off. Once a timing cycle is completed, pin 3 goes to a ground potential and is held there. With an IR signal present, the base of Q3 and pin 2 of U2 are repeatedly triggered by negative-going pulses from pin 7 of U1B. As a result, the timing cycle of

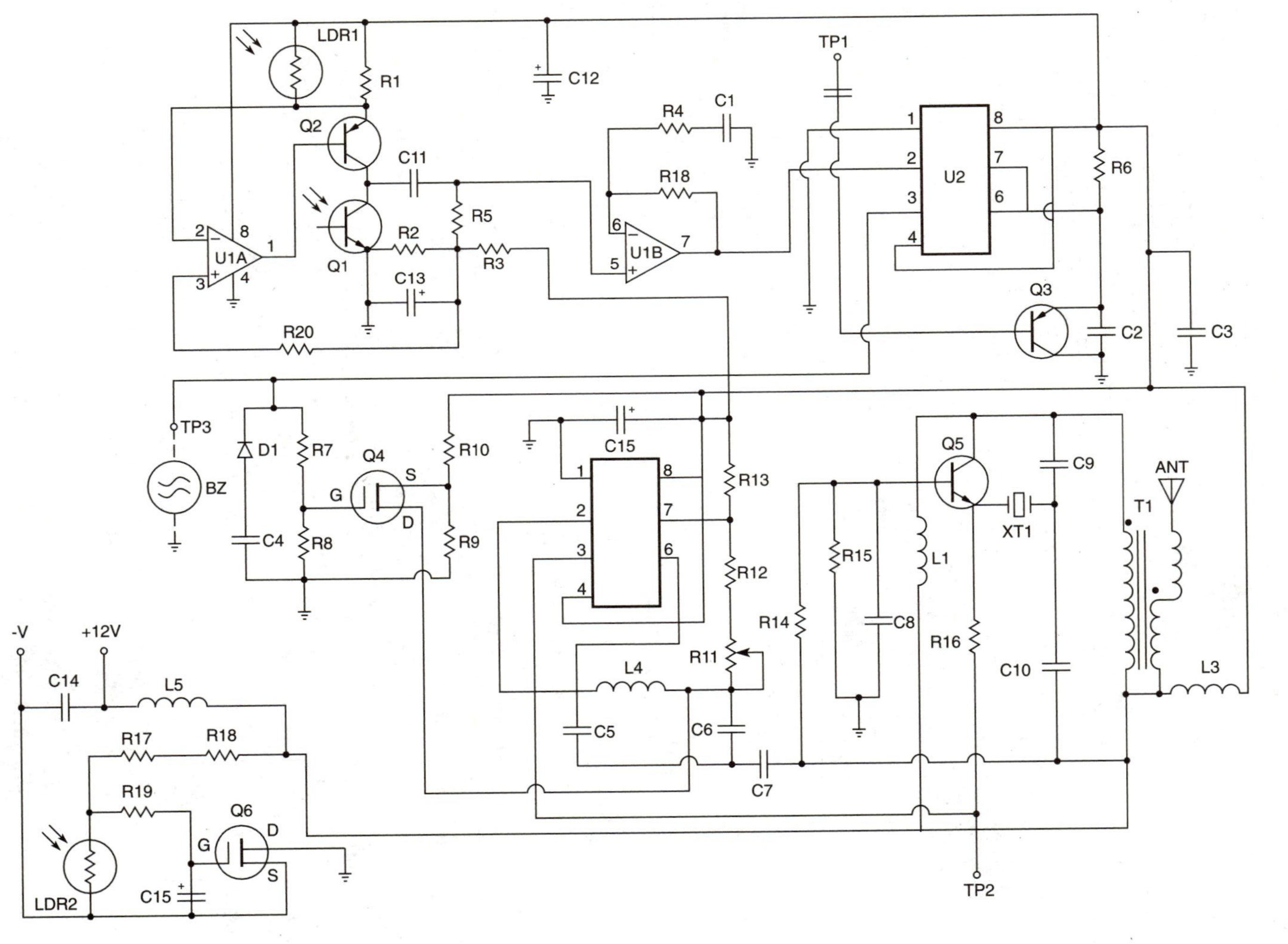

7-17 Wireless IR security system—IR receiver/RF transmitter. Copyright Radio Electronics, 1989. Reprinted with permission.

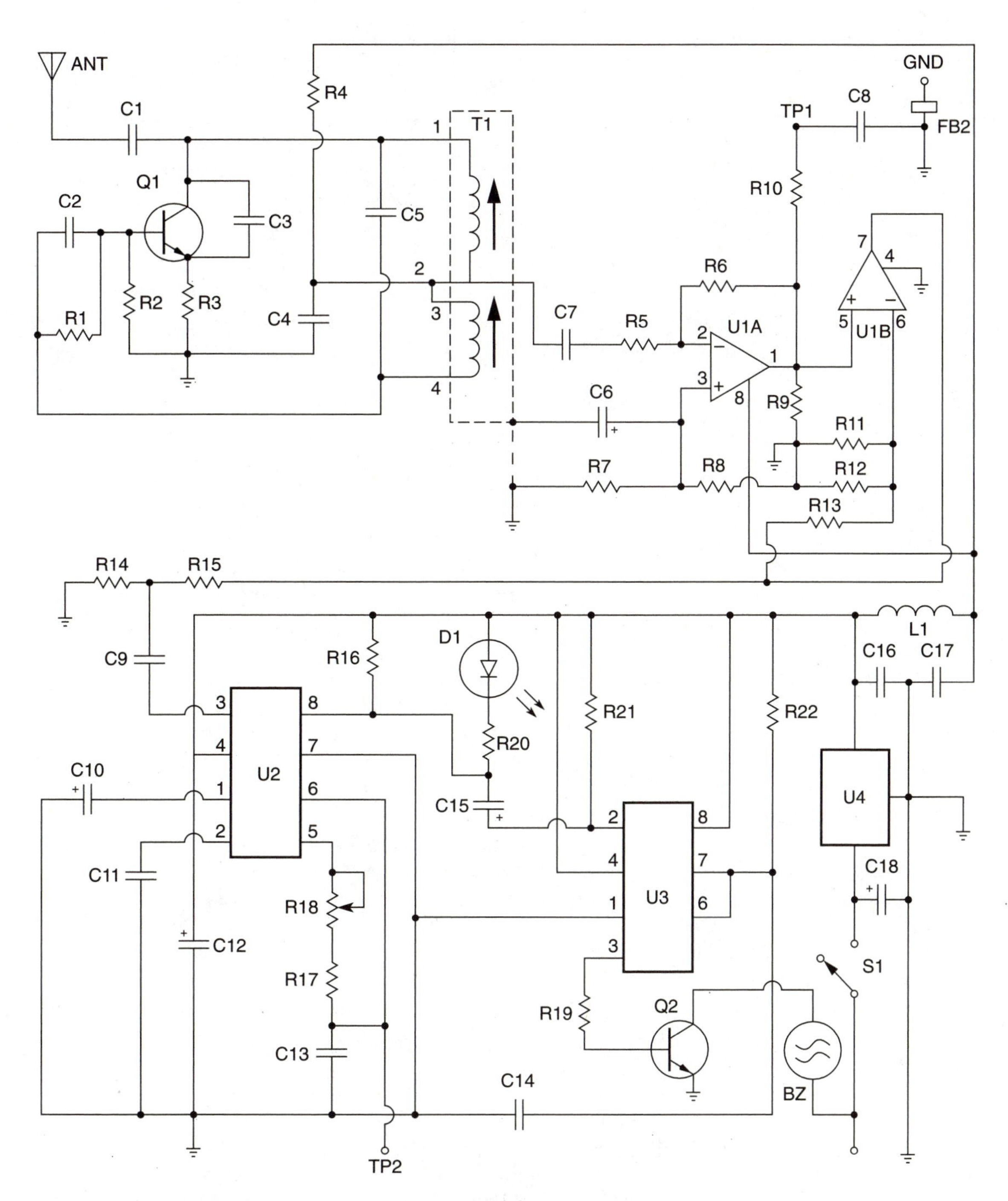

7-18 Wireless IR security system—RF receiver/alarm. Copyright Radio Electronics, 1989. Reprinted with permission.

U2 is continuously interrupted before it has a chance to complete one time-delay cycle. This causes pin 3 to remain high. With pin 3 high, the RF transmitter is off; when pin 3 is low, the transmitter is turned on.

The RF transmitter consists of a crystal-controlled oscillator and an audio-tone generator. Resistors R14 through R16 set up a dc bias of about 7 mA. The transmitter is tuned to 49.890 MHz by coils L1, C9, and C10. Coil T1 provides an impedance match between the oscillator's output and the antenna. The antenna-loading coil L2 tunes a one-meter whip antenna to 49.890 MHz. Amplitude modulation is obtained by connecting R16 to pin 3 of U3. Resistors R11 through R13 and capacitor C6 set the IC's frequency to 490 Hz. Pin 3 of U3 goes low once during each 490-Hz cycle, which turns transistor Q5 on, which in turn activates the RF transmitter. Since the audio frequency is set to 490 Hz, a 490-Hz AM RF carrier is generated by the circuit at Q5. The RF transmitter is turned on and off by Q4, which is controlled by pin 3 of U2. Capacitor C4 charges through R7 when pin 3 goes high. The voltage across C4 and Q4's gate-source biases Q4 when it reaches 3 volts. Now when pin 3 goes low, C4 discharges through D1, biasing Q4 off.

The voltage divider formed by R9 and R10 biases Q4's source terminal to about .7 volts, and that voltage in combination with R8 provides a several-second delay, which turns Q4 from on to off. The RC time-constant circuit activates the RF transmitter for several seconds. Transistor Q6 and the surrounding components function as an on/off switch that is controlled by the intensity of the ambient light. LDR2's resistance is low when the device is exposed to light, thereby forcing Q6's gate-source voltage below the "turn on" threshold. Capacitor C15 charges through resistors R17 to R19 when LDR2 is not illuminated. When transistor Q6 is on, the negative terminal of the power supply is connected to the circuit's ground, thereby applying power to the RF transmitter. The IR receiver/RF transmitter is powered from a 12-volt power source.

The third portion of the wireless security system is the RF receiver/alarm unit, shown in Figure 7-18. The RF receiver is of the super-regenerative type. The AM RF carrier is coupled from the antenna through capacitor C1 to T1 and then to the base of Q1. Resistors R1 through R4 provide a bias to the base of Q1. Capacitor C4 bypasses the RF signal, and capacitor C2 is selected in this case to cause self-oscillation, a requirement of a super-regenerative receiver. Transistor Q1 oscillates at 49.890 MHz at a repetition rate of 450 kHz. During each 450-kHz cycle and just before Q1 breaks into oscillation, the circuit functions as a very high gain RF amplifier. The average emitter current of Q1 goes up and down according to the amplitude of the RF signal. Since the RF signal is AM-modulated by the 490-Hz tone, a 490-Hz voltage appears at Q1's collector and at R4. Capacitor C7 couples the 490-Hz signal from the receiver to U1A, which provides about a 10-dB gain. U1B amplifies the signal once again and clips the signal, shaping it into square-wave pulses. Resistors R7 through R9, R11, and R12 allow U1 to be powered from a single power supply. The square-wave pulses from U1B are fed to the tone decoder at U2 through R15 and C9. The tone decoder's input-signal voltage is reduced by R14, thus increasing the immunity to false triggering. Note that R16 is optional and will reduce the system's range. Capacitors C10 and C11 set the decoder's bandwidth. Resistors R17 and R18 along with C13 determine the frequency to which U2 will respond. Pin 8 of U2 goes low with a 490-Hz signal. A high-to-low transition at pin 8 is coupled to U3's trigger input at pin 2 through C15. Pull-up resistor R21 keeps pin 2 high at all other times. Once triggered, U3's output on pin 3 goes high, biasing Q2 to switch on. Transistor Q2 drives the alerting buzzer. Resistor R22 and capacitor C14 determine the length of time the buzzer sounds. The RF receiver/alarm circuit can be powered from a 6-volt power source, as is the IR transmitter.

Each of three components that make up the wireless beam-break security system should be mounted in metal chassis boxes to ensure a minimum of outside interference. Install phototransitor Q1 into the enclosure of the IR receiver/RF transmitter. Mount LDR1 adjacent to Q1 so both devices are exposed to the same light intensity. Mount LDR2 in a location with a maximum exposure to ambient light. You might need a light shield at Q1, depending on the amount of ambient light in the area the system is used.

The alignment of the wireless security system is somewhat involved. You need a frequency counter that is capable of audio frequencies and an audio amplifier/speaker. Cover LDR2 with some black tape. Apply power to the IR receiver/RF transmitter and allow a minute for the circuit to stabilize. Then adjust L1's core until it protrudes to $\frac{1}{32}$ inch above the coil form. Attach the frequency counter to TP2 and adjust R1 for a 490-Hz tone. Disconnect the frequency counter and attach the audio amplifier and speaker to TP1, which should produce a 490-Hz tone. While listening to the audio at TP1, place the IR transmitter in line with Q1, in the receiver unit. A higher-pitch tone of about 1500 Hz should now be produced. After a few seconds, the 490-Hz tone should stop, leaving only the 1500-Hz tone. Momentarily interrupt the IR beam, and the 490-Hz tone should reappear. Now apply power to the RF receiver/alarm unit. Connect the frequency counter to TP2 and adjust R18 for 490 Hz. Then connect the audio amplifier/speaker to TP1. Adjust T1's core until the top of the core is level with the top of the coil form. Adjust T1 and L1 in both the IR receiver/RF transmitter and the RF receiver/alarm unit for the loudest 490-Hz signal. Extend the antennas, place the transmitter apart from the receiver unit, and adjust for the best reception.

The wireless IR beam-break system could prove to be very useful if you want to protect a remote building, a detached garage, a pool area, or outside walkways from burglars. Why not let the wireless IR security system protect you?

Wireless security system, IR transmitter parts list

R1,R2	2.2-ohm, $\frac{1}{4}$-watt resistor
R3	33-kilohm, $\frac{1}{4}$-watt resistor
R4	1.3-kilohm, $\frac{1}{4}$-watt resistor
R5	10-kilohm, $\frac{1}{4}$-watt resistor
R6	82-kilohm, $\frac{1}{4}$-watt resistor
R7	3.3-kilohm, $\frac{1}{4}$-watt resistor
C1	100-µF, 25-volt electrolytic capacitor
C2	.01-µF, 25-volt disc capacitor
D1,D2	VT1261 IR LED or equivalent
Q1	2N4403 transistor
Q2	TIP-110 transistor or equivalent
U1	7555 CMOS timer
S1	SPST switch

Wireless security system, IR receiver/RF transmitter parts list

R1,R5,R20	1-megohm, $\frac{1}{4}$-watt resistor
R2	200-kilohm, $\frac{1}{4}$-watt resistor
R3,R6,R10	100-kilohm, $\frac{1}{4}$-watt resistor
R7,R8,R17,R18,R19	10-megohm, $\frac{1}{4}$-watt resistor
R9,R13	10-kilohm, $\frac{1}{4}$-watt resistor
R11	20-kilohm potentiometer
R12	53.6-kilohm, 1-percent, $\frac{1}{4}$-watt resistor
R14	6.2-kilohm, $\frac{1}{4}$-watt resistor
R15	3.3-kilohm, $\frac{1}{4}$4-watt resistor
R16	470-ohm, $\frac{1}{4}$-watt resistor
C1,C2,C11	.01-µF, 25-volt disc capacitor
C3,C7,C8,C14	.001-µF, 25-volt disc capacitor
C9	27-pF, 25-volt Mylar capacitor
C10	180-pF, 25-volt Mylar capacitor
C4,C12,C13,C15	10-µF, 25-volt electrolytic capacitor
LDR1,LDR2	VT835 light-dependent resistor (Vactec)
Q1	VT1314 phototransistor (Vactec)
Q2,Q3	2N4403 transistor
Q4,Q6	BS170 FET transistor
Q5	MPSH11 transistor
U1	LM358 op-amp
U2,U3	7555 CMOS timer
Xtal	49.890-MHz series resonant crystal
L1	.47-µHz coil (Toko 7KM series)
L2	.6- to 50-µHz miniature choke
T1	RF transformer TS2343-18-5 (Time Space Scientific)
BZ	Piezo buzzer
ANT	One-meter whip antenna

Wireless security system, RF receiver/alarm parts list

R1,R18,R19,R21	10-kilohm, $\frac{1}{4}$-watt resistor
R2	2.2-kilohm, $\frac{1}{4}$-watt resistor
R3	47-ohm, $\frac{1}{4}$-watt resistor
R4	2-kilohm, $\frac{1}{4}$-watt resistor
R5	4.7-kilohm, $\frac{1}{4}$- watt resistor
R6	470-kilohm, $\frac{1}{4}$-watt resistor
R7,R8,R22	100-kilohm, $\frac{1}{4}$-watt resistor
R9	6.2-kilohm, $\frac{1}{4}$-watt resistor
R10	56-kilohm, $\frac{1}{4}$-watt resistor
R11	47-kilohm, $\frac{1}{4}$-watt resistor
R12	33-kilohm, $\frac{1}{4}$-watt resistor
R13	10-megohm, $\frac{1}{4}$-watt resistor
R14	130-ohm, $\frac{1}{4}$-watt resistor
R15	6.8-kilohm, $\frac{1}{4}$-watt resistor

Wireless security system, RF receiver/alarm parts list

R16	20-kilohm, $\frac{1}{4}$-watt resistor
R17	15-kilohm, $\frac{1}{4}$-watt resistor
R20	1-kilohm, $\frac{1}{4}$-watt resistor
C1	5-pF, 25-volt mica capacitor
C2,C4	.002-µF, 25-volt disc capacitor
C3	24-pF, 25-volt mica capacitor
C5	18-pF, 25-volt mica capacitor
C6,C14,C18	10-µF, 25-volt electrolytic capacitor
C7	.039-µF, 25-volt disc capacitor
C8,C16,C17	.01-µF, 25-volt disc capacitor
C9,C13	.1-µF, 25-volt disc capacitor
C10,C11	4.7-µF, 25-volt electrolytic capacitor
C12	100-µF, 25-volt electrolytic capacitor
C15	1-µF, 25-volt electrolytic capacitor
D1	LED
Q1	MPSH11 transistor
Q2	2N3904 transistor
U1	LM358 op-amp
U2	LM567 tone decoder
U3	7555 CMOS timer
U4	78L05 regulator
L1	50-MHz coil
T1	RF transformer (primary 18 turns #28-gauge wire and secondary 5 turns #24-gauge wire on .23-inch-diameter #43 ferrite core)
BZ	piezo buzzer
ANT	One-meter whip antenna

Portalarm: wireless pyroelectric security system

The Portalarm wireless pyroelectric security system, illustrated in Figures 7-19 through 7-22, can protect your home while you are away, and it reports directly to a friend or neighbor's house. Now you can feel secure in knowing that your property is being monitored, whether you are away for a few hours or many months. The Portalarm wireless alarm system monitors your home with an advanced passive pyroelectric sensor, which can provide coverage up to 50 feet. The pyroelectric sensor activates a logic circuit that turns an FM transmitter on, sending a "warble" tone over the air for 15 seconds to alert your neighbor. A sensitive microphone is then switched into the circuit to allow your neighbors to "listen in" to your home for about four minutes, at which time they can call the police or investigate personally.

You can use the Portalarm system with just about any FM transmitter: either an FM broadcast-band transmitter or a VHF/UHF public-service-band transmitter operating on one of the itinerant FM frequencies of 154.570, 151.625, or 154.600 MHz. The neighbor or friend monitoring your home would then need only an FM receiver or scanner tuned to the Portalarm's transmitter frequency.

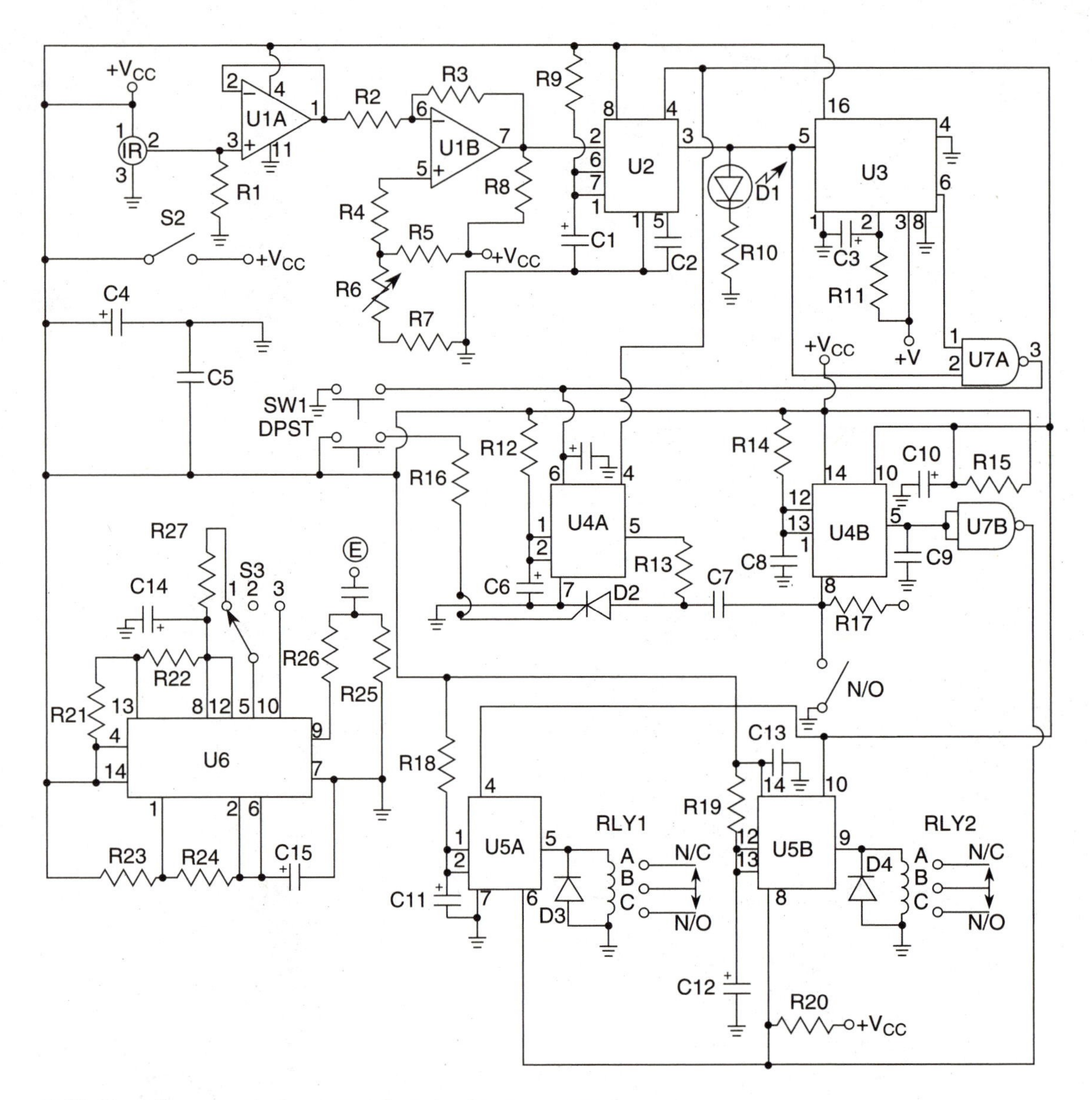

7-19 Portalarm—wireless pyroelectric alarm system.

The Portalarm's main logic board is shown in Figure 7-19 and Photo 7-7. The front end of the circuit begins with an Eltec 5192 dual element pyroelectric sensor. The pyroelectric sensor is constructed from a thin film of lithium tantalate, with conductive electrodes attached to both ends of the sensing elements. The pyroelectric material produces an internal electrical field that is collected via the end electrodes. Monitoring the charge across the lithium tantalate crystal requires a high-impedance amplifier. Therefore, most sensors of this variety are packaged with an internal FET amplifier, load resistor, and impedance converter. These pyroelectric sensors are very

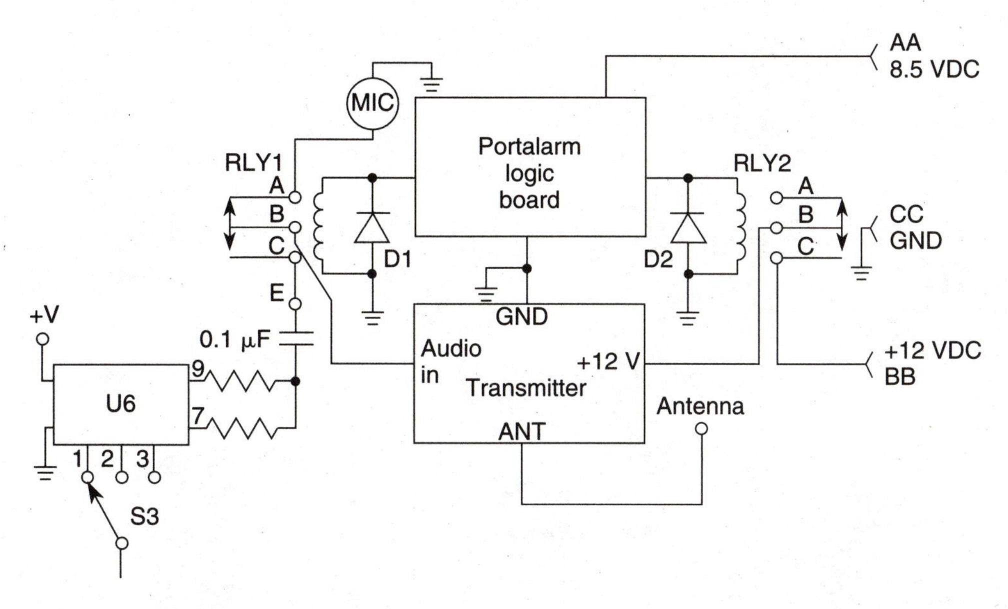

7-20 Portalarm—interconnection diagram.

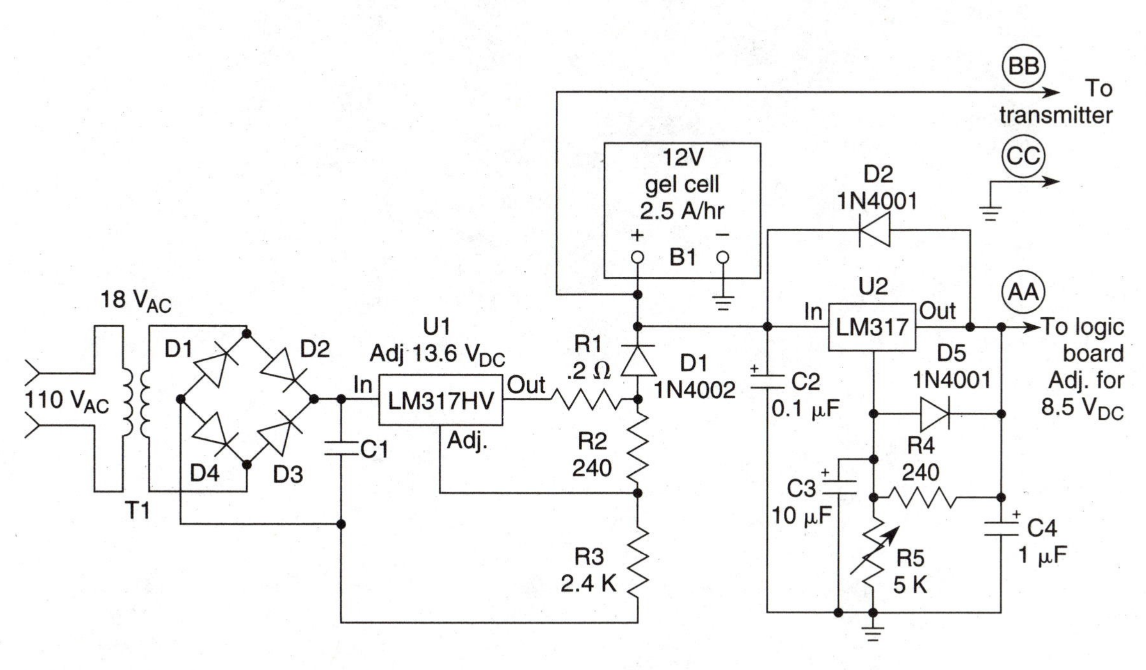

7-21 Portalarm—power supply.

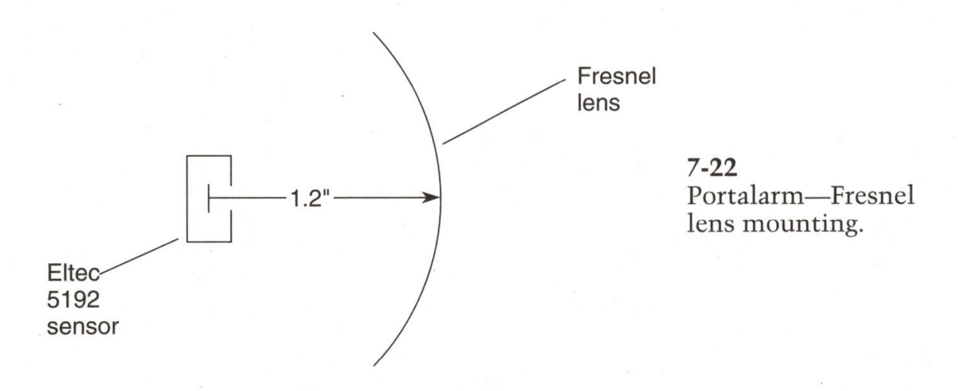

7-22
Portalarm—Fresnel lens mounting.

Photo 7-7 Portalarm system board.

sensitive and respond quickly to changes in infrared intensity. Both sensing elements are mounted in a TO-5 transistor-type package, with a square aperture window fitted with an infrared filter. The dual element detector is wired parallel-opposed to help eliminate the effects of temperature variations and to prevent false alarms.

A person walking in the field of view of the sensor elements causes two distinct pulses, positive and negative. The detector produces no output when both sensing elements receive equal amounts of IR radiation. As soon as one of the elements receives slightly more radiation than the other, an output is produced. In practical applications, it is necessary to focus the IR radiation onto the sensing elements. A

Fresnel lens is generally placed in front of the sensor to increase the range and shape the pattern or coverage of the sensor. Low-cost Fresnel lenses come in many different configurations. Long-range lenses can sense up to 50 feet away, a narrow-beam-width lens only a few degrees, and wide-angle lenses out to 12 or so feet, with a wide angle of coverage.

The Eltec 5192 pyroelectric sensor is a three-lead device. Pin 1 is the power input, pin 2 is the signal output lead, and pin 3 is the case or ground lead. The Portalarm circuit begins with a 47-kilohm resistor at R1, which draws a small amount of current for the sensor elements. The signal output from the pyroelectric sensor on pin 2 is fed directly to an LM324 op-amp. Op-amp U1A isolates the sensors using a simple voltage follower. The signal is then passed on to U1B, which acts as a signal amplifier. R3 sets the overall gain of the amplifier section. The network connected to pin 5 of U1B is used as a threshold control and is adjusted via R6. The output from pin 7 of U1B is passed on to an LM555 timer, set up as a multistable multivibrator, which produces a time pulse about one second in duration. The LED at D1 assists in the setup and adjustment of the pyroelectric sensor, which is done via R6. The signal from pin 3 of U2 feeds U3, a CD4528 that acts as a time window discriminator circuit. The time period of the window is set by the C3/R11 combination for about 10 seconds, to allow passage of a few detector pulses before resetting. This window detector helps to prevent false alarms.

The signal is then passed on to the Portalarm's main timers. The first set of timers begins with U4, a LM556 dual timer. The first half of U4 is the exit/entry timer, which allows you to leave the room when initially setting the alarm. This delay timer is also used for reentering the protected space to deactivate the alarm. The exit/entry timer is set for about 20 to 25 seconds, which is ample time to leave the protected space. You can readjust the exit/entry timer to suit. The C6/R12 combination sets the exit/entry period. Once the exit/entry timer has been started via S1, the SCR fires, preventing U4B from being triggered during the time-delay period. Note that multiple sensors can be wired directly to point AA, pin 8 of U4B. Any normally open type of sensor can be connected to this point. The time period of U4B is set by the C8/R14 pair. The output of U4B is a positive-going pulse, and it must be inverted in order to trigger U5. U5A and U5B are two independent timers. The first timer at U5A switches between the warble-tone generator and the microphone used to "listen in" to the protected area. The C11/R12 combination is set for about 15 seconds. Once the alarm is triggered, the warble tone is transmitted for about 15 seconds to get your neighbor's attention before switching over to the microphone.

At the instant U5A is triggered, U5B also begins its timing period and power is applied to the FM transmitter module. The timer at U5B is configured to allow the transmitter to send a voice-modulated signal for about four minutes, after which time the transmitter is turned off. Figure 7-20 illustrates the relay connections for the warble-tone generator/microphone hook-up and for the transmitter power-on relay. A power reset circuit is formed by C10/R15, which resets all the timers upon power-up.

The warble-tone generator is comprised of U6, a dual-timer IC that is configured in the astable mode. Switch S3 selects the three-tone output combinations. The

C14/R22 and the R24/C15 pairs determine the actual frequencies generated by the warble-tone generator. The output of the sound generator at point E is coupled via C16 to the audio input of the transmitter module, as shown.

The next portion of the Portalarm is the FM transmitter. Since building and testing RF transmitters is not trivial and usually requires special equipment for testing and adjustment, I recommend that you purchase an FM exciter or transmitter kit. There are, however, a few alternatives. You can obtain a VHF walkie-talkie, or perhaps a discarded cordless phone. If you decide to purchase an FM transmitter (suppliers are listed in the Appendix), Ramsey Electronics offers an FM broadcast-band transmitter with a range of up to a half mile with an external antenna. The FM-4 kit can readily be used for this application. More powerful transmitter kits, such as the 5-watt model shown in Photo 7-8, are available from Hamtronics. I suggest FM transmitters for the Portalarm since the companion receiver or scanner can be used to eliminate noise problems when no signal is present.

The power-supply circuitry for the Portalarm is shown in Figure 7-21. Power is provided from a 12-volt, 1.2-amp/hour gel-cell battery that is constantly charged from a battery-charger circuit. The logic board is powered from the LM317 to provide 8.5 volts dc, while the transmitter section is powered directly from the 12-volt battery. In this configuration, the Portalarm remains wireless and will not be affected by power outages or tampering.

Constructing the Portalarm is somewhat more involved than most projects in this book. The transmitter board should be placed in a metal or shielded enclosure of its own inside the Portalarm cabinet or system box. The logic board and power supply are mounted in the system chassis box side by side. The prototype was housed in a 8 × 10-inch metal chassis box. Perhaps the most difficult part of constructing the Portalarm involves mounting the pyroelectric sensor. You can use either a small chassis box or tuna can to house the sensor atop the main system enclosure. The pyroelectric sensor is generally mounted behind the Fresnel lens, as shown in Figure 7-22. In the diagram, the focal distance is shown as 1.2 inches, but this might vary depending on the

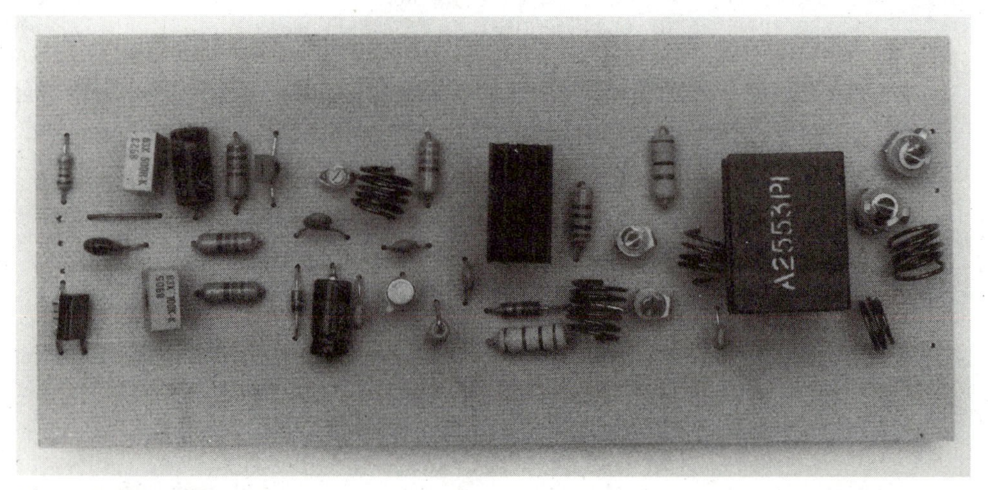

Photo 7-8 Portalarm transmitter board.

actual lenses used. Two main lens types are available. A long-range lens allows you to see out to 50 feet, but with a narrow beamwidth; with a wide-angle lens you can see out to about 12 feet, but at a wider angle. One approach is to cut a slot in the side of a tuna can and wrap the lens around the interior of the slot. The sensor can then be mounted just behind the lens. Remember to keep the leads as short as possible and to use shielded wire between the sensor and the electronics.

Operating the Portalarm is straightforward, once everything is wired and mounted. First, select the warble tone of choice via S3, and attach your antenna to the transmitter. You can construct your own half-wavelength dipole antenna using the following formula:

$$\text{length of antenna} = 492 / \text{frequency (MHz)}$$

or you could use a whip antenna with reduced range. An outdoor antenna or dipole placed in front of a window works great. Any number of additional open-circuit sensors such as floor mats, door switches, and vibration sensors can be connected to point AA on the main logic board. The Portalarm should be placed on a counter or table with the longest possible nonobstructed view, to allow the sensor to "see" through your home. Never point a pyroelectric sensor toward a fire, sunlight, or heating vents.

Locate a companion receiver or scanner. Turn on the receiver, which is tuned to the Portalarm transmitter. Turn on the Portalarm and adjust R6 to a point just below triggering the pyroelectric sensor. Wave your hand in front of the sensor and D1 should light up. The transmitter should begin sending the warble tone once the exit/entry timers are activated. After 15 seconds, the transmitter should begin sending audio to the receiver. To test the exit/entry function, turn off the Portalarm for a few seconds, then turn on S2 and press S1. Quickly walk away from the protected area. Wait 30 to 40 seconds and enter the protected area; the alarm should trigger.

The Portalarm is now ready to protect your home from potential burglars. You can feel secure in knowing that the Portalarm is standing guard 24 hours a day!

Portalarm wireless pyroelectric alarm system parts list

FM transmitter	See text
R1,R5	47-kilohm, ¼-watt resistor
R2,R8,R20	10-kilohm, ¼-watt resistor
R3	500-kilohm, ¼-watt resistor
R4	33-kilohm, ¼-watt resistor
R6	2-kilohm trim potentiometer
R7	2.5-kilohm, ¼-watt resistor
R10,R13	30-kilohm, ¼-watt resistor
R11,R12,R24	100-kilohm, ¼-watt resistor
R14	220-kilohm, ¼-watt resistor
R15,R16,R17	24-kilohm, ¼-watt resistor
C1	22-µF, 25-volt electrolytic capacitor
C2	.05-µF, 25-volt disc capacitor
C3	100-µF, 25-volt electrolytic capacitor
C4	200-µF, 25-volt electrolytic capacitor
C5,C7,C13,C14,C16	.1-µF, 25-volt disc capacitor

Portalarm wireless pyroelectric alarm system parts list continued

C6,C12	47-µF, 25-volt electrolytic capacitor
C8,C11	10-µF, 25-volt electrolytic capacitor
C10	1-µF, 25-volt electrolytic capacitor
C15	3.3-µF, 25-volt electrolytic capacitor
D1	LED
D2	NTE 5404 SCR
D3,D4	1N4002 silicon diode
U1	LM324 op-amp
U2	LM555 IC timer
U3	CD4528 one-shot
U4,U5,U6	LM556 dual timer
U7	LM7400 Nand gate
IR	5192 dual-opposed pyroelectric sensor (Eltec)
RlY1,RLY2	5-volt, low-current relay (RS275-243)
SW1	DPST push-button switch
SW2	SPST on/off switch
Misc	IR lens, circuit board, and chassis boxes

Portalarm power supply parts list

R1	.2-ohm, ¼-watt resistor
R2	240-ohm, ¼-watt resistor
R3	2.4-kilohm, ¼-watt resistor
R4	240-ohm, ¼-watt resistor
R5	5-kilohm potentiometer
C1	470-µF, 25-volt electrolytic capacitor
C2	.1-µF, 25-volt disc capacitor
C3	10-µF, 25-volt electrolytic capacitor
C4	1-µF, 25-volt elecrolytic capacitor
D1-D5	1N4001 silicon diode
U1	LM317HV regulator
U2	LM317 regulator
T1	110-volt to 18-volt (ac) transformer
B1	12-volt gel-cell battery
Misc	Circuit board, chassis box, and power cord

Long-range IR laser perimeter alarm system

The last project in this chapter is a long-range laser beam-break alarm system that can protect driveways, walkways, and the entire perimeter of your home, shop, or office. The long-range beam-break perimeter alarm is shown in Figures 7-23 through 7-26. The laser perimeter alarm consists of an IR FM laser transmitter and an FM IR phase-locked loop receiver/alarm unit.

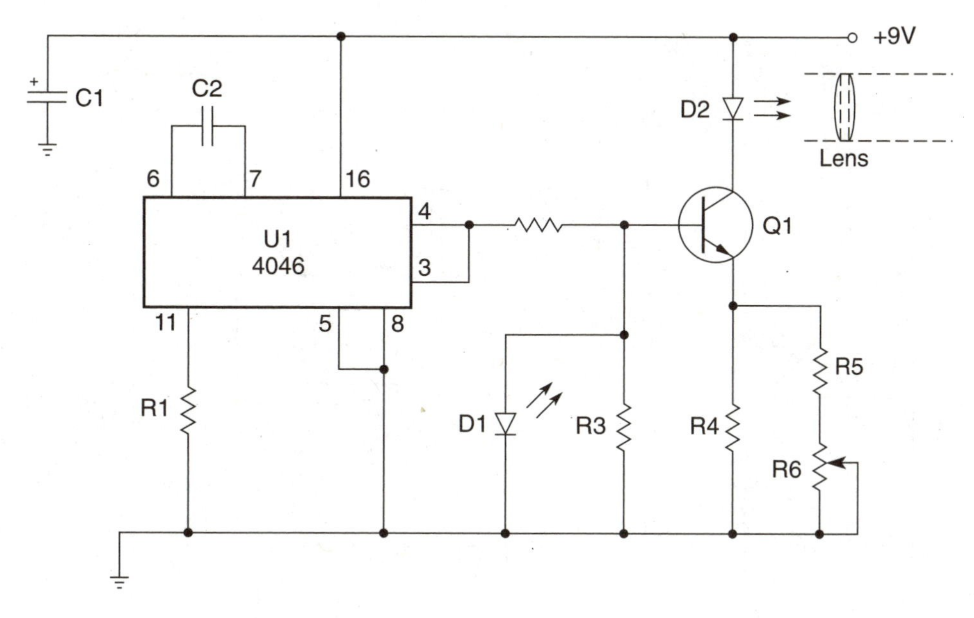

7-23 Long-range IR laser perimeter alarm transmitter.

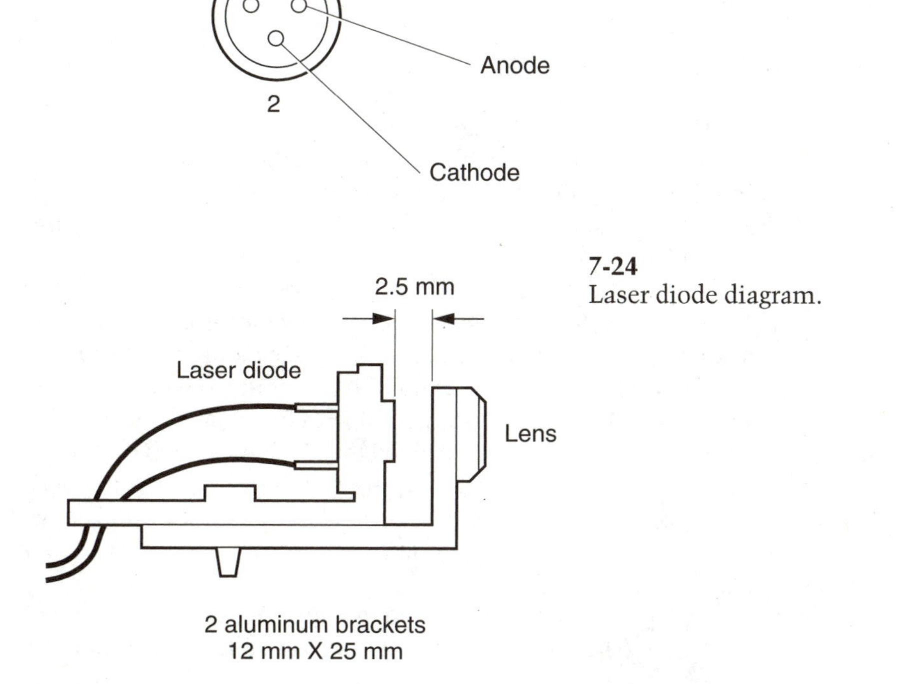

7-24
Laser diode diagram.

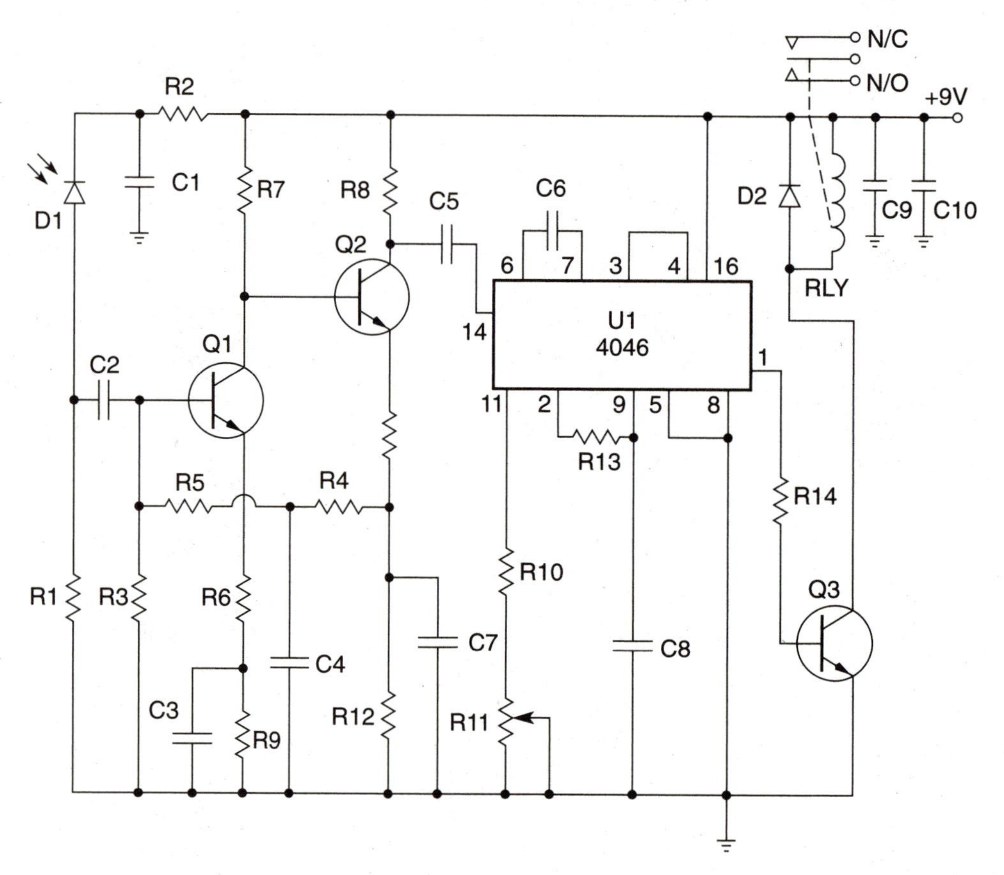

7-25 Long-range IR laser perimeter alarm receiver.

The transmitter portion of the laser perimeter alarm is shown in Figure 7-23. The heart of the transmitter is the 4046 phase-locked loop (PLL) chip, which oscillates at about 200 kHz. The PLL oscillator frequency is controlled by C2 and R1. The output of the PLL at pin 4 is fed back into the phase comparator at pin 3. The output signal of the PLL is coupled through R2 to Q1. This square-wave output signal at 200 kHz has a peak-to-peak output of about 6 volts. Transistor Q1 drives the laser diode. The laser diode used in the laser perimeter alarm transmitter is the LT026 IR laser, which has a maximum forward current of 100 mA. Exceeding the 100-mA current will ultimately destroy the laser, so be careful. The current to the laser diode is controlled by R5 and R6, which limit the current to 65 mA for normal operation. The current to the laser diode is stabilized or controlled by D1. The LED at D1 holds the base voltage of Q1 to about 1.8 volts, thus clipping the signal voltage applied to the base of Q1. The voltage across R4, therefore, is limited to 1.2 volts and voltage to the emitter is held at 1.2 volts. The quiescent current through Q1 measures about half the recommended value, or about 34 mA.

The laser diode has three leads, as shown in Figure 7-24, but only leads 1 and 2 are used in this circuit. The laser diode mounting assembly used to hold the laser

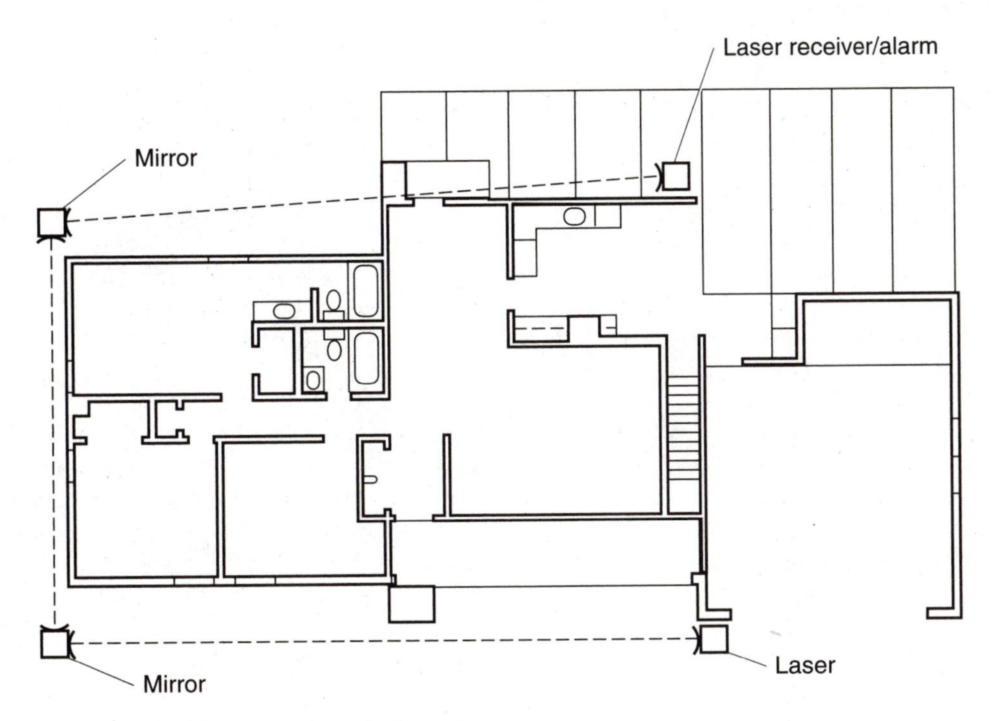

7-26 Typical laser perimeter alarm layout.

diode can be fabricated from two small aluminum angle brackets. To ensure that the lens and laser diode are exactly aligned, clamp the two brackets together, one over the other, as shown, and drill a pilot hole through both brackets. A single machine screw was used to hold one bracket over the other, as shown. The lens bracket is 2.5 mm in front of the laser diode. The distance between the lens and laser diode depends on the actual lens used, so you might have to find this focal distance from the manufacturer's specifications or by experimenting.

The laser perimeter alarm receiver section is illustrated in Figure 7-25. The IR detector diode is first amplified by a high-gain dc-coupled amplifier formed around transistors Q1 and Q2. The output of this amplifier is then coupled to U1 via C5. The frequency of the internal voltage-controlled oscillator (VCO) in the PLL is determined by C6, R10, and R11. A low-pass filter is formed by R13 and C8. The low-pass filter connects the output of the internal phase comparator to the input of the VCO. The output pin of U1 is connected to Q3, which drives the alarm relay. The output at pin 1 goes high when the internal oscillator is locked onto the input signal at the input stage of the PLL, which is derived from the remote transmitter. Under normal operating conditions, the relay is pulled in when the receiver PLL is locked onto the transmitter signal. When the laser beam is broken, the interrupted beam causes the relay to drop out, setting off an external alarm siren. Diode D2 limits the back EMF from the relay coil.

Both the laser transmitter and receiver sections can be powered from 9-volt power supplies or batteries. When power is applied to the laser transmitter, the LED at D1 should light up, indicating an output from the PLL. The relay on the receiver should pull in once the transmitter and receiver are pointed at and aligned with each other. If the receiver does not lock onto the transmitter, then you might have to adjust R11 on the receiver to ensure that both units are operating at the same frequency.

Since the laser beam from the transmitter is in the infrared region of the spectrum, you might have some difficulty in aligning the transmitter and receiver sections. To assist you in aligning the system units, you might want to obtain an IR test card from Radio Shack. This test card is about the size of a credit card and can readily be used in system alignment. First energize the IR test card from either sunlight or an incandescent lamp. Once energized, the card is ready to detect an IR beam. Hold the IR test card in front of the IR laser beam and a red spot should appear on it.

The diagram shown in Figure 7-26 illustrates a typical perimeter alarm configuration. The laser transmitter can be mounted in a waterproof box fastened to the side of your house, about 2½ feet above ground. Measure the same distance up from the ground and mount the waterproof receiver at a second location along the perimeter of your house. If you adjust the height of the transmitter and receiver high enough, you can protect most of your house. To bounce the laser beam around the corners of your house, use a few surplus front-surface mirrors because they are much more efficient than conventional mirrors.

You could also use the laser perimeter alarm in other alarm configurations. The perimeter alarm could be used indoors for interior perimeter protection. If you have long hallways or many windows in a row, you could easily protect large interior spaces by bouncing the beam around the interior walls. You can use the laser perimeter alarm to protect outdoor walkways and driveways at your home. The laser perimeter alarm is a low-cost but highly effective alarm component. A few laser perimeter alarms could be combined with an alarm control box to form an entire alarm system.

Long-range IR laser perimeter alarm transmitter parts list

R1	22-kilohm, ¼-watt resistor
R2	3.3-kilohm, ¼-watt resistor
R3	2.2-kilohm, ¼-watt resistor
R4,R5	22-ohm, ¼-watt resistor
R6	200-ohm trim potentiometer
C1	1000-µF, 25-volt electrolytic capacitor
C2	100-pF, 25-volt disc capacitor
D1	LED
D2	IR laser diode LT026 or equivalent
U1	4046 PLL
Q1	BC337 transistor
Misc	Chassis box, batteries, lens, etc.

Long-range IR laser perimeter alarm receiver parts list

R1	10-kilohm, ¼-watt resistor
R2	1-kilohm, ¼-watt resistor
R3	82-kilohm, ¼-watt resistor
R4,R14	22-kilohm, ¼-watt resistor
R5,R7,R10	12-kilohm, ¼- watt resistor
R6,R9	220-ohm, ¼-watt resistor
R8	3.9-kilohm, ¼-watt resistor
R11	20-kilohm trim potentiometer (trim)
R12	2.2-kilohm, ¼-watt resistor
R13	120-kilohm, ¼-watt resistor
C1,C4	.47-µF, 25-volt disc capacitor
C2	470-pF, 25-volt Mylar capacitor
C3	10-nF, 25-volt Mylar capacitor
C5	68-pF, 25-volt ceramic capacitor
C6	100-pF, 25-volt ceramic capacitor
C7	.1-µF, 25-volt ceramic capacitor
C8	1-nF, 25-volt Mylar capacitor
C9	.1-µF, 25-volt ceramic disc capacitor
C10	1000-µF, 25-volt electrolytic capacitor
D1	BPW50 IR photodiode detector
D2	1N4007 silicon diode
Q1,Q2,Q3	BC548 transistor
U1	4046 PLL
Misc	Chassis box, lens, and hardware

8
CHAPTER
Optical communication

In this chapter we will explore light-wave listeners, which can "listen in" to natural phenomena as well as man-made light-modulated sources. I will discuss simplex and duplex "free space" or "through the air" light-wave communications systems, light-wave signaling systems, and AM/FM voice communication links. Then I will move on to a pulse-modulated data communication link, which can transfer computer data through the air. You will see a wireless or remote speaker system, with which you can quietly listen to radio or TV programs while others are sleeping, as well as a laser communication system that can be used to send voice or data through "free space." Lastly I will demonstrate how to use touch tones to remotely control functions via a laser beam.

Modulated light waves can be sent through either optical fibers or the air (free space). The advantage of free-space light-wave communications is that it is license-free, private, and jam-proof. Short-range signaling and communication up to 10 or 12 feet away is very easy to design and align, while long-range systems require lenses, filters, collimators, and tripods for stable operation.

A simplex light-wave link consists of a sending unit or transmitter and a receiver unit, as shown in Figure 8-1. A two-way (duplex) communication system consists of a transmit-and-receive unit, or transceiver, at each end of the system, as shown in Figure 8-2. A typical light-wave sending unit would contain a microphone and a modulator that drives an LED or laser diode light source. The microphone is used with the modulator to vary the brightness of the transmitter's light source in response to the audio signal at the microphone.

A free-space light-wave receiving link typically consists of a detector such as a photodiode, phototransistor, or solar cell, which is coupled to an electronics circuit that amplifies and demodulates the incoming light-wave signals, translating them back to an amplified audio signal to drive a speaker. The reconstructed audio is representative of the original audio going into the transmitter unit. Light-wave communication systems can use incandescent light bulbs, LEDs, infrared LEDs, laser diodes, or helium-neon lasers to link a sending unit to a receiver unit. The best results are obtained when infrared LEDs or lasers are used in light-wave links. Signals are often modulated on an AM or FM carrier, which eliminates stray light

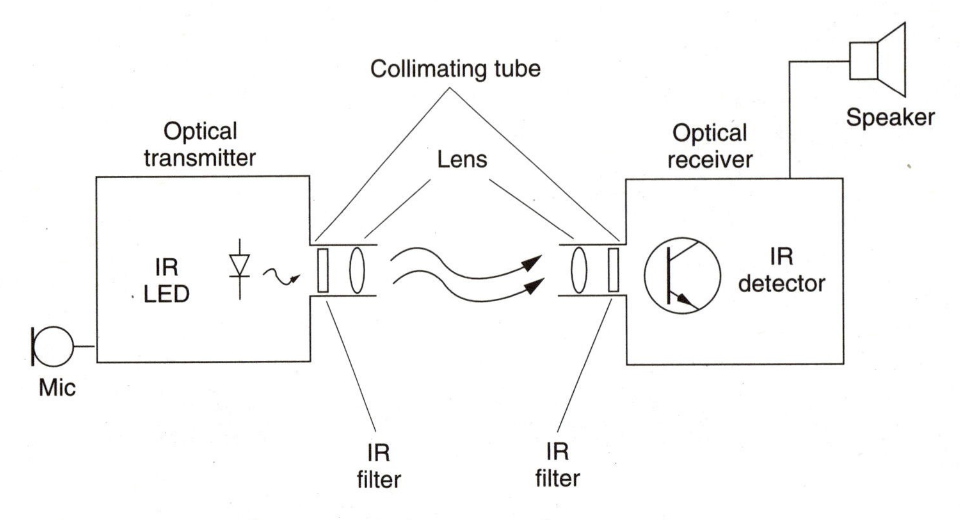

8-1 One-way simplex transmission.

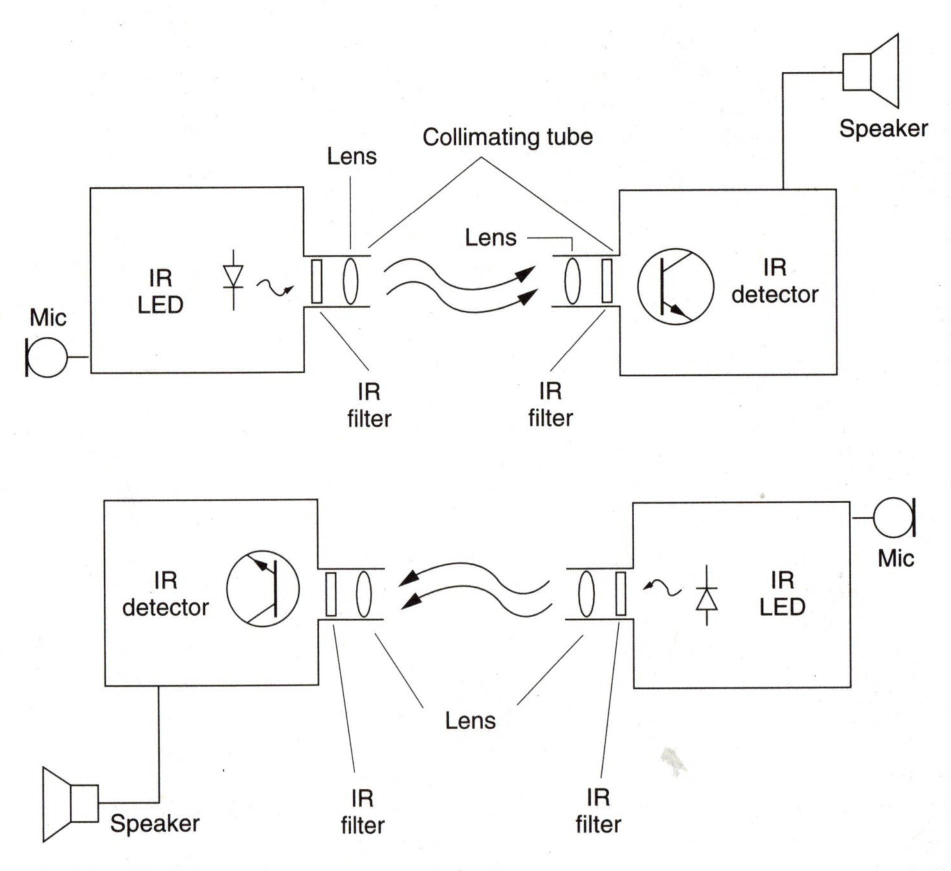

8-2 Duplex IR communications system.

and interference from the light-wave communication system. FM and PFM modulation techniques provide the best noise-free communication for audio systems. Often a form of pulse-code modulation is used to send data signals from the transmitting unit to a light-wave receiver.

In order to increase the range of a free-space communications link from 10 feet to a long-range system, you need a pair of lenses, one in front of the light-wave sending unit and another ahead of the detector. Often filters are used in conjunction with lenses for long-range infrared light-wave links. Lenses can range from plastic surplus lenses to gun bore-sight telescopes, depending on your application. Many filter and lens sources are listed in the Appendix.

For optimum long-range results, consider using a collimator or external light-shield tube ahead of the detector. A collimator consists of a hollow tube lined with black paper or painted inside with flat black paint. The collimator reduces unwanted light and reflections from reaching the detector assembly. Figure 8-3 illustrates a typical free-space optical link with lenses on both the transmitting and receiving units. Note the collimator at the receiving end ahead of the detector. The range of the free-space link is determined by the following formula:

$$R = \frac{P_o A_{rec}}{D_S \phi^2}$$

where:

P_O = LED power in milliwatts
A_{rec} = receiver lens area in meters
D_S = detector sensitivity in milliwatts
ϕ = LED beam divergence in radians

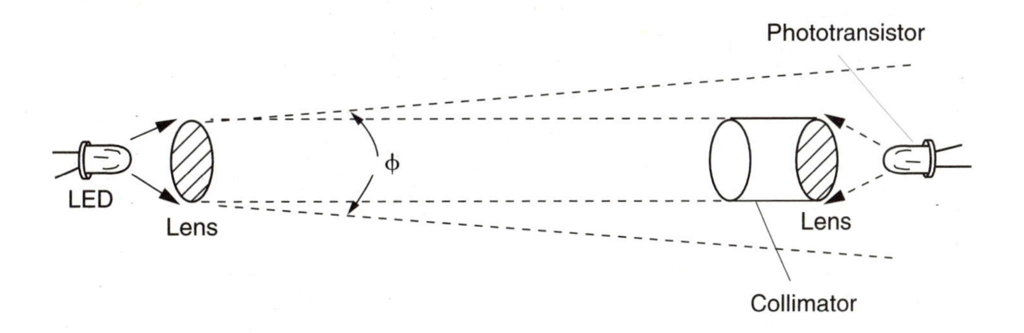

8-3 Lens and collimator.

The best long-range free-space communication links consist of infrared (IR) lasers at the sending unit and an IR phototransistor at the receiver link, with an IR filter and lens ahead of the detector with a collimator, modulated by an FM or PFM modulator.

Focusing and aligning an IR free-space communications link can be a bit tricky. Both transmitter and receiver need to be mounted on a flat table top or tripod with an unobstructed view between the sending and receiving units. A quick method of aligning an IR communications link is to obtain a Radio Shack IR detector card and use it to test the IR TV remote controls. This detector will greatly aid you in aligning your IR light-wave link.

Sensitive light-wave receiver

A very sensitive light-wave listener or light-wave receiver is shown in Figure 8-4. This light-wave receiver consists of two high-gain stages of amplification. The light-wave listener can transform pulsating or modulated light that the eye cannot discern into sounds that your ear can readily hear. The light-wave listener can be used indoors as well as outdoors to "listen in" to both natural and man-made sounds.

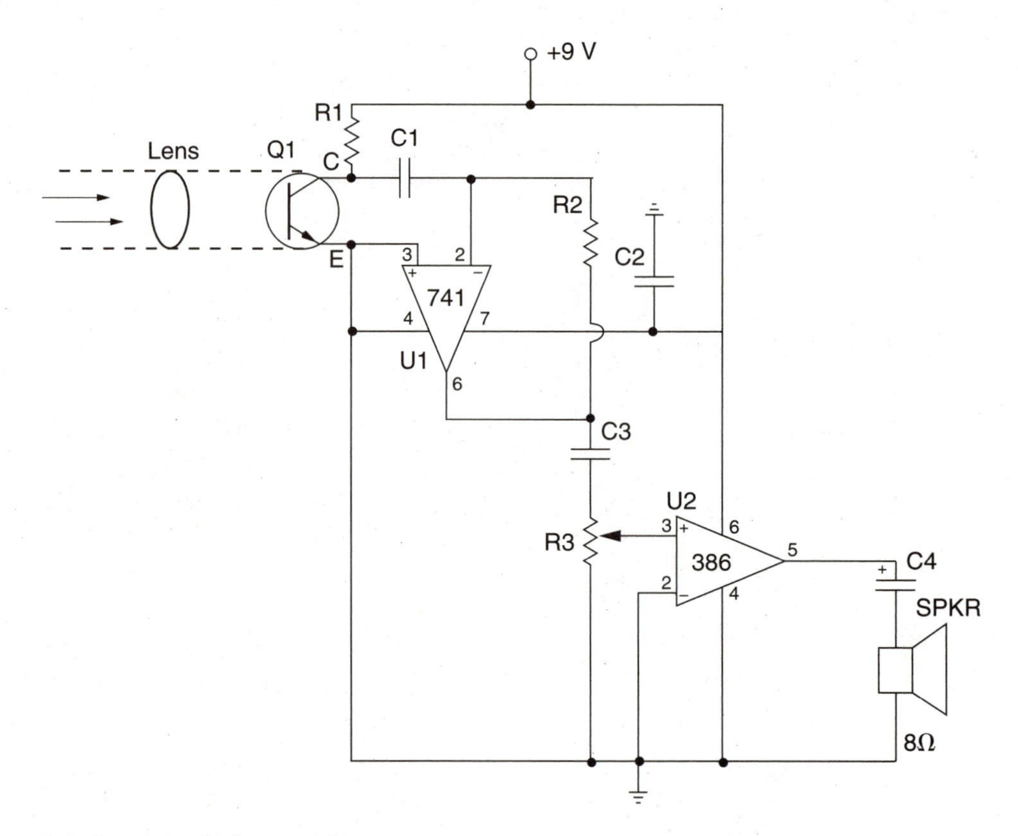

8-4 Sensitive light-wave listener.

The light-wave listener can detect lightning flashes and produce pops and clicks in response; it can even detect some lightning missed by the human eye. Try the following experiment. Light a candle in front of your light listener and you will hear various interesting sounds. When the surrounding air is still, a soft rushing sound will be produced. When the flame is disturbed from moving air, you will hear crackles and pops from the speaker.

For an interesting and novel experiment, use your light listener (with a small collimator and lens) to listen to flying insects or a bird's wing beats. If you place your light listener beneath an insect or hummingbird, in line with sunlight or a light source, you will easily hear when the insect or bird's wing is reflecting sunlight into the detector because the speaker will produce a distinct buzz or hum. Take your light listener outside at dusk and nearby fireflies will produce soft clicks for each flash of light.

You can also use the light listener to detect man-made modulated light sources. Sweep a flashlight beam in front of your light listener and a soft swishing sound will be produced. Then, with a flashlight aimed at your listener, tap the flashlight with a pencil; you will hear ringing sound as the lamp filament vibrates. The headlights of cars will produce a distinct ringing when the vehicle is driven down a bumpy road. Electric displays or clocks will produce a hum or buzz when brought near the light listener. TV or computer screens will produce a buzz, while a camera flash will produce a loud pop. The light listener could also be placed in the path of a rotating fan blade or propeller, allowing you to detect the speed change with your ears. The light-wave listener could make an interesting science-fair project.

The heart of this sensitive light-wave receiver is the phototransistor Q1. Resistor R1 is used to bias the phototransistor. The output of the phototransistor is coupled to the input of U1 via capacitor C1. The 100-kilohm resistor R2 is used to set up the overall gain of the op-amp at U1. The capacitor at C3 couples the op-amp to the audio amplifier at U2, an LM386. Potentiometer R3 controls the level of signal reaching U2. Capacitor C2 is a bypass capacitor that prevents oscillation in the circuit. The output of U2 is fed to an 8-ohm speaker or headphone via C4. The light-wave listener is powered by a 9-volt transistor radio battery. The light-wave receiver in Figure 8-4 can be used with the AM light-wave transmitter in Figure 8-5 or in a number of other interesting projects.

Sensitive light-wave receiver parts list

R1,R2,R3	100-kilohm, ¼-watt resistor
C1,C2,C3	.1-µF, 25-volt disc capacitor
C4	100-µF, 25-volt electrolytic capacitor
U1	LM741CN op-amp
U2	LM386 audio amplifier
Q1	Phototransistor (Motorola MRD-300)
SPKR	8-ohm speaker or headphone
B	9-volt battery
Misc	Enclosure, battery clip, and collimator

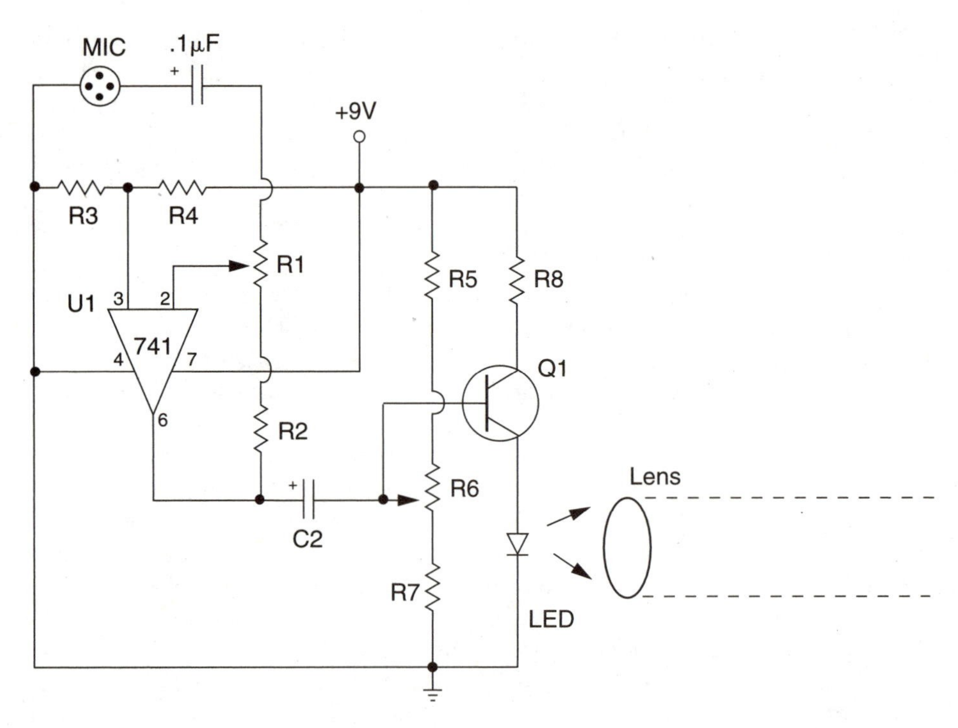

8-5 Light-wave voice transmitter.

AM light-wave voice transmitter

The light-wave receiver shown in Figure 8-4 can be readily combined with the light-wave transmitter in Figure 8-5 to form a medium-range light-wave communications link. The light-wave transmitter consists of an electret microphone coupled to an LM741C op-amp, which modulates the LED via transistor Q1. The overall gain is controlled by resistors R1 and R2, while R6 is adjusted for best sound quality in the receiver unit. The op-amp is coupled to the LED driver via C2. The LED shown can be a conventional, super-bright, or IR LED. Using a very bright IR LED with lenses at both ends of the link will allow communications up to hundreds of feet. You will obtain the best results by placing a lens and collimator at the receiver input and using tripods to stabilize both the transmitter and receiver.

AM light-wave transmitter parts list

R1,R6	50-kilohm potentiometer
R2	1-megohm, ¼-watt resistor
R3,R4	5.6-kilohm, ¼-watt resistor
R5,R7	1-kilohm, ¼-watt resistor
R8	220-ohm, ¼-watt resistor
C1	.1-μF, 25- volt disc capacitor
C2	10-mF, 25-volt electrolytic capacitor
U1	LM741CN op-amp

AM light-wave transmitter parts list continued

Q1	2N2222 transistor
LED	Bright LED
MIC	Electret microphone
B	9-volt battery
Misc	Enclosure, battery clip, and lens

Short- to medium-range signaling system

Figures 8-6 and 8-7 illustrate a short- to medium-range tone-signaling or Morse-code light-wave link. The tone transmitter consists of an LM555 timer IC, which forms an oscillator that modulates the transmitter's LED sender. The LM555's pulse rate is controlled by the 100-kilohm potentiometer at R1. The tone oscillator is set up for a 50-percent duty cycle. Use the code key, push-button switch, or relay contacts to activate the tone transmitter unit. The tone transmitter is powered by a 9-volt transistor radio battery.

The companion light-wave receiver consists of a phototransistor that receives the incoming modulated light-wave signals and an electronic receiver that contains two transistors, used to drive a sounder. Resistor R1 biases transistor Q1. The modulated light-wave signal from the phototransistor is coupled to the second transistor

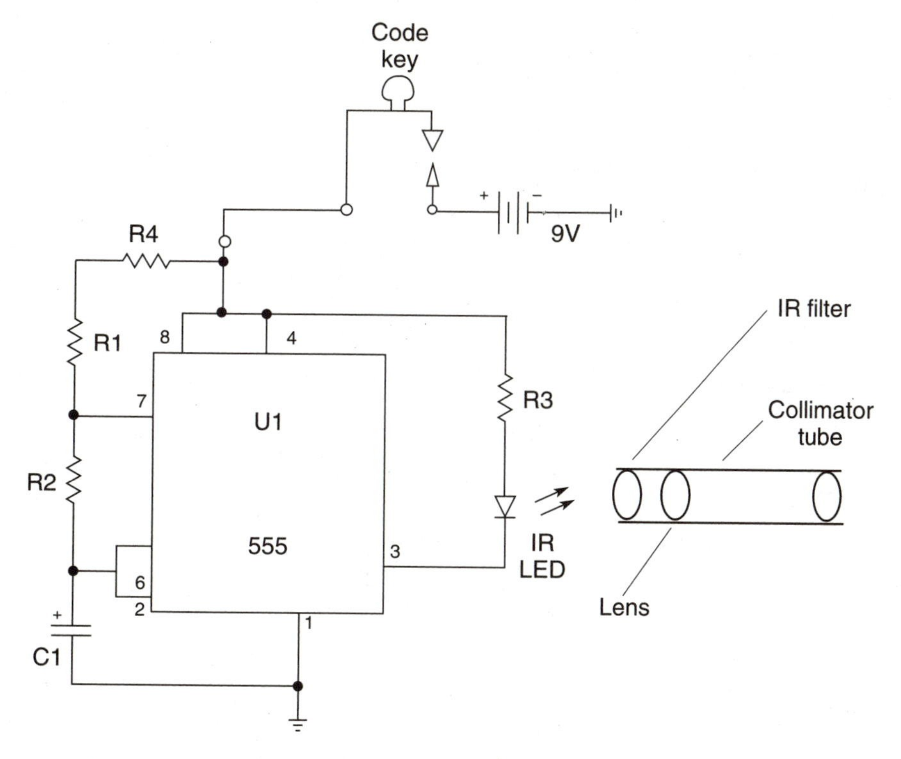

8-6 Short/medium-range signaling transmitter.

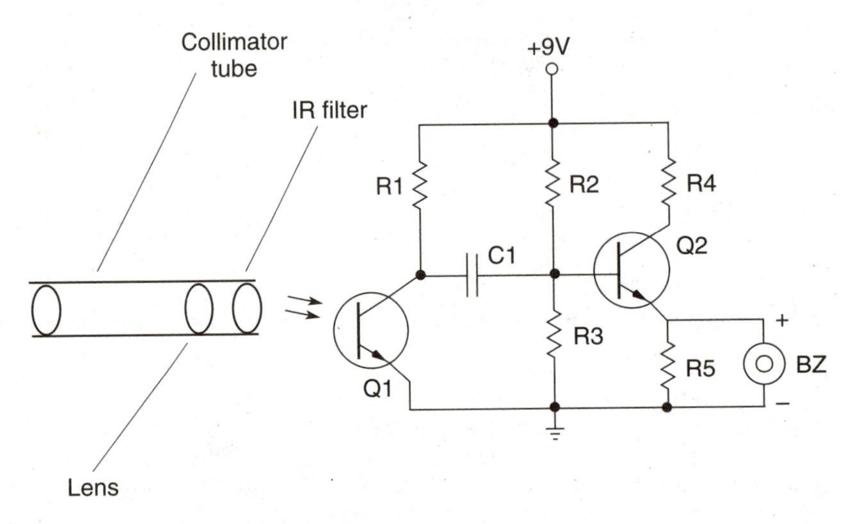

8-7 Short/medium-range signaling receiver.

Q2 via capacitor C1. Transistor Q2 drives the piezo buzzer to reproduce the code sent from the transmitter unit. Both the tone transmitter and receiver units are powered by the ubiquitous 9-volt transistor radio battery. The simple tone transmitter and companion receiver would also make a good demonstration project for a science fair.

Signaling system transmitter parts list

R1	100-kilohm potentiometer
R2	10-kilohm, ¼-watt resistor
R3	220-ohm, ¼-watt resistor
R4	5-kilohm, ¼-watt resistor
C1	.01-mF, 25-volt disc capacitor
D1	LED
U1	LM555 timer IC
B	9-volt battery
CK	Code key
Misc	Enclosure and battery clip

Signaling system receiver parts list

R1	47-kilohm, ¼-watt resistor
R2,R3,R5	4.7-kilohm, ¼-watt resistor
R4	22-kilohm, ¼-watt resistor
Q1	Phototransistor
Q2	2N2222 transistor
BZ	Piezo buzzer
B	9-volt battery
Misc	Enclosure, battery clip, collimator, and lens

30-kHz IR AM-modulated light-wave communications link

This versatile infrared AM-modulated light-beam communications system is shown in Figures 8-8 and 8-9 and Photo 8-1. The system has a ¼-mile range when used with IR filters and a simple lens system. The infrared light-beam communicator is available in kit form from Ramsey Electronics; it consists of a transmitter and receiver on a single detachable PC board. If the two boards are detached, you can use them as a wireless IR TV link from your TV to your easy chair for quiet listening. You could also use the IR transmitter/receiver pair as a TV remote-control repeater by connecting the receiver's output section to the input of the IR transmitter via a coupling capacitor. For operation, aim your existing TV remote control at your infrared light-beam receiver. The receiver will then be coupled to the IR transmitter to rebroadcast or repeat the signal, thus increasing the range of your existing TV remote control, even around corners.

If you elect not to separate the transmitter from the receiver unit, you can mount the pair of boards in a plastic enclosure, which would form one end of a duplex light-beam communicator. By mounting a complete transceiver on a camera tripod separated by ¼ mile, you would have a complete IR communications system. The IR light-beam communicator uses a 30-kHz carrier wave onto which an AM-modulated carrier signal is applied. This modulated carrier approach provides more reliable communication with less light interference than the simpler light-wave systems described earlier.

The AM-modulated IR transmitter shown in Figure 8-8 takes a microphone or audio input signal and applies it to transistor Q3. The microphone gain is controlled

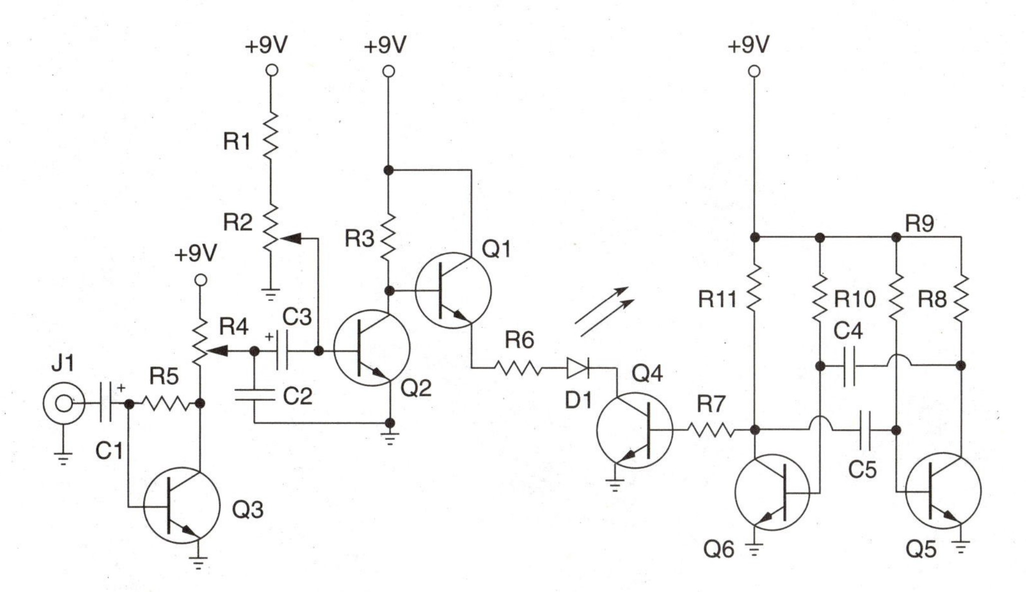

8-8 30-kHz IR AM voice transmitter. Ramsey Electronics.

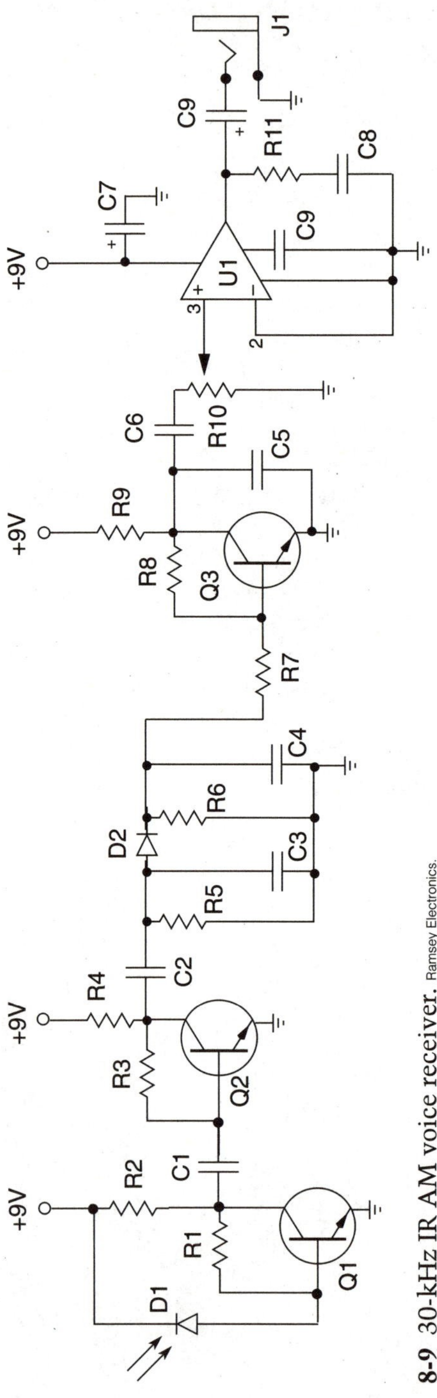

8-9 30-kHz IR AM voice receiver. Ramsey Electronics.

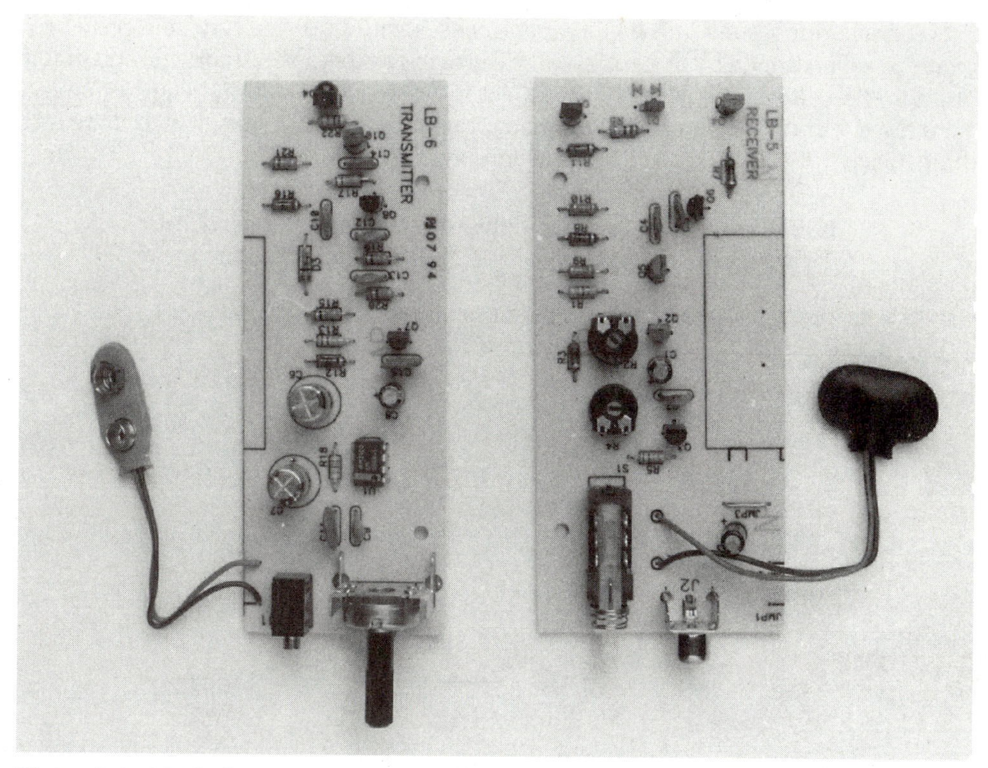

Photo 8-1 AM light-wave transmitter and receiver.

by R4 as needed and the resultant output from Q3 is coupled to Q2. The signal from Q2 is then applied to Q1, which modulates the IR LED. This circuit is unique in that Q1, Q2, and Q3 apply audio to D1 while transistors Q4, Q5, and Q6 form the 30-kHz carrier oscillator that is applied to the opposite lead of D1. The IR 30-kHz light-wave transmitter is powered from a 9-volt radio battery.

The infrared light-beam receiver shown in Figure 8-9 uses an IR photodiode at D4 to receive the incoming modulated light beam from the 30-kHz transmitter. The output of D4 is first amplified by Q10 followed by Q8, which amplifies the incoming signal. The 30-kHz AM detector consists of the components clustered around detector diode D3. The output of the detector is then coupled to Q7, which boosts the derived demodulated audio signal recovered from the carrier wave. The derived audio is further amplified by the LM386 audio power amplifier. Potentiometer R14 is used as a gain control ahead of the power amplifier. The output of the LM386 on pin 5 is coupled via C7 to the audio output jack at J1. Then you could use headphones to listen to the audio output at J1. If you are using the receiver unit as a TV remote repeater, then the output would be coupled, via an 8-ohm to 1-kilohm transformer, to the input of the transmitter unit. A coupling capacitor would be used at both sides of the transformer to couple the audio signals.

For superior results, use IR filters and lenses with both the transmitter and receiver. A collimator on the receiver end will go a long way toward reducing stray light or interference from reaching the IR detector diode. The AM light-beam communicator used without lenses and filters has a range of 10 to 12 feet, which increases dramatically with IR filters and a simple lens system.

30-kHz IR AM-modulated light-wave transmitter parts list

R1	4.7-kilohm, ¼-watt resistor
R2,R4	2.2-kilohm potentiometer
R3,R7,R8,R11	1-kilohm, ¼-watt resistor
R5	100-kilohm, ¼-watt resistor
R6	51-ohm, ¼-watt resistor
R9,R10	47-kilohm, ¼-watt resistor
C1,C3	2.2-µF, 25-volt electrolytic capacitor
C2	.01-µF, 25-volt disc capacitor
C4,C5	.001-µF, 25-volt disc capacitor
D1	IR LED
Q1-Q8,Q10	2N3904 transistor
J1	⅛-inch phono jack
Misc	Enclosure, battery clip, lens, and filter

30-kHz IR AM-modulated light-wave receiver parts list

R1	1-megohm, ¼-watt resistor
R2,R5,R6,R7	10-kilohm, ¼-watt resistor
R3,R8	100-kilohm, ¼-watt resistor
R4,R9	1-kilohm, ¼-watt resistor
R10	10-kilohm trim pot
R11	2-ohm, ¼-watt resistor
C1	.001-mF, 16-volt disc capacitor
C2,C3,C4,C9	.01-mF, 16-volt disc capacitor
C5,C8	.1-mF, 16-volt disc capacitor
C6	2.2-mF, 25-volt electrolytic capacitor
C7	220-mF, 25-volt electrolytic capacitor
Q1,Q2,Q3	2N3904 transistor
U1	LM386 audio amplifier IC
D1	IR photodiode detector
D2	1N270 detector diode
J1	⅛-inch audio jack
B	9-volt battery
Misc	Enclosure, battery clip, IR filter, lens, and collimator

Pulse frequency-modulated IR link

Amplitude-modulated communication links are ideal for simple light-wave links, but are quite susceptible to both artificial and natural noise. The transmission range of AM systems actually controls the volume of the receiver audio unless some type of automatic gain control is used. To overcome these problems, you could use pulse frequency modulation for designing a superior communications link.

The pulse frequency modulation (PFM) technique is ideal for constructing a free-space communications link. PFM resembles FM modulation in that the transmitter section emits a steady train of pulses, called a *carrier*. The audio information is then superimposed on the carrier by changing the carrier's frequency. Detection of the PFM signal is quite straightforward since no clock signal is required for synchronization. PFM pulses have uniform duration and intensity.

The PFM infrared transmitter is shown in Figure 8-10 and is designed around the ubiquitous LM555/7555 timer IC. In operation, audio is taken from the electret microphone and fed to a 100-kilohm potentiometer. The 100-kilohm potentiometer acts as a gain control; if you want additional gain, substitute a 1-megohm potentiometer. The output from R2 is immediately fed to the noninverting input of an LM741 op-amp for amplification. The gain path for the system is formed by R2 and R4. The audio signal from U1 is then passed on to U2 via capacitor C1. The LM555 oscillates at a center frequency determined by R6 and C2. The center frequency is usually about 40 kHz. The signal from U1 enters U2 via the modulation input of the LM555 and alters the LM555's oscillation frequency with the input audio. Potentiometer R5 adjusts the actual carrier frequency. The IR-transmitting LED can be a GaAS, AlGa, As, or GaAs:Si type LED.

For long-range applications, use a super-bright LED such as the Xciton XC880 series IR diode or a GE F51D1/F5E1 series LED. The peak pulse current through a conventional LED is about 50 to 60 mA, with a 5-microsecond pulse. The entire PFM IR transmitter link can be powered from a 9-volt battery, but if you are using a high-power LED driver, then you should power the system from a dc or dc wall-cube power supply. The PFM IR transmitter should be mounted in a light-tight enclosure with the IR LED facing outward into a collimator. Use an IR filter in long-range applications, placed between the LED and the collimator.

The companion PFM IR receiver is depicted in Figure 8-11. The PFM receiver is centered around an LM565 phased-locked loop (PLL) chip. The PLL is tuned by R4 and C2/C3 to the approximate center frequency of 40 kHz. Capacitor C3 is used to fine-tune the center frequency. Incoming optical signals enter the PFM receiver via phototransistor Q1. The resulting photocurrent is converted to a voltage by resistor R1. The output of Q1 is then coupled to capacitor C1, and the signal from Q1 is amplified 1000 times by U1. The gain path of U1 is set by resistors R2 and R3, and the output of U1 is coupled directly to the PLL at U2. The phase comparator in the PLL generates an error voltage proportional to the difference between the PLL's on-chip voltage-controlled oscillator (VCO) and the instantaneous transmitter frequency. The error voltage is fed back to the VCO in a feedback loop. This causes the VCO to

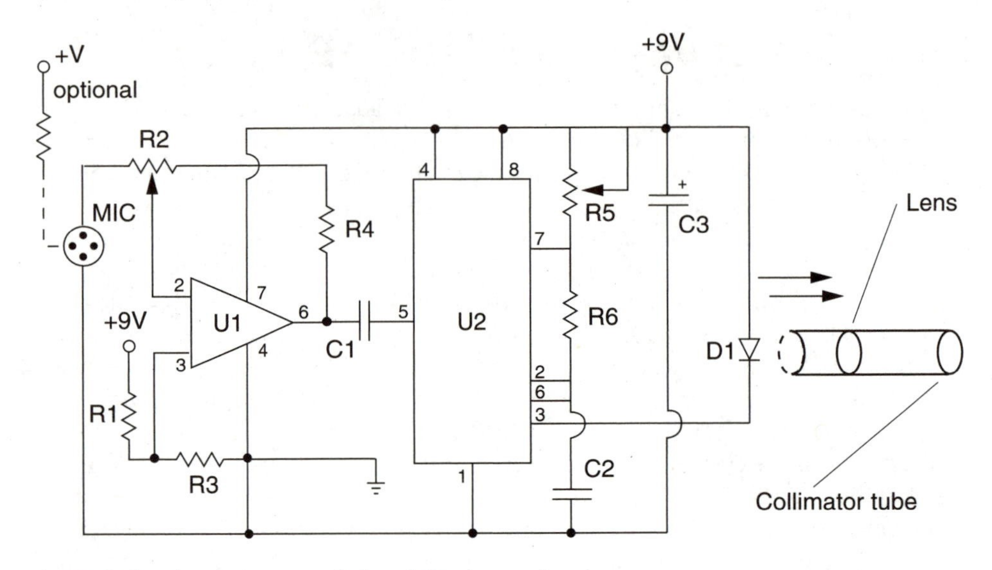

8-10 Pulse frequency-modulated IR transmitter.

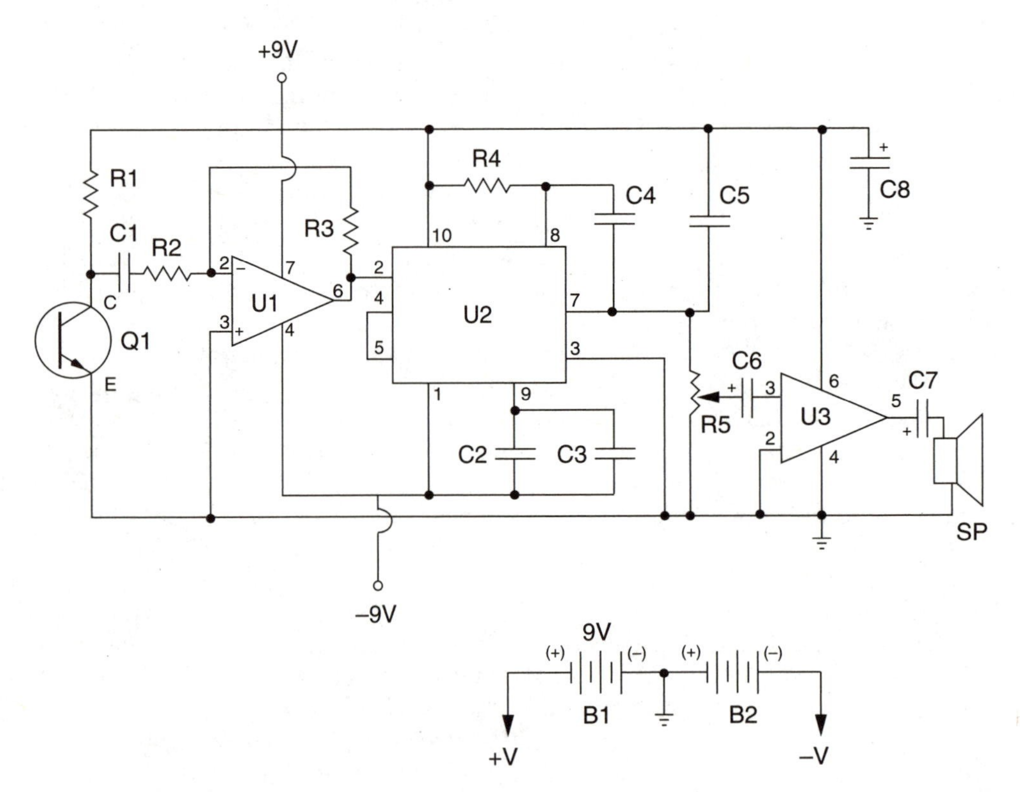

8-11 Pulse frequency-modulated IR receiver.

track the transmitter's output frequency. The error voltage represents the demodulated analog of the transmitter's audio signal. The demodulated audio output at pin 7 of U2 is sent to R5 and then on to the power amplifier at U3. Potentiometer R5 controls the audio into the output speaker.

The PFM receiver is powered from a ± 9-volt supply, i.e., two 9-volt batteries or power supplies. Be sure to keep the leads from the phototransistor to the receiver electronics as short as possible. For best overall results, shield Q1 with a collimator tube to keep external or ambient light from reaching the detector. For long-range applications, use an IR filter and lens arrangement ahead of the phototransistor.

Begin testing and operating the PFM communications link by checking the circuits over carefully before applying power, to ensure that there are no omissions, shorts, or wiring errors. Place collimator tubes and lenses on both the IR LED and the receiving phototransistor. For initial tests, you could replace the microphone with a transistor radio to allow you to make adjustments and still have a free hand. Connect the output of a radio headphone jack via a 1-µF capacitor to the input potentiometer on the transmitter unit, with the minus lead of the capacitor toward R2. Turn the radio on to allow volume, select a station, and apply power to both the transmitter and the receiver. Next, aim the transmitter toward the receiver and listen to the the IR PFM receiver's audio output. Set R2 on the transmitter midway and be sure to adjust the audio amplifier control R5 at the receiver to avoid extremely loud volume. Your PFM IR communications link is now ready to serve you.

Pulse frequency-modulated IR link transmitter parts list

R1,R3	5.6-kilohm, ¼-watt resistor
R2,R4	100-kilohm potentiometer (trim)
R5	100-kilohm, ¼-watt resistor
R6	10-kilohm, ¼-watt resistor
C1	.1-mF, 25-volt disc capacitor
C2	470-pF, 25-volt Mylar capacitor
C3	10-mF, 25-volt electrolytic capacitor
D1	IR LED (RS 276-143)
U1	LM741CN op-amp
U2	LM555Cn IC timer
MIC	Electret microphone
Misc	Enclosure, chassis, battery, wire, and hardware

Pulse frequency-modulated IR link receiver parts list

R1	220-kilohm, ¼-watt resistor
R2	1-kilohm, ¼-watt resistor
R3	1-megohm, ¼-watt resistor
R4	3.9-kilohm, ¼-watt resistor
R5	10-kilohm potentiometer
C1	.1-mF, 25-volt disc capacitor

Pulse frequency-modulated IR link receiver parts list continued

C2,C3,C4	.001-mF, 25-volt disc capacitor
C5	.047-mF, 25-volt disc capacitor
C6	10-mF, 25-volt electrolytic capacitor
C7	100-mF, 25-volt electrolytic capacitor
C8	2.2-mF, 25-volt electrolytic capacitor
Q1	IR phototransistor (RS 276-145)
U1	LM741CN op-amp
U2	LM565 phased-locked loop
U3	LM386 audio amplifier
SP	8-ohm speaker
Misc	Enclosure, chassis, batteries, wire, and hardware

Wireless speaker system

Late-night television viewing can often result in arguments, especially if the night-owl viewer is hard of hearing. Loud TV often disturbs others, but you can easily solve this problem with the IR wireless speaker system shown in Figures 8-12 and 8-13.

The IR wireless transmitter shown in Figure 8-12 is a straightforward design using three transistors and a NE566 function generator chip configured as an FM modulator. The eight-pin chip is actually a general-purpose voltage-controlled oscillator designed for highly linear frequency modulation. Audio from a TV, radio, or stereo is fed to the circuit input at R1. An input network consists of C1 and R2, which are coupled via C2 to the input amplifier stage at Q1. Audio input levels are adjusted with R5, a 5-kilohm potentiometer. The audio signal is then coupled to U1 through capacitor C5. The frequency of the FM carrier is controlled by C8 (see Table 8-1). You can fine-tune the modulator by adjusting potentiometer R9. The FM carrier signal and the audio modulation exit the NE566 on pin 3 of the chip. The square-wave output is then coupled to the LED drivers via C9 and R11. Transistor Q2 in turn drives the four IR LEDs. Make sure to select current-limiting resistors R15 and R16 to ensure correct current values through the selected LEDs. While two LEDs would work in this application, the four IR LEDs give a higher output with greater dispersion.

Table 8-1 IR wireless speaker frequency components.

Carrier frequency	C8 value
30 kHz	.0022-µF, 25-volt disc
100 kHz	.001-µF, 25-volt disc
200 kHz	470-pF, 25-volt Mylar

The IR speaker system receiver is shown in Figure 8-13. The modulated carrier wave is picked up by the photodetector at D1, and the varying light levels are coupled via C1 to Q1. The tuned circuit formed by C3 and L1 determines the actual

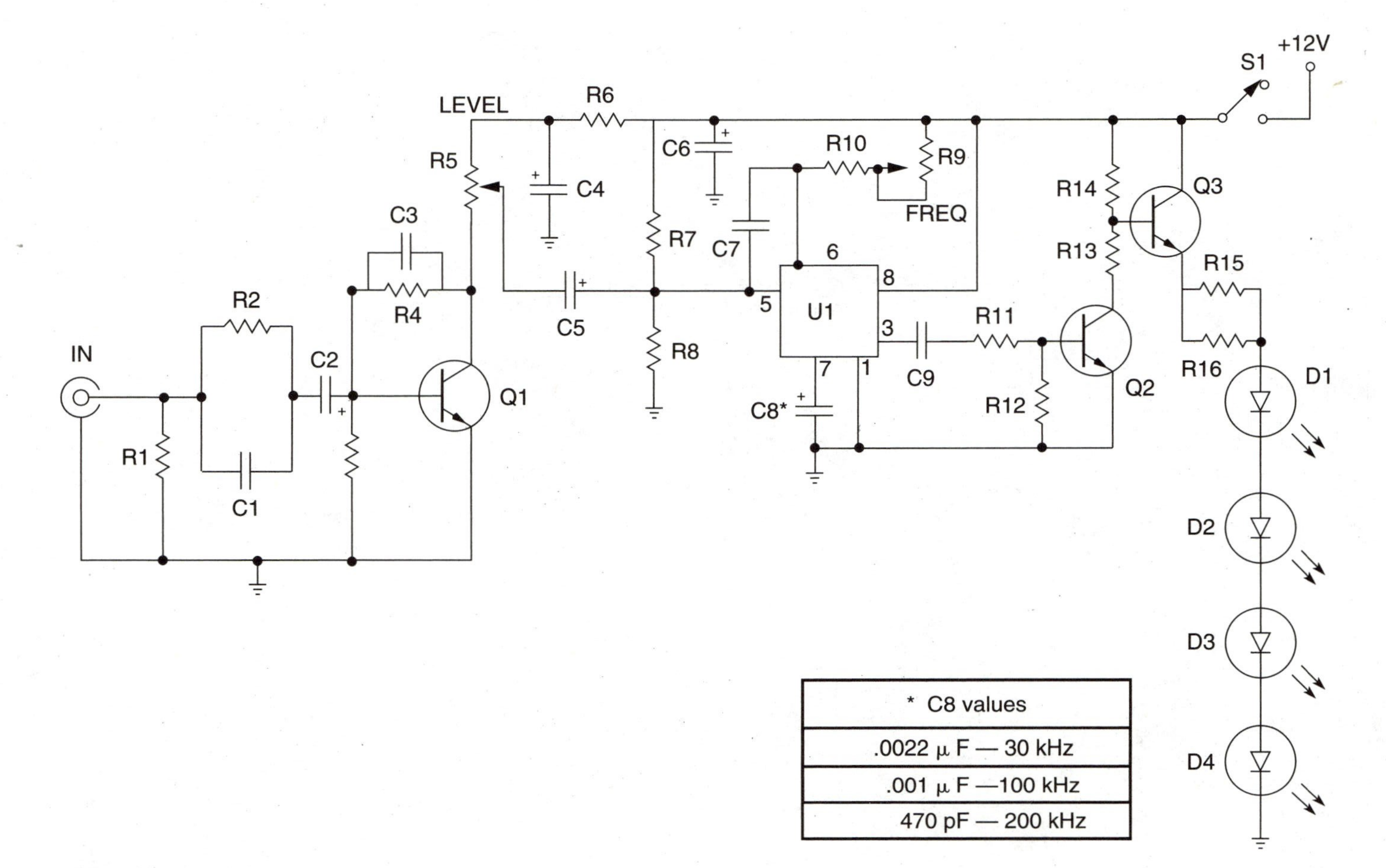

8-12 IR wireless speaker system—transmitter. Copyright Radio Electronics, 1988. Reprinted with permission.

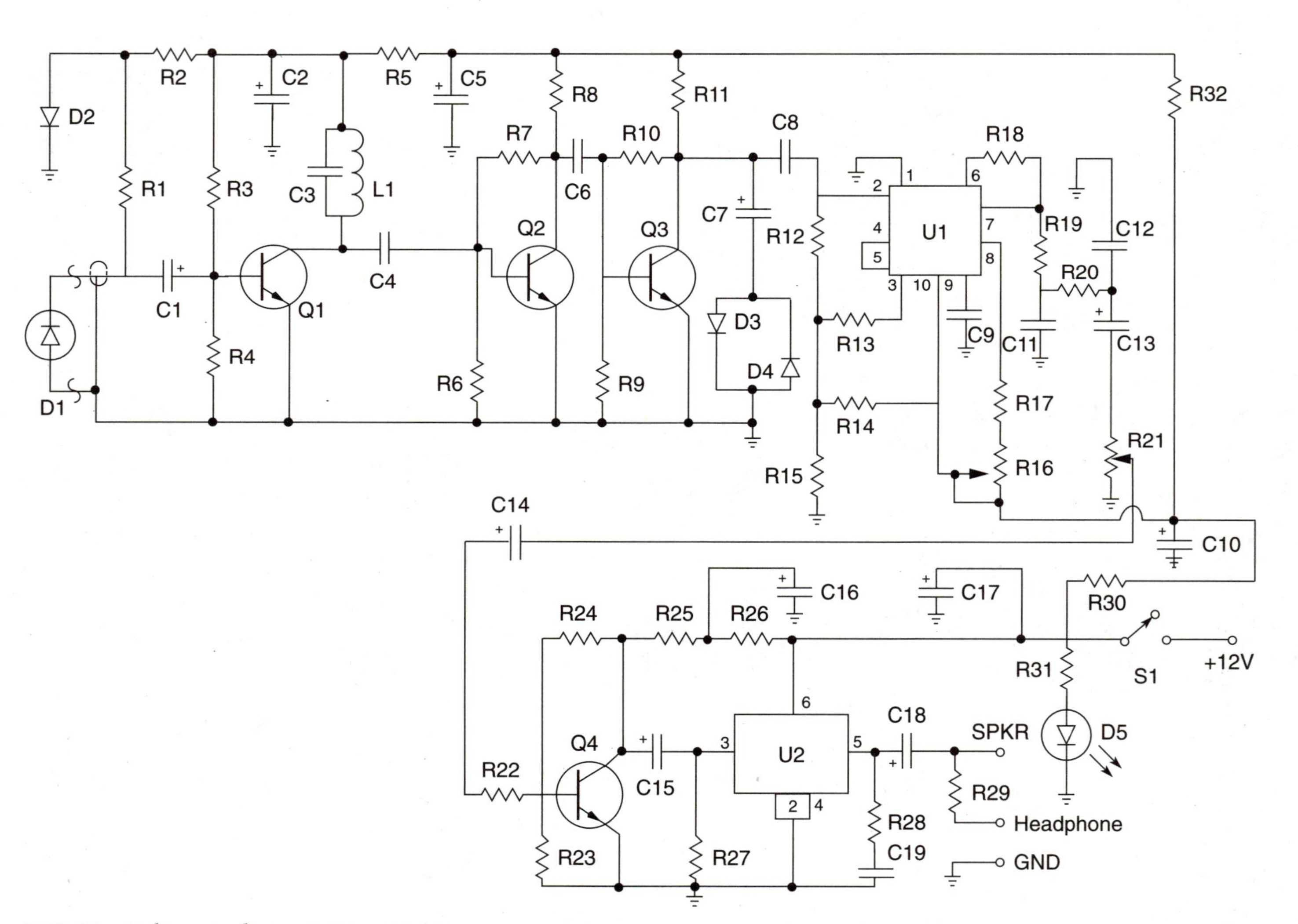

8-13 IR wireless speaker system—receiver. Copyright Radio Electronics, 1988. Reprinted with permission.

frequency (see Table 8-2). The resultant input signal is then amplified via Q2 and Q3. The output of the amplified signal is fed to an NE565 phased-locked loop IC at U1. The phase-locked loop essentially demodulates the FM carrier wave, which has been detected, amplified, and conditioned. The derived audio signal is then passed on to the final amplification stages at Q4 and U2. The high-gain audio amplifier at U2 boosts the recovered audio in order to drive a speaker or headphone for private listening. Any 8-ohm speaker or headphone could be used for private nighttime listening. The wireless speaker system receiver is powered from a 12-volt power supply or battery. Power to the receiver is switched by S1, and LED D5 is used as a power indicator.

Table 8-2 IR wireless speaker component chart.

Frequency	L1	C3	C9
31 kHz	4.7 mh	.0047 µF	220 pF
40 kHz	4.7 mh	.0033 µF	220 pF
50 kHz	4.7 mh	.0022 µF	220 pF
60 kHz	4.7 mh	.0015 µF	220 pF
75 kHz	4.7 mh	.001 µF	220 pF
75 kHz	1.0 mh	.0047 µF	100 pF
108 kHz	1.0 mh	.0033 µF	100 pF
130 kHz	1.0 mh	.0022 µF	100 pF
160 kHz	1.0 mh	.0015 µF	100 pF
200 kHz	1.0 mh	.001 µF	100 pF

In order to test and use your IR wireless speaker system, you need to ensure that both transmitter and receiver are tuned to the same operating frequency (refer back to Tables 8-1 and 8-2). Once the correct values are chosen, it is simply a matter of adjusting the frequency control in the transmitter and receiver to match each other. If you have an oscilloscope or frequency counter, the task will go much more quickly.

For best results and optimum range, use IR filters for both the transmitter and receiver input.

Wireless speaker system transmitter parts list

R1	4.7-kilohm, ¼-watt resistor
R2	22-kilohm, ¼-watt resistor
R3	10-kilohm, ¼-watt resistor
R4	100-kilohm, ¼-watt resistor
R5	5-kilohm potentiometer (trim)
R6	1-kilohm, ¼-watt resistor
R7	6.8-kilohm, ¼-watt resistor
R8	47-kilohm, ¼-watt resistor
R9	10-kilohm potentiometer (trim)
R10	2.2-kilohm, ¼-watt resistor

Wireless speaker system transmitter parts list continued

R11,R12,R13	1-kilohm, ¼-watt resistor
R14	100-ohm, ¼-watt resistor
R15,R16	See text
C1	.0033-mF, 25-volt disc capacitor
C2,C4,C5,C6	10-mF, 25-volt electrolytic capacitor
C3	47-pF, 25-volt mica capacitor
C7	.001-mF, 25-volt disc capacitor
C8	See Table 8-1
C9	.01-mF, 25-volt disc capacitor
Q1	2N3565 transistor
Q2	2N3904 transistor
Q3	2N3906 transistor
U1	NE566 IC
D1-D4	IR155 IR LED

Wireless speaker system receiver parts list

R1	1-megohm, ¼-watt resistor
R2,R3,R7,R10,R24	100-kilohm, ¼-watt resistor
R4,R6,R9,R14,R18,R23	10-kilohm, ¼-watt resistor
R5,R8,R11,R13	4.7-kilohm, ¼-watt resistor
R15,R20,R25	4.7-kilohm, ¼-watt resistor
R16	10-kilohm potentiometer
R17,R31	2.2-kilohm, ¼-watt resistor
R19,R26	1-kilohm, ¼-watt resistor
R21	50-kilohm potentiometer
R22	22-kilohm, ¼-watt resistor
R27	47-kilohm, ¼-watt resistor
R28	10-ohm, ¼-watt resistor
R29	100-ohm, ¼-watt resistor
R30	47-ohm, ¼-watt resistor
R32	470-ohm, ¼-watt resistor
C1,C2	1-mF, 25-volt electrolytic capacitor
C3	See Table 8-2
C4,C6,C8,C11,C12	.01-mF, 25-volt disc capacitor
C5,C7,C10,C13,C15,C16	10-mF, 25-volt electrolytic capacitor
C9	See Table 8-2
C17,C18	470-mF, 25-volt electrolytic capacitor
C19	.1-mF, 25-volt disc capacitor
D1	PD600 silicon photodiode
D2,D3,D4,D5	1N914B silicon diode
Q1,Q2,Q3,Q4	2N3565 transistor
U1	NE565 phased-locked loop IC
U2	LM386N audio amplifier IC

IR PLL laser communications system

The feature project in this chapter is the laser communications system shown in Figures 8-14 and 8-15 and Photo 8-2. The laser communicator provides a true high-quality FM free-space optical link. The FM laser communications link is based on a 4046 phase-locked loop or PLL IC operating at 200 kHz at both the transmitter and receiver sections. The laser communicator can serve as a wireless communication or signaling system, as well as a long-range TV listening link. For simultaneous or duplex communication, you would need both a transmitter and a receiver at both ends of the communications link.

Let's take a tour around the laser transmitter section (see Figure 8-14). The microphone input is immediately amplified by U1. The system gain is controlled by resistor R14 and ranges from unity to 20. The output of U1 is coupled to the phase-locked loop at U2 through resistor R7 into the VCO input on the 4066 IC. Note that digital data can be input at pin 9 of U2 in the transmitter section. The frequency of the PLL is determined by C5 and R9 at approximately 200 kHz. The output of the PLL at pin 4 is also fed back into pin 3, the phase comparator input. The output of U2 is a frequency-modulated 200-kHz signal that connects to transistor Q1 via R10. The output frequency is essentially a square wave with a peak-to-peak signal of 6 volts. Transistor Q1 drives an IR laser diode or an IR LED if shorter distances are to be spanned.

The laser diode chosen for this project is an LT026, which has a maximum forward bias of 100 mA (be careful not to exceed the current ratings). The average

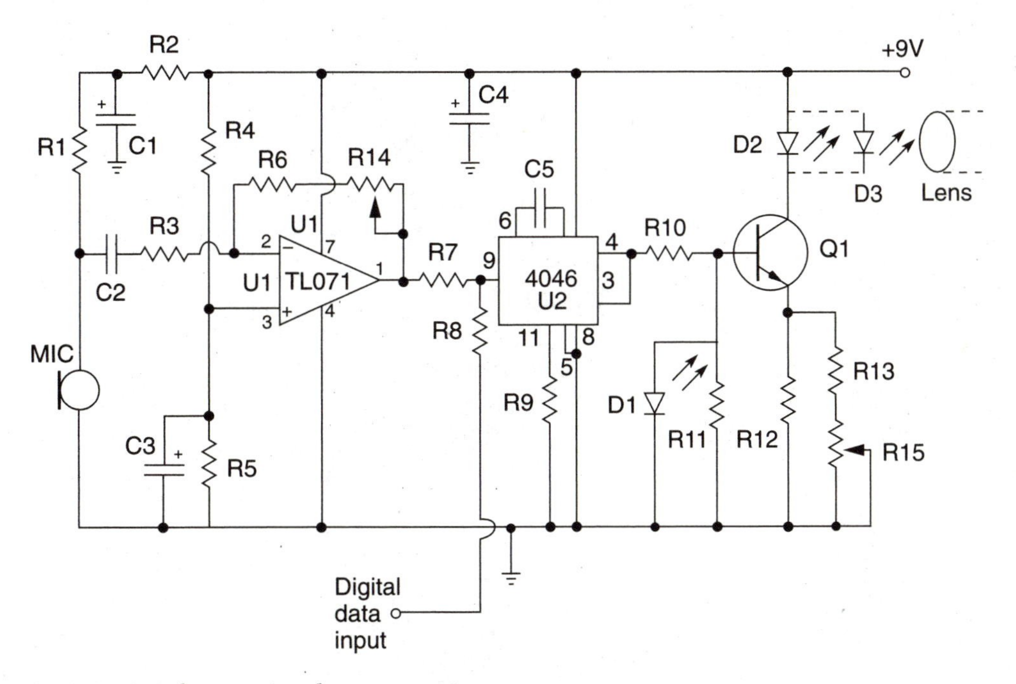

8-14 IR PLL laser voice data transmitter.

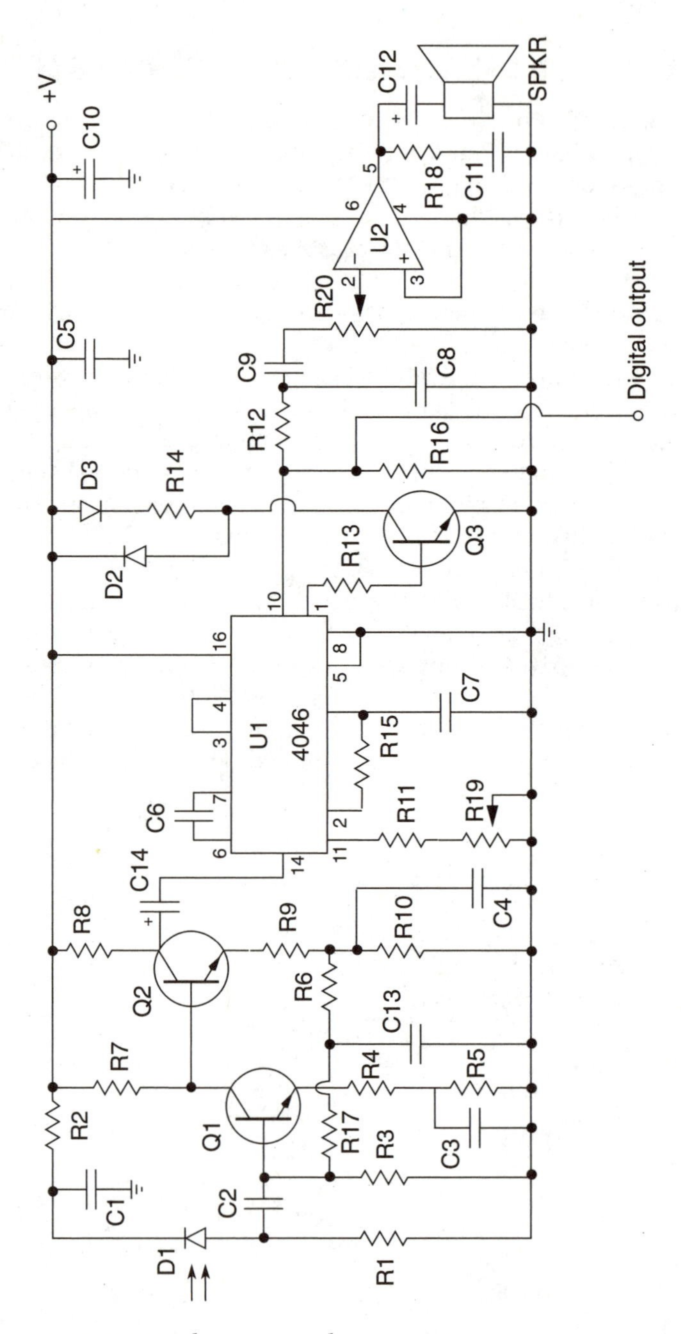

8-15 IR PLL laser voice data receiver.

current through the laser in this project is about 65 mA, and the bias to the IR laser is controlled by R15. The LED diode LED1 or D1 holds or stabilizes the laser diode current through the base of Q1 at 1.8 volts. Since the duty cycle of the output signal averages 50 percent, the quiescent current through Q1 should measure half the rec-

Photo 8-2 Laser communications link.

ommended value. For the laser diode shown, therefore, you would measure a current of about 34 mA.

The receiver section of the laser communications link is shown in Figure 8-15. The PLL once again is at the heart of the link. The input signal is received via the IR detector, and the IR detector is in turn amplified by a high-gain dc amplifier at Q1 and Q2. The output of the dc amplifier is coupled to the PLL via C14. The PLL's internal frequency of operation is determined by C6 and R11 along with potentiometer R19. Potentiometer R19 adjusts the receiver's frequency to the PLL frequency (200 kHz) of the transmitter section. A low-pass filter is formed by R15 and C7 and is connected between the internal phase comparator and the input of the VCO.

The output of the PLL on pin 1 drives Q3, which in turn powers the LED "lock" indicator. The output on pin 1 goes high when the internal oscillator is locked to the input signal at pin 14. When the lock lamp is on, the receiver is synchronized with the transmitter unit. The demodulated signal from pin 10 of the PLL is fed to the low-pass filter formed by R12 and C8. The output of the low-pass filter is then coupled via C9 to R20, which controls the final audio amplifier gain. If you choose to send digital data instead of audio signals, data output is provided on pin 10 of the PLL chip.

The laser diode used in this project is an LT026. Only pins 1 and 2 (the anode and cathode) are used. The laser diode must be used with a collimating lens in order to focus the beam to a sharp point. Without the lens, the laser beam would spread out and lose its intensity. The laser diode and lens brackets are made from two pieces of aluminum, as shown in Figure 8-16, and the lens bracket is slotted and

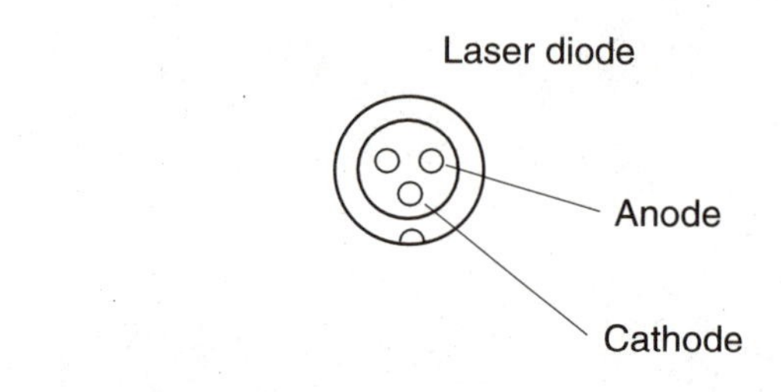

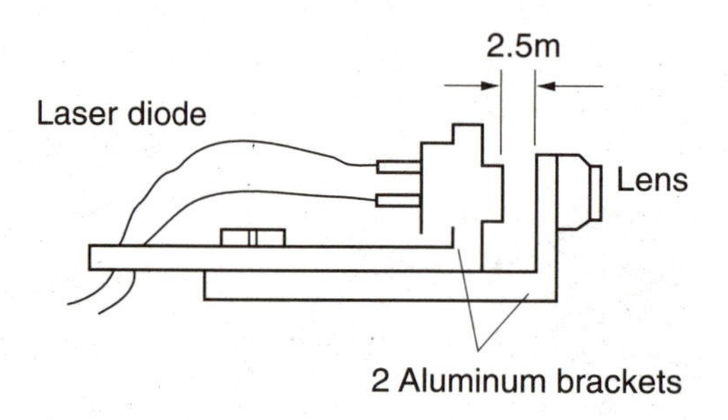

8-16 Laser diode mounting.

placed on top of the laser diode bracket with a sheet metal screw holding the two brackets together. The focal distance between the laser diode and the lens should be about 2.5 mm, and is determined by the particular lens chosen. The spacing between the laser and the lens is crucial and should be adjusted carefully.

When constructing the laser communicator, be certain to carefully observe the polarity of the capacitors and diodes, as well as the electret microphone and transistors, to ensure the project works when power is applied. Both the transmitter and receiver units can be powered from a 9-volt battery, but for long-term operation use a dc wall-cube supply.

When power is applied to the transmitter, LED1 should light up, indicating that the laser is actually on and sending. When the transmitter is powered and properly aimed at the companion receiver, the PLL lock LED should light up, indicating an established link. Note that the output of the laser diode is invisible, so be sure not to look directly into the beam when aligning the system. An inexpensive method to assist in aligning the communications link is a Radio Shack TV remote-control IR test card.First sensitize the card with visible light from a light bulb or sunlight, and then place the card in front of the laser transmitter. The card will glow when the beam hits the test card.

Place both the laser transmitter and receiver on a clean, level bench top to check for primary alignment. Set the gain of the transmitter to maximum, and then adjust the volume control on the receiver to $\frac{1}{4}$ turn. When the two units are aligned properly, the lock lamp should light up. Adjust potentiometer R20 on the receiver to provide the cleanest audio. If all works, you can extend the range between the transmitter and receiver. Camera tripods are an excellent means to elevate and align a long-range communications link.

Once you have tested your laser communications link, you can find a host of applications for it, from sending voice or data between two points to controlling numerous functions via touch-tone control over long distances.

IR PLL laser communications system transmitter parts list

R1,R8	10-kilohm, ¼-watt resistor
R2,R7,R10	3.3-kilohm, ¼-watt resistor
R3,R6	27-kilohm, ¼-watt resistor
R9,R12,R13	22-kilohm, ¼-watt resistor
R11	2.2-kilohm, ¼-watt resistor
R14	1-megohm potentiometer (trim)
R15	200-ohm potentiometer (trim)
C1,C3,C4	100-mF, 25-volt electrolytic capacitor
C2	.1-mF, 25-volt disc capacitor
C5	100-pF, 25-volt ceramic disc capacitor
LED1	LED
Q1	BC337 NPN transistor
U1	TL-071 op-amp
U2	4046 PLL IC
LASER	LT026 IR laser diode or equivalent
B	9-volt battery
MIC	Electret microphone
Misc	Lens, chassis box, brackets, battery clip, and switches

IR PLL laser communications system receiver parts list

R1	10-kilohm, ¼-watt resistor
R2	1-kilohm, ¼-watt resistor
R3	82-kilohm, ¼-watt resistor
R4,R5	220-kilohm, ¼-watt resistor
R6,R13,R16	22-kilohm, ¼- watt resistor
R7,R11,R17	12-kilohm, ¼-watt resistor
R8,R14	3.9-kilohm, ¼-watt resistor
R9	100-ohm, ¼-watt resistor
R10	2.2-kilohm, ¼-watt resistor
R12	6.8-kilohm, ¼-watt resistor
R15	120-kilohm, ¼- watt resistor
R18	4.7-ohm, ¼-watt resistor
R19	20-kilohm potentiometer (trim)
R20	100-kilohm panel potentiometer
C1,C13	.47-mF, 25-volt tantalum capacitor
C2	470-pF, 25-volt ceramic capacitor
C3,C11	10-nF, 25-volt polyester capacitor

IR PLL laser communications system receiver parts list continued

C4,C5,C9	100-pF, 25-volt ceramic capacitor
C7	1-nF, 25-volt ceramic capacitor
C8	3.3-nF, 25-volt ceramic capacitor
C10	1000-mF, 25-volt electrolytic capacitor
C12	100-mF, 25-volt electrolytic capacitor
C14	68-pF, 25-volt ceramic capacitor
Q1,Q2,Q3	BC548 NPN transistor
U1	4046 PLL IC
U2	LM386 audio amplifier IC
D1	LED
D2	IR detector BPW50 or equivalent
B	9-volt battery
Misc	Chassis box, battery clip/holder, and switches

Touch-tone encoder/decoder

You could use the laser communications link or AM light-wave communicator shown earlier to remotely control the functions of a number of devices. One of the most flexible methods of remotely controlling devices, however, is using dual-tone multifrequency signals, or DTMF. A DTMF generator consists of two tone generators with a keypad to synchronize them. When a keypad button is pressed, a tone from each generator combines to make a two-tone sound (see Table 8-3). Each key produces two of eight tone combinations, as does your touch-tone phone.

Table 8-3 Touch-tone frequencies.

Signal digit	Low tone	High tone	Binary output	2 of 8 output
1	697	1209	0001	0000
2	697	1336	0010	0001
3	697	1477	0011	0010
4	770	1209	0100	0100
5	770	1336	0101	0101
6	770	1477	0110	0110
7	852	1209	0111	1000
8	852	1336	1000	1001
9	852	1477	1001	1010
10	941	1336	1010	1101
*	941	1209	1011	1100
#	941	1477	1100	1110
A	697	1633	1101	0011
B	770	1633	1110	0111
C	852	1633	1111	1011
D	941	1633	0000	1111

The diagram in Figure 8-17 illustrates two LM555 timer chips configured in the astable mode to produce DTMF coding. The keypad allows you to select various resistances, so U1 is capable producing four tones in rows, while U2 produces tones for the columns. Potentiometers R10 and R11 each control a bank of tones to the correct frequency range. The outputs from each tone generator are combined and capacitor C1 couples the tone generators to an amplifier formed by U3, a 741CN op-amp. The output of the op-amp is varied by R17 and then fed to C4, which couples the DTMF tone generators to the microphone input of the IR laser transmitter shown earlier in the chapter. Another approach to generating DTMF would be to modify a touch-tone generator such as the Radio Shack pocket tone dialer (43-145). With this approach, you would need a coupling capacitor to feed the output of the tone dialer to the microphone input of the laser communications link.

In order to decode the tone functions at the opposite end of the laser communications link, you would need a DTMF decoder coupled to the speaker or audio output of the laser communications receiver, shown in Figure 8-18. Teltone Corporation produces a number of DTMF decoders. The M-957 DTMF decoder is used in this system; it is a 22-pin chip that uses a 3.85-MHz color-burst crystal for timing.

The DTMF signal from the audio output of the laser communicator's receiver is coupled to pin 12 of the M-957 decoder via capacitor C1. A 3.58-MHz crystal and a 1-megohm resistor are connected across pins 14 and 15 of the M-957 decoder. When pin 16 is a logic 1, the internal crystal oscillator is selected. The M-957 chip has four output pins (1, 20, 21, and 22) grouped at one end of the chip. These pins provide two kinds of binary bit patterns, which correspond to the detected DTMF output signal (see Table 8-3). Several control and output pins are also featured. Pin 5 on the M-957 selects between two ranges of decoded signals. When pin 5 is at logic 1, 12 tones are selected. When pin 5 is at a logic 0, then four additions of tones are decoded, other than the normal 12 tone groups used for special functions. The A and B input pins (8 and 9) control the input sensitivity of the tone decoder chip. Applying various combinations of logic to these two pins adjusts the sensitivity of the chip in steps to a maximum of –31 dB. The output enable (OE) input on pin 3 controls whether the output pins are enabled or not. When logic 1 is applied to pin 3, the outputs are enabled. The hex input on pin 2 of the M-957 controls the output format. When pin 2 is at logic 1, the output pins provide a standard 4-bit binary pattern; when pin 2 is at logic 0, the output pins provide a 2 of 8 binary code (refer back to Table 8-3). The strobe output on pin 18 indicates that a valid frequency pair is present.

The DTMF decoder circuit shown back in Figure 8-18 uses four optional LEDs connected at the outputs of the tone decoder to provide a visual indication of the received signal. The four outputs of the tone decoder are connected to a 74C154, a 1 of 16 decoder. The 1 of 16 decoder takes the binary output from the tone decoder and converts it to 1 of 16 outputs, which can then control up to 16 different relays or control functions (lights, motors, fans, motion displays, etc.). The output of each of the 16 outputs is coupled to a 1-kilohm resistor that feeds a 2N2222 transistor that controls a small, low-current relay. The diode D1 absorbs the reverse voltage generated by the collapsing field of the relay. The relay is normally

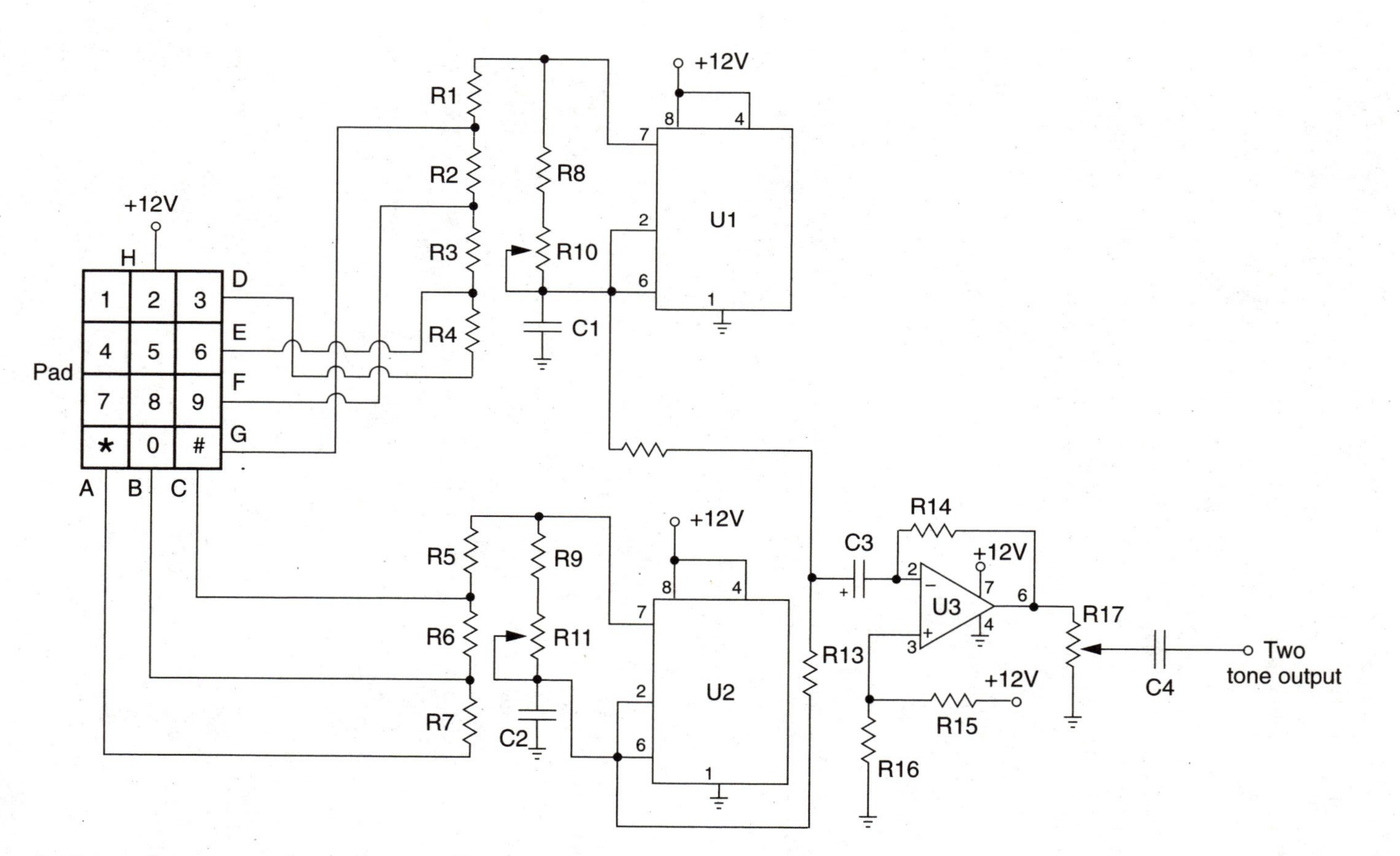

8-17 Touch-tone signal encoder.

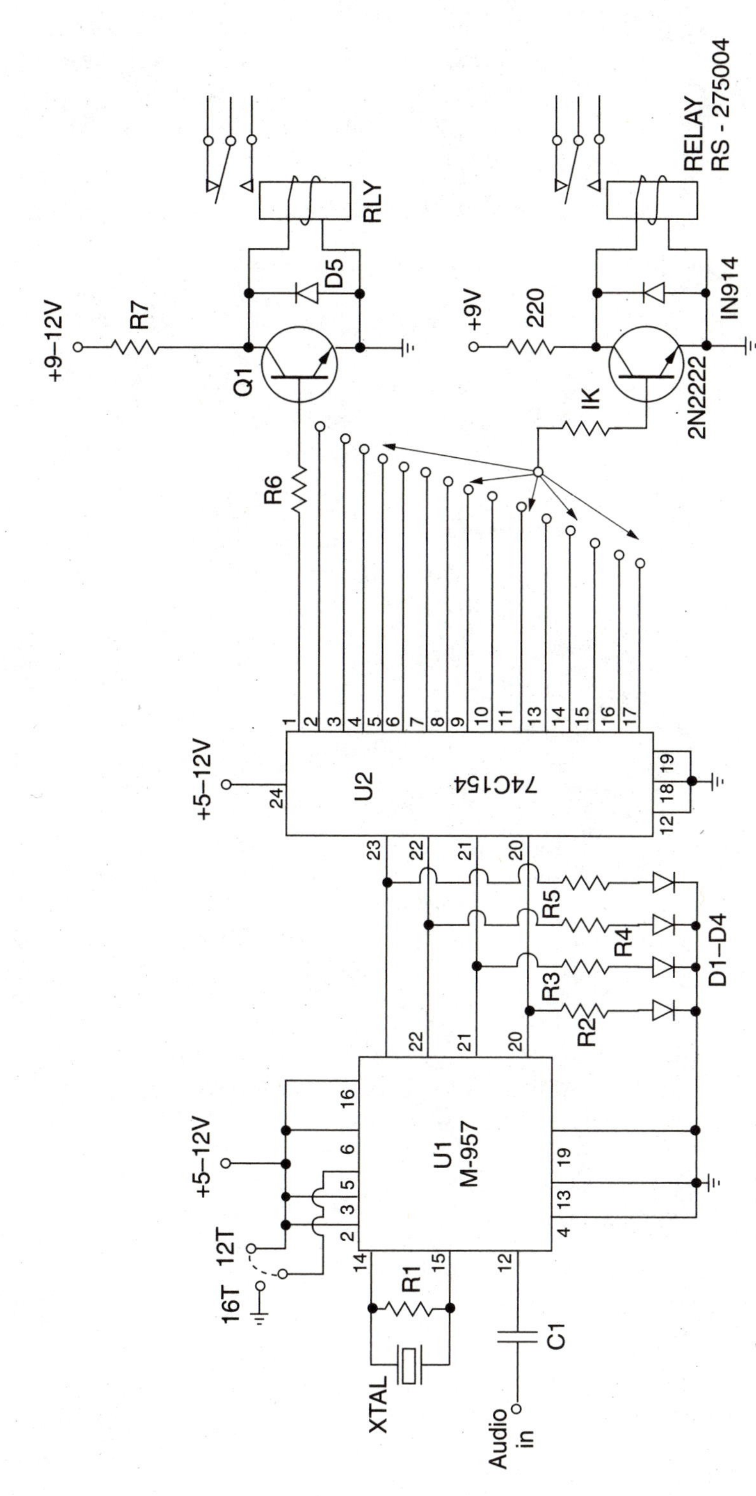

8-18 Touch-tone signal decoder.

de-energized until a valid decoded output is present. The DTMF tone decoder circuit can be powered from 5 to 12 volts dc. The M-957 tone decoder is a CMOS chip, so observe proper antistatic handling procedures when touching the component.

The DTMF control circuit opens up a whole new realm of possibilities for the IR laser communications link, allowing you to control the world remotely.

Touch-tone encoder parts list

R1,R4	4.3-kilohm, ¼-watt resistor
R2	3.3-kilohm, ¼-watt resistor
R3,R5	3.9-kilohm, ¼-watt resistor
R6	2.2-kilohm, ¼-watt resistor
R7	2.4-kilohm, ¼-watt resistor
R8	12-kilohm, ¼-watt resistor
R9	5.6-kilohm, ¼-watt resistor
R10,R11	5-kilohm trim potentiometer
R12,R13,R14	100-kilohm, ¼-watt resistor
R15,R16	10-kilohm, ¼-watt resistor
R17	10-kilohm potentiometer (trim)
C1,C2	.047-mF, 25-volt disc capacitor
C3	20-mF, 25-volt electrolytic capacitor
C4	.005-mF, 25-volt disc capacitor
U1,U2	LM555 timer IC
U3	LM741CN op-amp
KEYPAD	Chromerics ER-21623

Touch-tone decoder parts list

R1	1-megohm, ¼-watt resistor
R2,R3,R4,R5,R6	1-kilohm, ¼-watt resistor
R7	220-ohm, ¼-watt resistor
D1,D2,D3,D4	Red LED
D5	1N914 silicon diode
C1	.01-mF, 25-volt disc capacitor
U1	Teltone M-957 touch-tone decoder
U2	74C154 1 of 16 decoder
Q1	2N2222 transistor
XTAL	3.58-MHz crystal
RLY	500-ohm, low-current relay
Misc	Switches and a power supply

9
CHAPTER
Fiber optics

Optical fibers are most often associated with communications applications, and are also widely used for numerous sensing applications, such as liquid-level, vibration, pressure, and seismic sensors, gyroscopes, medical fiberscopes, and art sculptures. Optical fibers are currently used in high-end audio systems to prevent digital signals from interfering with analog signals.

Over 100 years ago, British physicist John Tyndall demonstrated how a beam of light was internally reflected through a stream of water flowing from a tank, and in 1934, the first light-pipe patent was issued to Bell Labs. Then serious research began. Light was first transmitted over short distances in the 1950s when American Optical, Inc. developed the first glass fibers. Corning Glass developed the first true long-distance optical fibers in the early 1970s.

Optical fibers are of two major types, either glass or plastic, and they are generally composed of three components, as shown in Figure 9-1. The core at the center—either of plastic or glass—is the actual light path or guide. The secondary layer is the cladding material around the core. This is a reflecting medium that bounces light from edge to edge. The diagram in Figure 9-2 illustrates light bouncing from edge to edge inside an optical fiber's cladding. An optical fiber might also contain an outer jacket or protective sheath of opaque insulation or plastic, used primarily for protection and to keep light from leaking out.

If an optical fiber were perfectly straight, light would pass through the medium as if it were passing through a pane of glass. If, on the other hand, the fiber were bent slightly, light would eventually strike the outer edge of the fiber, as shown. Optical fibers transmit light by total internal reflection (TIR). If the angle of incidence is greater than the critical angle, light is reflected internally and continues its path through the fiber. If the fiber bend is large, then the angle of incidence is small and light is passed through the fiber and lost.

Optical fibers are fabricated when a strand of glass or plastic is pulled through a small orifice or extrusion. The process of pulling is repeated over and over again until the strand of fiber is a few hundredths of a micrometer or less in diameter. While

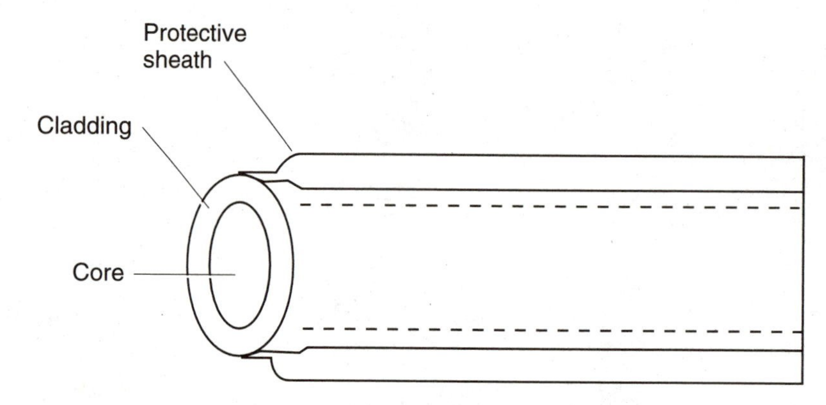

9-1 Optic fiber cross section.

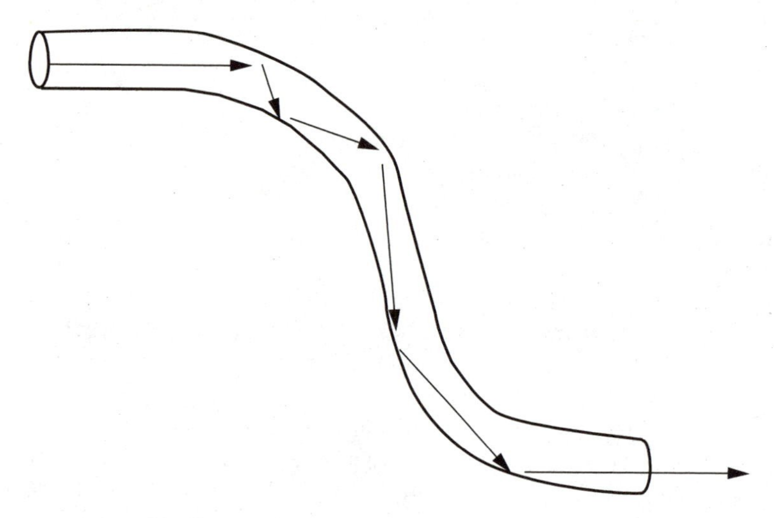

9-2 Optic fiber internal reflection.

single strands of optical fiber are used in special applications, they are often bundled or fused together to form a larger-diameter light guide.

Optical fibers are classified into two categories. *Coherent* fibers of glass are most commonly used for medical applications, because they allow an image to be transmitted from one end of the fiber to the opposite end. These coherent fibers allow doctors to view inside human bodies to find blockages or tumors. As you might imagine, this type of fiber is the most expensive. *Incoherent* optical fibers are used for most other applications, such as sensing and communication, and are much less expensive than the coherent type.

Optical fibers fall into yet another classification. They can be either the *graded index* or *stepped index* type, as shown back in Chapter 2, Figure 2-16. The cladding layer determines whether the fiber falls into the stepped index or graded index category. Stepped index fiber provides a distinct boundary between the core's more dense region and the less dense region of the cladding. Stepped index fibers are less expensive and easier to manufacture than graded index fibers, but suffer loss of coherence from one end of the fiber to the other. Graded index fiber has no discernible boundary between the core and cladding, allowing light to be reflected and refracted more evenly at any angle of incidence. Most low-cost sensing applications and demonstration systems use stepped index fibers, while high-grade telephone circuits and medical applications use the more expensive glass graded index fiber bundles.

In order to use an optical fiber, you must first prepare the fiber by cutting and polishing the ends. Optical fibers are generally easy to work with and can be readily cut with wire cutters or a knife. Before cutting glass fibers, however, you should put on a pair of safety glasses and gloves since glass fibers can produce flying shards of glass. One simple method of cutting a glass fiber is to gently score or nick the fiber with a sharp knife or razor. Score the entire diameter in one curved motion, then snap the fiber in two. Once an optical fiber is cut, it then needs to be polished before it can be spliced or placed into an optical connector. The fiber ends are polished with extra-fine aluminum oxide or 330-grit sandpaper. The sandpaper is usually wetted and placed on a flat surface or table, and the fiber rubbed in a circular motion perpendicular to the sandpaper. Use a high-power magnifying lens to carefully inspect the fiber end during the polishing procedure. When you feel the fiber is polished sufficiently, shine a light from one end of the fiber to the opposite end. The magnified end of the fiber should display a bright round spot of light onto a piece of paper. Be sure the spot is uniformly bright around the diameter before splicing or fastening connectors.

There are a wide variety of available fiber-optic connectors that can be quickly attached to fiber-optic cable, and they are readily available from a number of sources, including Radio Shack, Jerryco, or Circuit Specialists (listed in the Appendix). One of the most common fiber-optic connectors is the FLCS package, shown in Figures 9-3 and 9-4. Figure 9-3 provides a cross section of the low-cost plastic FLCS connector. The LED chip is placed to the rear of the connector, with a molded lens placed ahead of the LED chip. The threads at the opposite end of the package allow the fiber cable to be quickly connected or disconnected from the LED or phototransistor housing. The LED or phototransistor inside the FLCS package can be readily soldered to a PC board. To mate the fiber cable to the FLCS connector, remove ½ inch of the outer jacket, place the mating screw assembly over the fiber, and then crimp the ferule onto the cable. The ferrule locks the cable onto the cable's screw assembly to prevent the cable from coming apart.

The diagram in Figure 9-4 depicts a GE FLCS fiber-optic connector assembly. The GE GFOE1A1 is an inexpensive gallium arsenide LED package that generates a near-infrared light source at about 940 nm, with a 30- to 50-mA forward current. GE also makes two low-cost photo-optic detectors. The GFOD1A1 is an NPN phototransistor in an FLCS package, while the GFOD1B1 is a photo-Darlington transistor. The response of the two GE detectors peaks at about 80-percent efficiency. Motorola and Hewlett-Packard have also entered the low-cost fiber-optic connector market. Serious fiber-optic experimenters should obtain the HP or Motorola optoelectric data books.

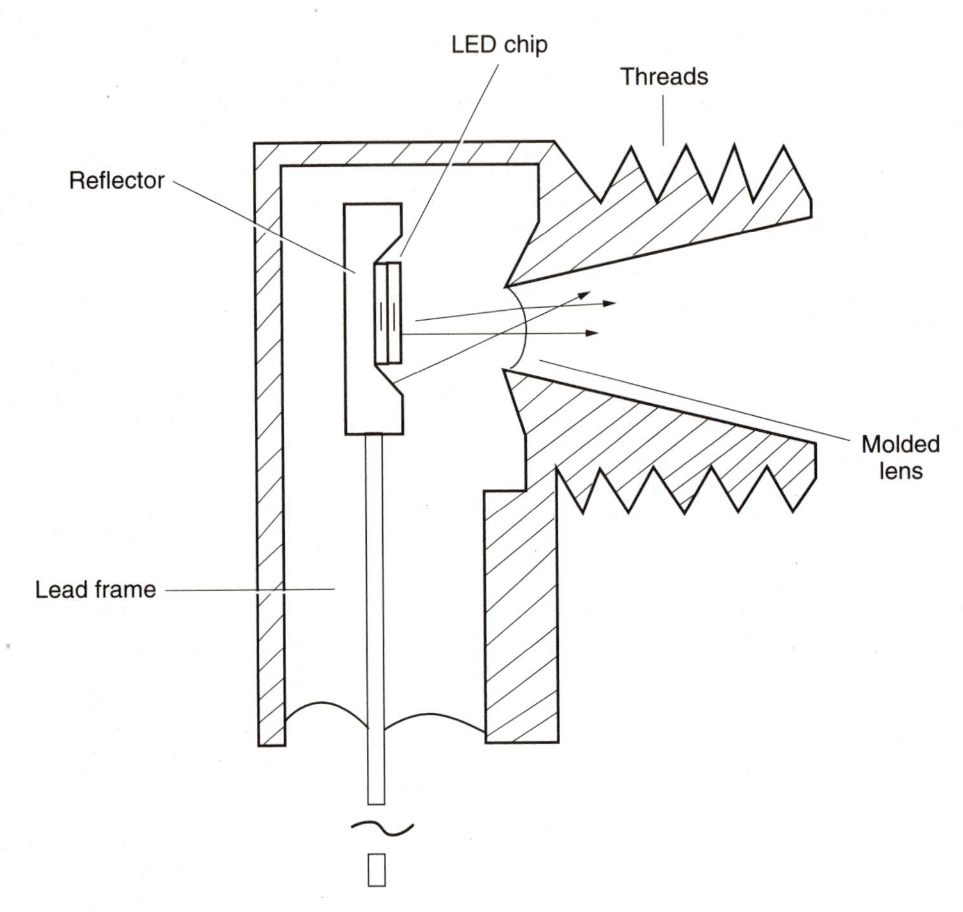

9-3 FLCS fiber-optic connector cross section.

An even lower-cost alternative is to create you own optical connections. One simple method of interfacing a small-diameter fiber-optic cable to an LED or phototransistor is to drill a small hole into the front rounded end of the LED or phototransistor, as shown in Figure 9-5. Be careful to drill only a short distance into the LED or phototransistor to avoid damaging the device. You can place a dab of epoxy to the fiber and LED housing to secure the fiber to the LED or phototransistor case. Once the fiber is secured to the LED or phototransistor package, place a length of heat-shrink tubing over the assembly and heat it very briefly.

Another low-cost connector is shown in Figure 9-6. This simple do-it-yourself connector can be used to couple a plastic multimode fiber to a semiconductor laser. Cut the fiber at one end and polish it with ultra-fine sandpaper. If the fiber is jacketed, remove at least ¾ to 1 inch of the jacket. Trim a common plastic wall anchor to fit a solderless RG/59u connector, as shown. Insert the anchor into the connector and gently turn or screw the anchor into the connector. Now place the polished cable end into the wall anchor and gently twist the cable to secure it in the assembly. In-

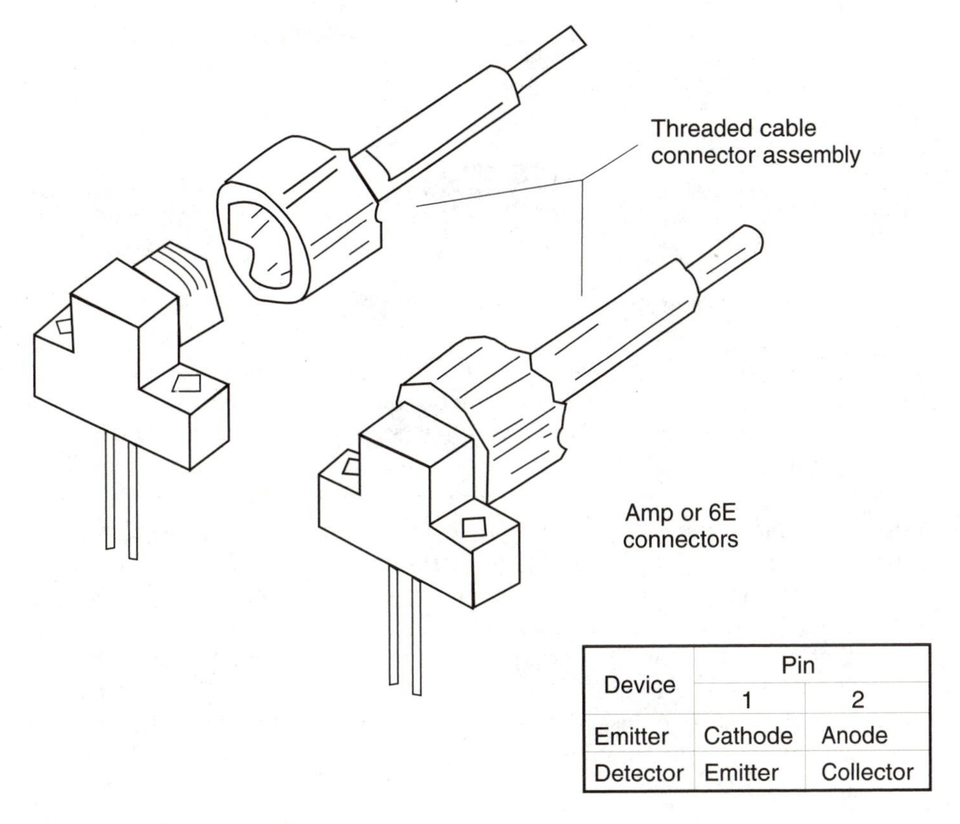

Device	Pin	
	1	2
Emitter	Cathode	Anode
Detector	Emitter	Collector

9-4 Typical FLCS package.

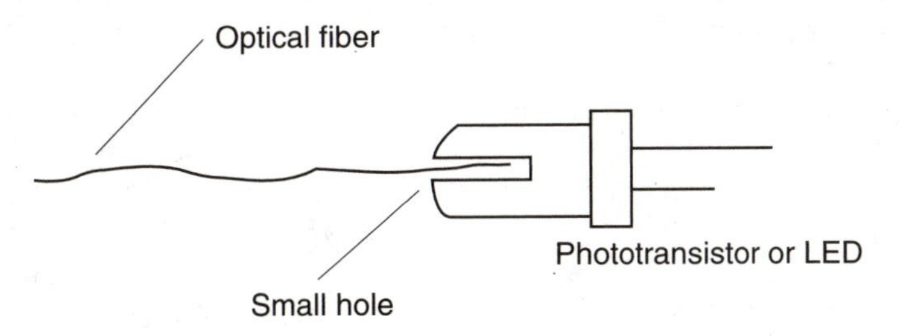

9-5 Do-it-yourself fiber-to-LED interface.

sert a laser diode into the free end of the RG/59u connector, its window just touching the end of the fiber.

You can easily construct your own optical light-wave splices. One simple method of splicing optical cables is to use a small-diameter piece of heat-shrink tubing. Just butt two polished fibers together, glue around the diameter, and couple them by placing the heat-shrink tubing over the two pieces of cable.

You can fashion fiber-optic couplers by drilling both ends of a small block of wood or dowel, forcing the fiber-optic bundles into each end of the dowel or block,

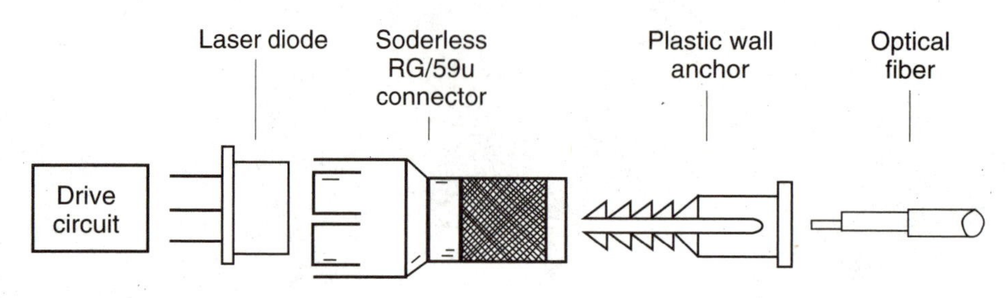

9-6 Do-it-yourself fiber-optic connector.

and then placing a dab of epoxy where the fiber bundle goes into the block or dowel. You can also use this method to couple a fiber cable to an LED or phototransistor. Drill a small hole at one end of the dowel or block to accept the fiber, and then drill a larger hole at the opposite end of the block to accept the LED or phototransistor case. Once seated, use a dab of epoxy to secure both the LED and optic fiber. You can use electronic spade lugs to secure optical fibers to a chassis by simply passing the fiber through the spade lug and gently crimping the lug. Then use the screw-down part of the lug to secure the fiber to the chassis. You could use two spade lugs to splice two sections of fiber-optic cable together. Just butt the two fiber-crimped lugs together, with heat-shrink tubing placed over the junction. This way the spade lugs would be secured to the chassis for further stability.When working with infrared sources, note that silica or glass fibers transmit this range of wavelengths more efficiently than plastic fibers. A number of mail-order and surplus vendors, listed in the Appendix, can provide you a variety of fiber-optic light-guide cables at very reasonable prices.

Optical fibers can be used to sense various phenomena, such as vibration, liquid level, and pressure, as you will see in this chapter. Optical fibers sense in two ways, indirectly (indirect-mode fiber) and directly (direct-mode fiber). The diagram in Figure 9-7 illustrates three basic types of indirect-mode fiber sensors. The *in-line sensor*, shown in the top figure of the diagram, is the simplest indirect-mode fiber sensor. It has many applications and can be easily assembled. In operation, a light source is coupled to one end of a fiber-optic bundle. The first fiber couples light into a second length of fiber, which is separated by a narrow gap. If pressure is applied to either length of fiber, the light beam will be either interrupted or diminished. This type of detector can be used in alarm applications where a door or window controls a light interrupter. It could also be used to sense pressure applied by coupling the detector to an op-amp and dc amplifier circuit. The second indirect-mode fiber sensor is the *dual reflective sensor*, shown in the middle diagram of Figure 9-7. This type of sensor or detector uses two separate optical fibers, with the detector and light source mounted at the same end. A reflecting surface or target is placed at the opposite end of the emitter/detector pair. These fibers can be bent or routed into hard-to-reach places for counting or alarm applications. The last indirect-mode fiber sensor is the *Y-fiber* or *bifurcated light-pipe sensor*. The bifurcated sensor consists of two light pipes, combined or fused at one end. Each of the separate fibers has either an LED or photodetector. The fused ends of the two fibers are polished and coupled at a 45-degree angle to each other, with a third optical fiber acting as a sensing probe. This type of detector uses reflected light for sensing. The probe end is often coupled to a tiny metal probe that can be bent to enter small or confined spaces. The bifurcated light-pipe sensor can be used as an infrared aid for blind people, or for counting and positioning applications. Light-pipe sensors of this type are ideal for harsh environments where neither electrical wires nor sensors can be used. The bifurcated light-pipe sensor is described in more detail in Chapter 5.

Direct-mode fiber sensors operate a bit differently. When an optical fiber is bent into a large radius, light travel through the fiber is relatively unaffected, but strange things happen when an optical fiber is bent into a small radius or circle. If tiny bends or microbends are present, light passing through the fiber might become attenuated. The phenomena occurs when some of the light passing through the core is coupled

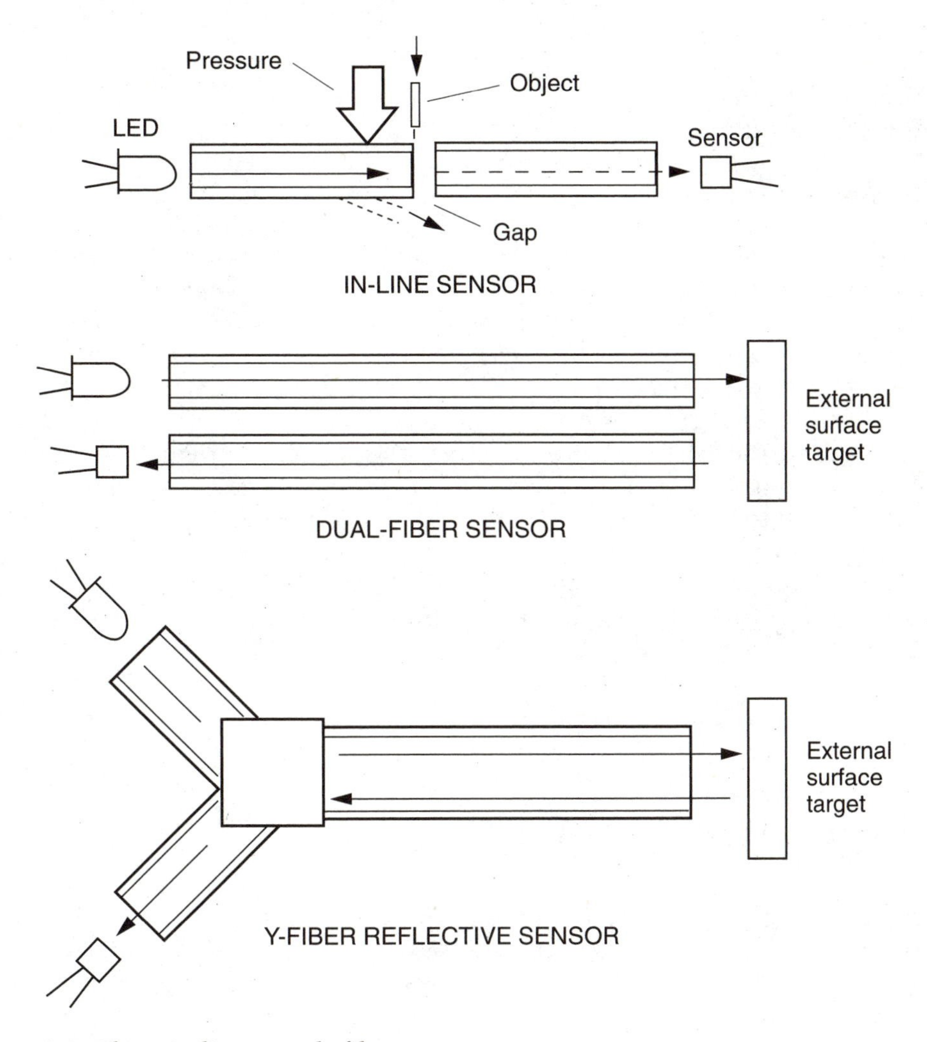

9-7 Three indirect-mode fiber sensors.

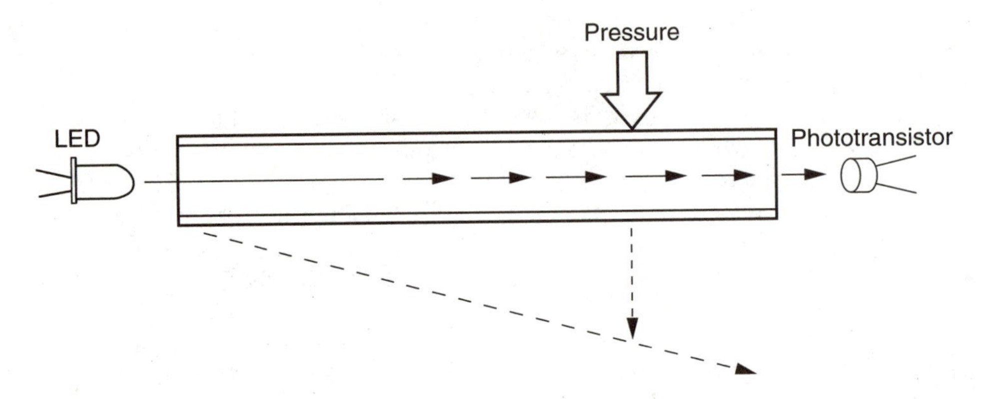

9-8 Pressure-sensing fiber.

out of the core and into the cladding. The diagram in Figure 9-8 shows a simple method for detecting pressure changes using an optical fiber in the direct sensing mode. Light from an LED is passed through an optical fiber bundle and detected by a phototransistor. When the fiber bundle is moved up or down, the light intensity at the phototransistor is modulated by the fiber movement. The resulting output from the photodetector can be coupled to an op-amp for amplification and then onto a dc amplifier to drive a meter, or sent to an I/O card in a personal computer.

The diagram in Figure 9-9 illustrates how an unclad fiber can be used to detect the presence of a liquid. When an unclad fiber is surrounded by water, some of the light is coupled out of the fiber and into the liquid. This occurs because the index of refraction of a liquid is higher than that of air. A detector such as a phototransistor can then sense the reduction in light level with the presence of a liquid. The diagram in Figure 9-10 depicts another type of liquid detector that uses two light pipes directly coupled to a prism from an LED and a phototransistor. The prism is painted black except for the lowest detection portion, which is angled at 45 degrees. When no liquid is present, light travels through the prism and total reflection allows light to be sent back up and out of the prism to the phototransistor. When the prism is lowered into water or some other type of liquid, light becomes lost in the liquid and the light level sent back up and out of the prism to the phototransistor is reduced. You could use the liquid monitor circuit in Figure 9-11 with either of the liquid detectors shown previously to provide a switch output signal.

Liquid monitor switch

The liquid monitor switch circuit in Figure 9-11 requires an eight-pin IC and a few discrete components to form an infrared optical interrupter. The heart of the liquid monitor switch centers around the LM567 phase-locked loop (PLL) tone decoder chip. The tone decoder chip contains a local oscillator, a PLL decoder, and a 100-mA output drive circuit. The local oscillator is tuned to 40 kHz by the R4/C5

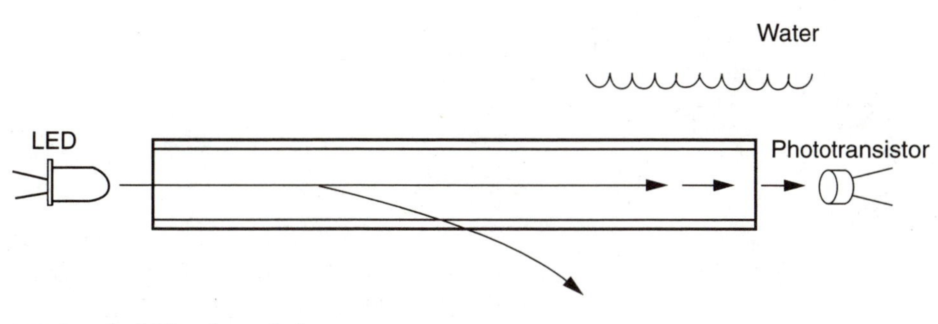

9-9 Unclad fiber liquid detector.

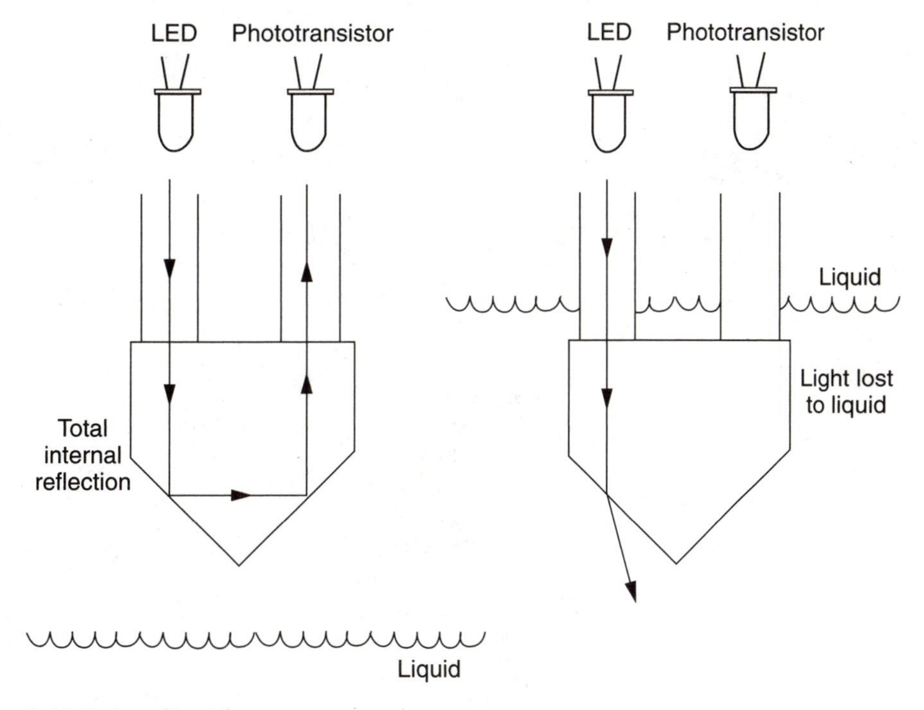

9-10 Prism liquid sensor.

combination, and drives transistor Q1. Transistor Q1 in turn drives the IR LED. The receiver or decoder portion of the IC centers around the PLL's input on pin 3. When photodetector Q2 detects the IR beam from the LED, the 40-kHz signal appears at pin 3 of the LM567. In this condition, the circuit locks in step to the 40-kHz signal and the output becomes high. So when the prism sensor is in air, the output is high; conversely, when the prism sensor is submerged in water, the path between the LED and Q2 is open and the IC output is low. A low-current, 5-volt miniature relay can be used at the IC's output at pin 8. The optical interrupter circuit can be powered from any 5- to 6-volt power source.

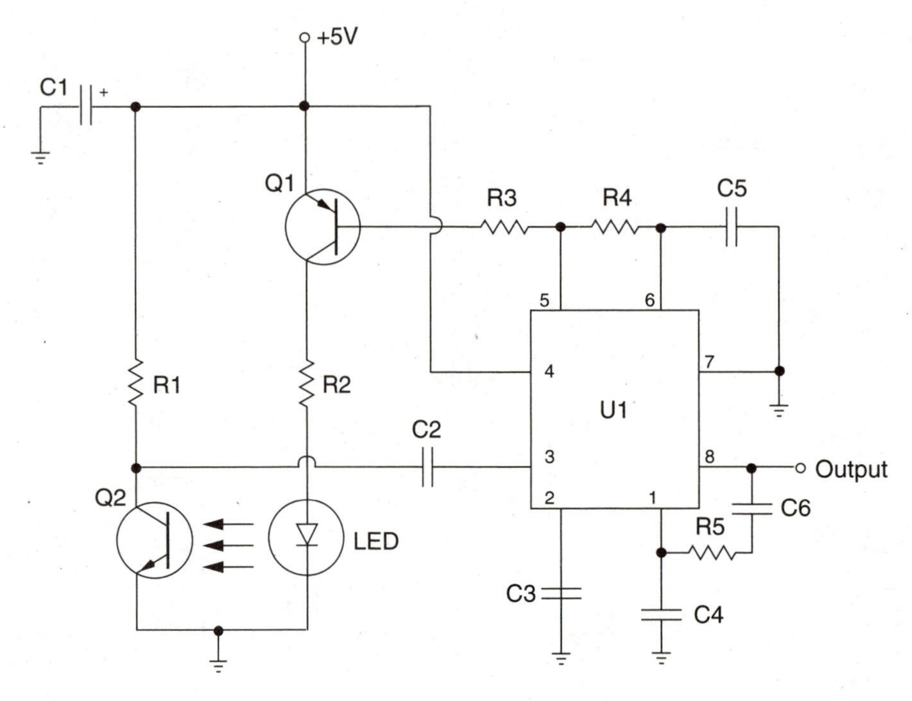

9-11 Liquid monitor switch.

Liquid monitor switch parts list

R1	1-kilohm, ¼-watt resistor
R2	560-ohm, ¼-watt resistor
R3,R4,R5	10-kilohm, ¼-watt resistor
C1	220-µF, 25-volt electrolytic capacitor
C2,C6	1-nF, 25-volt disc capacitor
C3,C4	33-nF, 25-volt disc capacitor
C5	2.7-nF, 25-volt disc capacitor
Q1	2N3906 PNP transistor
Q2	TIL78 IR phototransistor (Texas Instruments)
U1	LM567 tone decoder
LED	TIL32 IR LED (Texas Instruments)

Vibration detector

A simple vibration detector is shown in Figures 9-12 and 9-13. Begin by carefully drilling or slotting an LED to accept a short, rectangularly cut plastic arm with a micro-weight attached to the free end. Then mount a phototransistor near the free end of the plastic arm, as shown. Place the plastic arm on the center line of the phototransistor, as shown, in a light-tight box. In operation, light emitted by the LED is coupled to the photodetector. When the plastic arm is displaced by vibration, less light reaches the phototransistor. The signal from the phototransistor is coupled to an LM741 op-amp, which amplifies the light level from Q1. The output of the op-amp could be further amplified or fed to a display device. If vibrations exceed several cycles per second, connect a .1-µF capacitor between Q1 and R3.

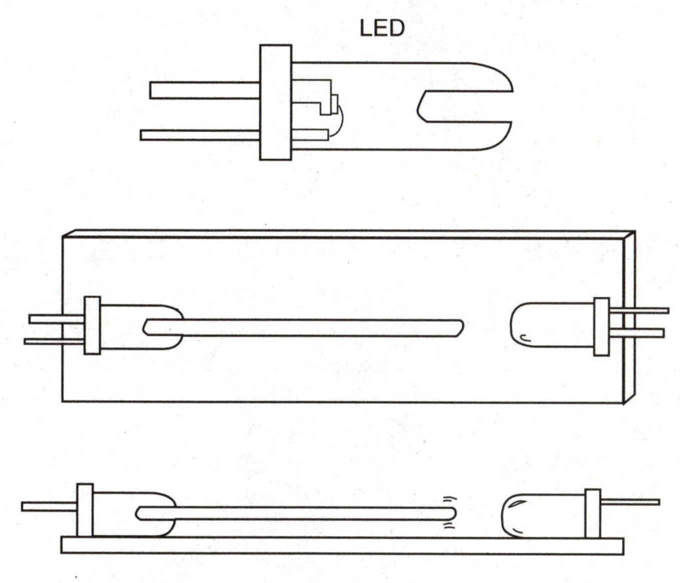

9-12 Fiber-optic vibration sensor.

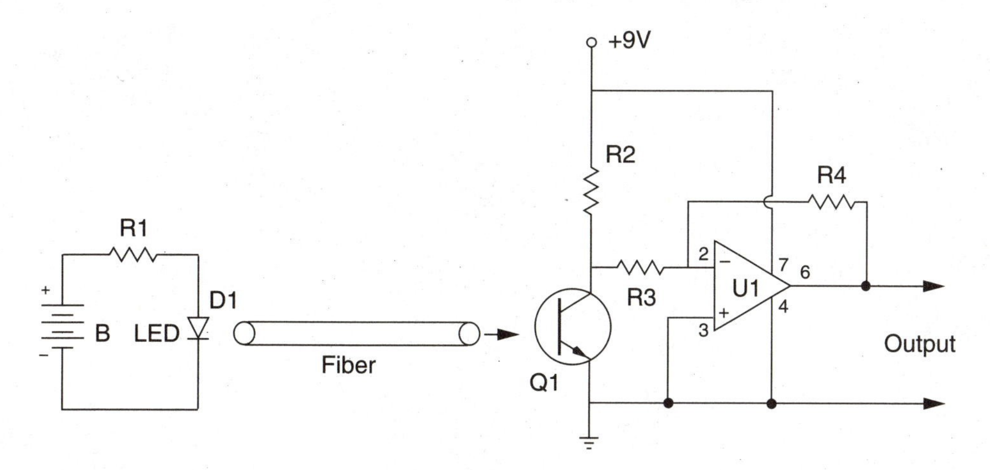

9-13 Fiber-optic vibration sensor electronics.

Vibration detector parts list

R1	500-ohm to 1-kilohm, ¼-watt resistor
R2	100-kilohm, ¼-watt resistor
R3	1-kilohm, ¼-watt resistor
R4	100-kilohm, ¼-watt resistor
D1	LED
Q1	Phototransistor
U1	LM741 op- amp
B	6- to 9-volt power source
Opt	.1-µF, 25-volt disc capacitor

Fiber-optic voice-communication system

The fiber-optic transmitter and receiver shown in Figures 9-14 and 9-15 and in Photo 9-1 is a great introduction to the fascinating world of fiber optics. The system allows you to send and receive voices and music well over 150 meters in electrically noisy environments, via a beam of light through a plastic fiber cable. You can combine two of these systems for full-duplex or two-way conversations. Let's begin with the simple fiber transmitter unit, illustrated in Figure 9-14. An electret microphone is fed to an LM380 audio amplifier chip via C1. The LM380 amplifier boosts the incoming audio level in order to drive the Motorola MF0T76 high-output LED assembly through the current-limiting resistor R2. Resistor R1 biases the electret microphone, and the entire transmitter is powered by a 9-volt battery. Once built, you can install the fiber-optic transmitter in a small metal chassis box, and use an SPST toggle switch to turn the transmitter off and on. Note that a number of LM380 pins are tied to ground.

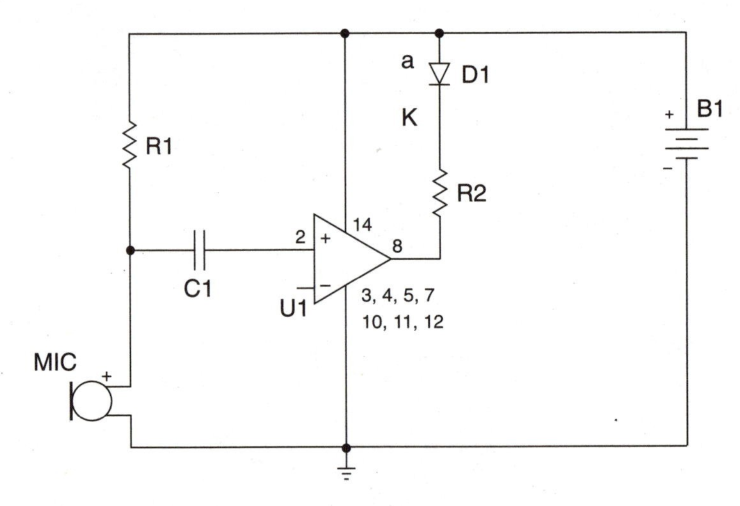

9-14 Voice fiber-optic transmitter link.

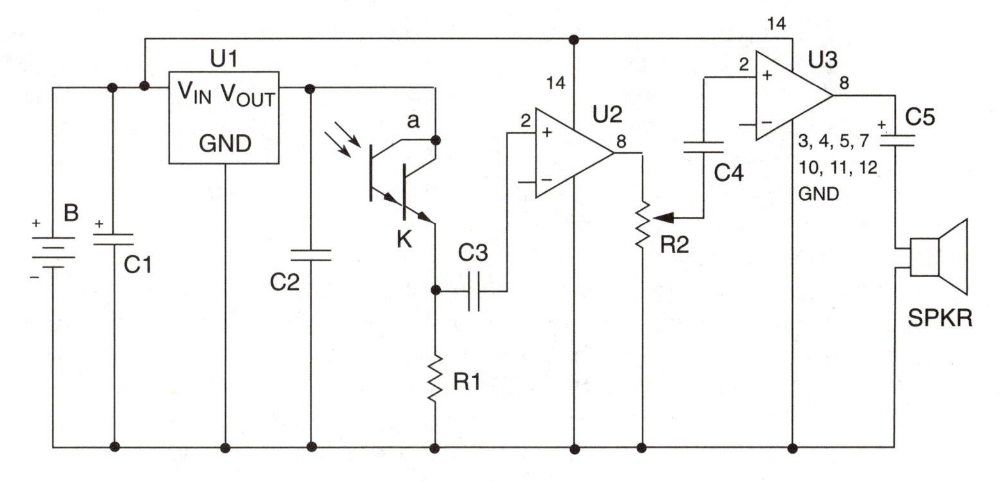

9-15 Voice fiber-optic receiver link.

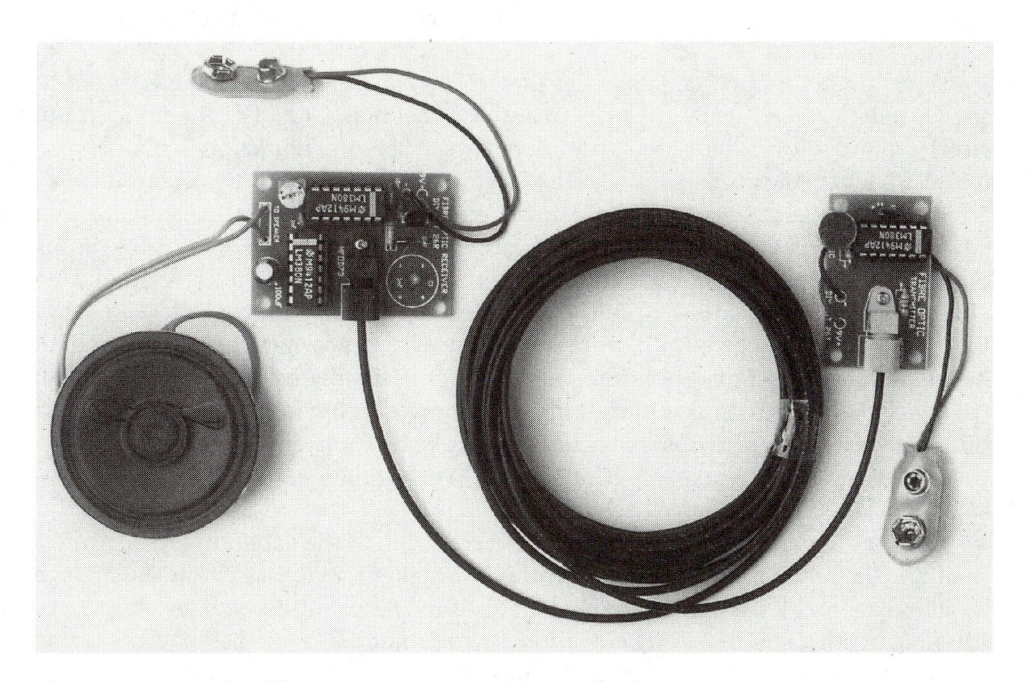

Photo 9-1 Simplex fiber-optic communications link.

The fiber-optic receiver, shown in Figure 9-15, is a bit more complicated than the transmitter, but it is straightforward once you look closely. The incoming light modulated by the transmitter is received by the MF0T73 Motorola photo-Darlington transistor assembly. The incoming signal is coupled to the first amplification stage at U1 via a 10-nF capacitor at C3. The amplified signal from the first stage is then sent to U2, a second audio gain stage, via a 2-kilohm potentiometer at U3. Note that many LM380 pins are connected to ground on both U2 and U3. The output of U2 is connected through capacitor C5 and in turn to the 8-ohm output speaker. The entire fiber-optic receiver is powered by a 9-volt battery. The battery voltage is sent to a 5-volt regulator, which provides a constant 5 volts to the photodetector. Once the fiber-optic receiver is constructed, locate a small chassis box and install the circuit, an SPST toggle switch, and a 9-volt battery, leaving an entry hole for the fiber-optic cable.

Now let's move on to preparing the fiber-optic cable for assembly. A 15-foot, 1-mm-diameter fiber-optic cable was used with the prototype. To obtain the highest transmission efficiency, it is important to squarely cut and polish the fiber-optic cable. Use a knife to score the end of the fiber and try to ensure a smooth, even cut at the end. Next, locate some extra-fine sandpaper and spread a few drops of water on it. Hold the fiber perpendicular to the sandpaper, which is placed on a flat table, and gently sand the end of the fiber in a circular motion. Once the cable ends are ready, simply fit them into the LED assembly at the transmitter end and then into the detector assembly on the receiver end. Turn the screw on the assembly to secure the cable.

Once the fiber-optic cables are secured, you can begin testing the entire system. Connect a 9-volt battery to the transmitter section. The LED should light up. Now connect a 9-volt battery to the receiver section and adjust the volume control at R2 for the desired volume. Place the transmitter unit in a nearby room with a radio playing next to the microphone. Then uncoil the fiber-optic cable, making sure there are no loop-backs or kinks. Try to ensure a minimum curvature radius of 2 cm for best results. Place the receiver in another room and you should be able to remotely listen to the radio placed next to the transmitter unit. Note that the fiber-optic connector at the receiver unit works best when the cable is pulled back from maximum insertion into the detector assembly by 1 to 2 mm. You can perform some interesting experiments with the system by pulling the fiber cable out of the assembly and observing how far back the cable can be from the detector before the audio signal is extinguished. Try waving a card in front of the unattached cable.

You can try cutting the black protective shielding of the cable while you are listening to the audio from the radio. For this experiment, make sure that the ends of the fiber are inserted into both the transmitting and receiving assemblies. As you are listening to the audio, gradually cut through the plastic cable and listen to the sound gradually being cut off. Hold the two cut ends close to each other and note how much audio signal jumps the air gap to the other cable. Take a cigarette lighter and melt the cables back together using the flame. You can try another experiment to see how important cable preparation really is. Take one end of the cable from the detector or LED assembly, rough it up, place it back into the light-guide connector, and observe the results. Have fun experimenting with your new fiber-optic communications system. Without much difficulty you will find a number of applications for this system around your home or shop.

Fiber-optic voice-communication transmitter parts list

R1	47-kilohm, ¼-watt resistor
R2	150-ohm, ¼-watt resistor
C1	10-nF, 25-volt Mylar capacitor
D1	Motorola MF0T76 LED fiber coupler assembly
U1	LM380 audio amplifier IC
M	Electret microphone
B	9-volt battery
Misc	Circuit board, chassis, battery holder, wire, and switch

Fiber-optic voice-communication receiver parts list

R1	220-kilohm, ¼-watt resistor
R2	2-kilohm potentiometer
C1	1-µF, 25-volt electrolytic capacitor
C2,C3	10-nF, 25-volt Mylar capacitor
C4	1-nF, 25-volt Mylar capacitor
C5	100-µF, 25-volt electrolytic capacitor
U1	LM78L05 5-volt regulator
U2,U3	LM380 audio amplifier chip
B	9-volt battery
SPKR	8-ohm speaker
Misc	Circuit board, chassis box, and battery holder

Communication or information can be transmitted in one of two basic ways: either analog or digital. The level of an analog signal varies continuously, while a digital signal can be at only a certain number of discrete levels. Analog technology has been used traditionally for audio and video communication and signaling because our ears detect continuous variations in sound levels, not just the presence or absence of sound. Our eyes also detect levels of brightness rather than simply the presence or absence of light. Telephone wires deliver a continuously varying signal to and from your handset, but digital signals are in fact more compatible with fiber optics. It's simpler and cheaper to design a circuit to detect whether a signal is at a high or low level (on and off) than to build one to accurately replicate a continuously varying signal. When an analog signal goes through a system that doesn't faithfully reproduce the signal, the result is a garbled signal that is often unintelligible, which is exactly what happens when you get a distorted voice on the telephone or radio. When a digital signal is not reproduced exactly, it is still possible to tell the "on" signals from the "off" signals, and the original message can still be recovered. If people really need analog signal information transfer but digital transmission works best, why not convert between the two? This is becoming increasingly common in both audio and telephone systems. Analog signals are converted to digital before transmission and then back to analog signals, since it is now economical with low-cost analog-to-digital (A/D) and digital-to-analog (D/A) converter integrated circuits.

Edulink fiber-optic data-transmission system

The Edulink fiber-optic transmitter and receiver pair shown in Figures 9-16 and 9-17 is ideally suited for accurate high-speed reconstruction of incoming data pulses. The Edulink transmitter is a TLL-compatible circuit that will convert an incoming logic signal to optical pulses. The transmitter also has a self-contained oscillator that provides a 1-kHz test/demonstration signal output. The fiber-optic transmitter consists of a CD4093 quad-2 input Nand Schmitt-trigger IC and a 2N4401 transistor driver that drives the Siemens GaAsP visible LED. Gates A and B of the CD4093 are connected as an astable oscillator that is enabled when the OSE and TXD inputs are high. The third gate, C, steers the signal from the built-in oscillator or from an external oscillator. Transistor Q1 drives the LED with resistor R1, which acts to limit the current through the LED to 40 mA. A glass or silica fiber-optic cable could be substituted for longer-distance operation in the infrared region.

The Edulink receiver depicted in Figure 9-17 and Photo 9-2 is optimized for operation with the Edulink transmitter. The incoming light pulses from the transmitter are received from D1, a high-speed PIN diode. Diode D1 is reverse- biased and connected to the first stage of U1, forming a voltage-to-current converter. The Edulink fiber-optic receiver is a unique design that is optimized to store the amplitude of incoming pulses. Amplifier U1B, together with Q1 and C1, forms a peak detector that stores the incoming pulses. This stored reference signal allows the circuit to sample the incoming signal at its point of minimum distortion, thereby reducing pulse-width distortion. The output from the peak detector is halved, relative to the voltage reference divider at R3/R4. This signal, along with the amplified signal from U1A, is then

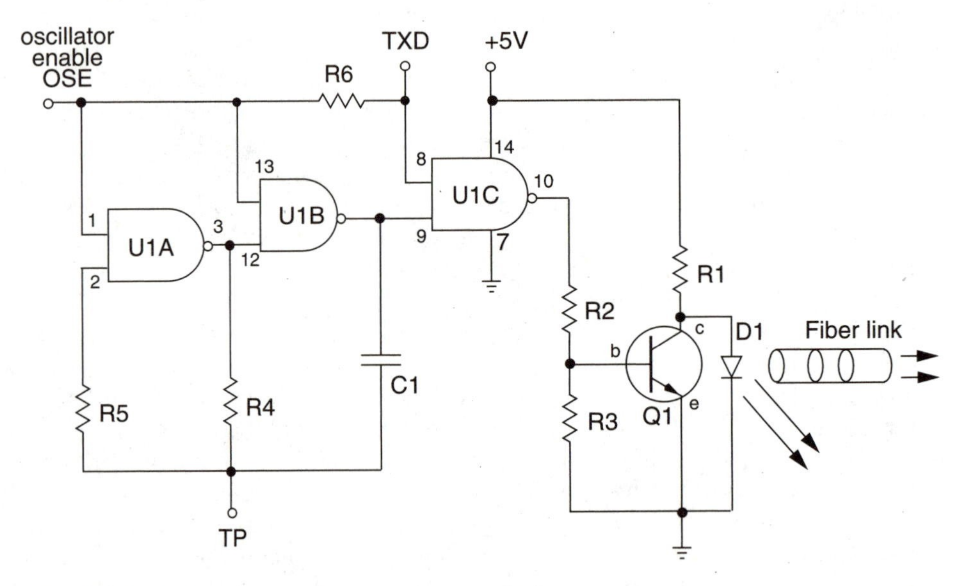

9-16 Edulink fiber-optic data transmitter link.

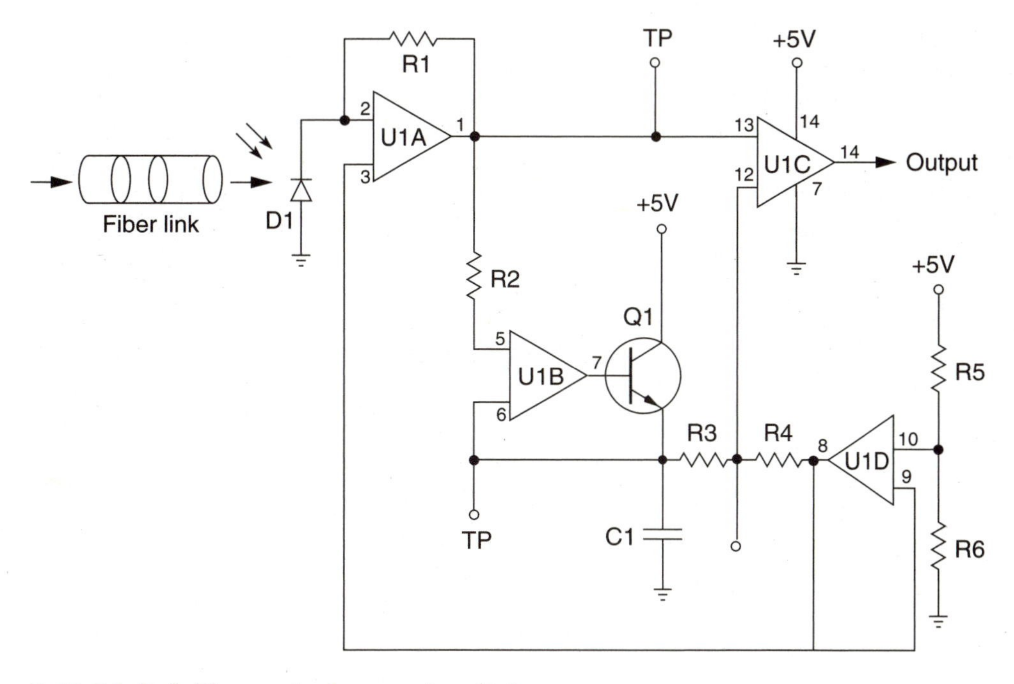

9-17 Edulink fiber-optic data receiver link.

applied to the output comparator at U1C. Amplifier U1D is configured as a unity gain follower that provides a buffered version of the voltage from the R5/R6 divider.

The Edulink transmitter and receiver pair is available in kit form from Edmund Scientific and is quite easy to set up and use. Both transmitter and receiver units use plastic fiber-optic connectors to couple the fiber cable, and these can be set up in a

Photo 9-2 Edulink fiber-optic data receiver.

few minutes. Remove ¼ inch of outer sheathing from both ends of the fiber, and then insert them into the plastic ferrules. Then cut the exposed fibers emerging from each ferrule with a knife, and insert the ferrules into each of the connectors on both the transmitter and receiver.

To test the Edulink fiber-optic data system, use a pair of 6-volt lantern batteries or battery packs made from C or D cells. In order to drop the voltage down to 5 volts, use a 1N4002 diode between the positive battery terminal and the circuit. You can readily test the system by connecting the output of the receiver to a small audio amplifier or an oscilloscope. Since this circuit uses a PIN diode as a detector, you can use the circuit to send higher data speeds than the previous TLL data systems. If silica fiber is used in place of the supplied plastic fiber, the system range can be increased significantly.

Edulink transmitter parts list

R1	75-ohm, ¼-watt resistor
R2	3.9-kilohm, ¼-watt resistor
R3,R4	30-kilohm, ¼-watt resistor
R5	300-kilohm, ¼-watt resistor
R6	4.7-kilohm, ¼-watt resistor
C1	.01-µF, 25- volt disc capacitor
D1	Red LED (see text)
Q1	2N4401 NPN transistor
U1	CD4093 IC
Misc	Circuit board, chassis box, battery, and battery holder

Edulink receiver parts list

R1	200-kilohm, ¼-watt resistor
R2,R5,R6	4.7-kilohm, ¼-watt resistor
R3,R4	47-kilohm, ¼-watt resistor
C1	.1-µF, 25-volt disc capacitor
Q1	2N4401 transistor
U1	LM324 op-amp
D1	High-speed PIN diode detector (Motorola MRD921)

High-speed digital transmission system

If you feel the need for speed, the fiber-optic link shown in Figures 9-18 and 9-19 should fill the bill. The 50-Mbps transmitter link illustrated in Figure 9-18 accepts data information via pins 1 and 2 on U1A. The transmit gate or enable input is gated high to send pulses through the transmitter to the LED. The pull-up transistor of the totem-pole output at U1B is used to turn the LED on, and the pull-down transistor is used to turn the LED off. The lower impedance and higher

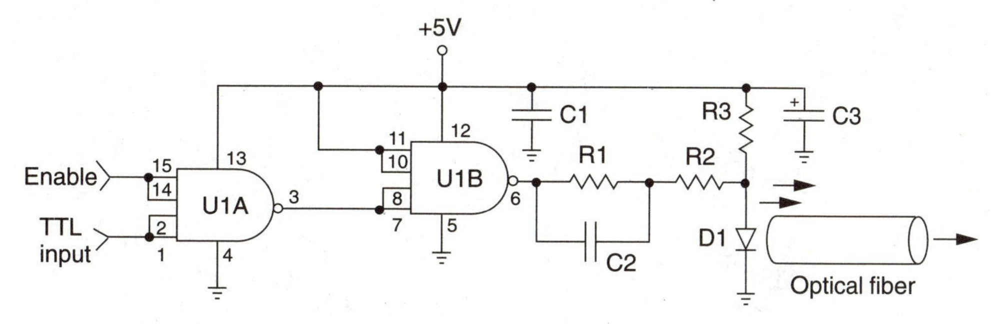

9-18 50 Mbps fiber-optic high-speed transmitter.

current-handling capability of the saturated pull-down transistor is an effective method of transferring the charge from the LED's anode to ground as its dynamic resistance increases during the turn-off period of the LED. The slightly higher output impedance of the pull-up stage ensures that the LED is not overpeaked during the less difficult turn-on transition. This asymmetrical current-handling capability of the output stage, with its variable impedance, greatly reduces the pulse-width distortion and long-tailed response. As the signal propagates through the two Nand gates, each transition passes through the high-to-low and low-to-high transition once, normalizing the total propagation delay through the transmitter circuit. This circuit is well suited to high-speed digital transmission over moderately long distances. The LED can be coupled to a length of 1-mm fiber-optic cable. The entire fiber-optic transmitter circuit can be powered from a 5-volt dc source.

The high-speed 50-Mbps fiber-optic companion receiver is shown in Figure 9-19. The optical signal from the fiber-optic transmitter is coupled to an Amperex BPF31 photodiode at pin 4 of U1. Current flowing in the diode also flows into the input of the NE5211 preamplifier, a fixed-gain block with a 28-kilohm differential transimpedance, and performs a single-ended differential conversion. With the signal in differential form, greater noise immunity is assured. The second-stage or postamplifier at U2 also includes a gain block, with an auto-zero detection and limiter circuit. The auto-zero circuit allows decoupling between U1 and U2, and cancels the signal-dependent offset because of the optical-to-electrical conversion process. The auto-zero capacitor must be 100 pF or greater for proper operation. The peak detector has an external threshold adjustment at pin 3 of U2, allowing you to tailor the threshold to specific needs. The receiver system also has provisions for hysteresis to minimize jitter introduced by the peak detector at pin 12 of U2. The output stage at pin 10 of U2 provides a single-ended TTL data-signal output with matched rise and fall times to minimize duty cycle distortion. The high-speed fiber-optic receiver is also powered from a 5-volt dc source. If your application requires a duplex transmission, you need to design each circuit board with both a transmitter and receiver integrated into a single PC board.

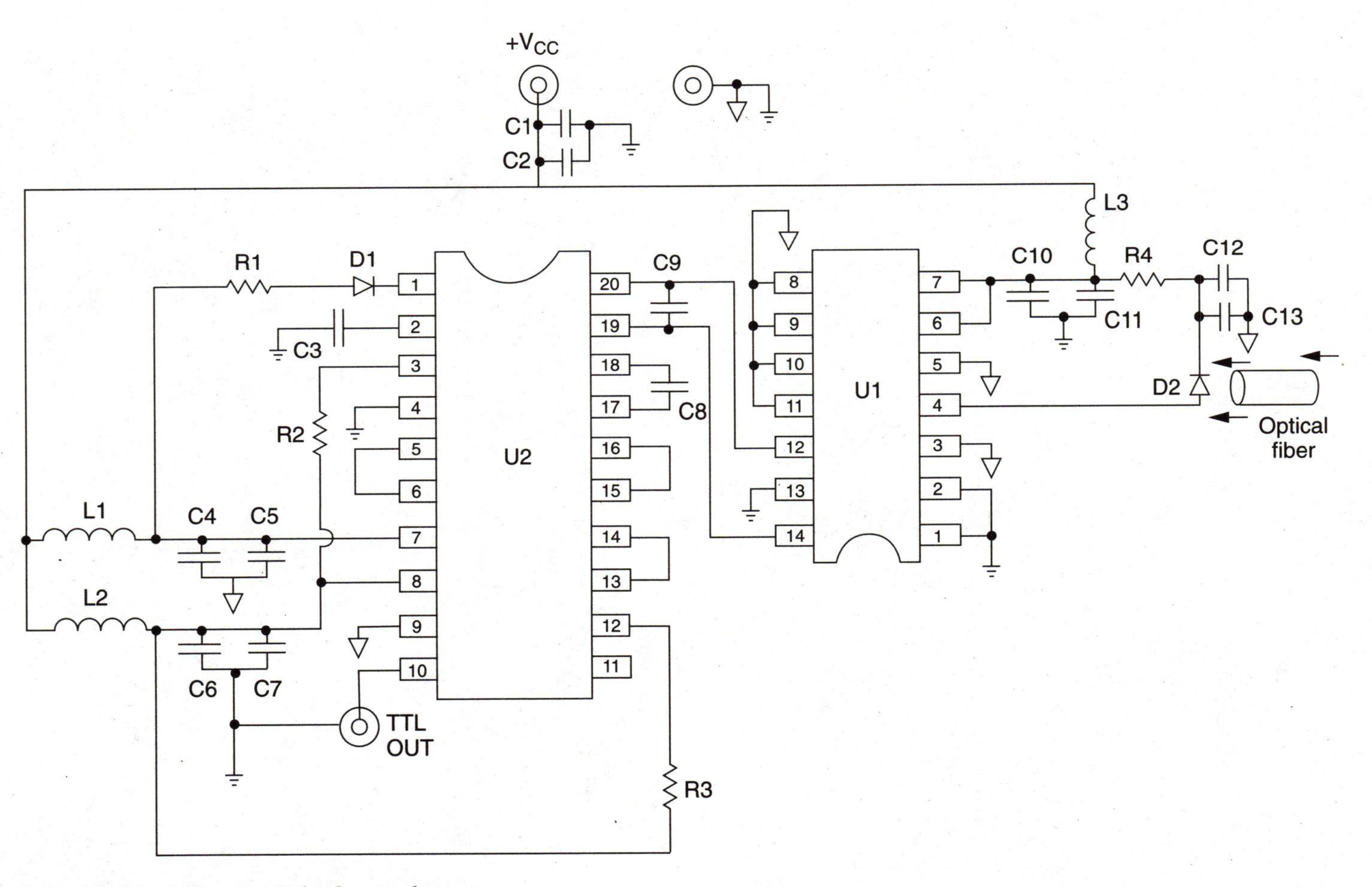

9-19 50 Mbps fiber-optic high-speed receiver.

High-speed digital transmitter parts list

R1	12-ohm, ¼-watt resistor
R2	4-ohm, ¼-watt resistor
R3	90-ohm, ¼-watt resistor
C1	.1-µF, 25-volt disc capacitor
C2	250-pF, 25-volt disc capacitor
C3	100-µF, 25- volt electrolytic capacitor
D1	CQF41 LED
U1	74F3040 Nand gate
Misc	Circuit board and chassis box

High-speed digital receiver parts list

R1	220-ohm, ¼-watt resistor
R2	47-kilohm, ¼-watt resistor
R3	5.1-kilohm, ¼-watt resistor
R4	100-ohm, ¼-watt resistor
C1	47-µF, 25-volt electrolytic capacitor
C2	.01-µF, 25-volt disc capacitor
C3	100-pF, 25-volt disc capacitor
C4,C6,C11	10-µF, 25-volt tantalum capacitor
C5,C7	.01-µF, 25-volt tantalum capacitor
C8	.1-µF, 25-volt disc capacitor
C9	100-pF, 25-volt Mylar capacitor
C10,C13	.01-µF, 25-volt Mylar capacitor
C12	1.0-µF, 25-volt tantalum capacitor
U1	NE5211
U2	NE5214
Misc	Circuit board and chassis box

Future fiber-optic trends

The key advantages of fiber optics are wide-bandwidth signals and greatly increased speed. Novel laser diodes and intense integrated packaging now allow multiple transmit and receive channels in a single IC package. As file sizes grow and data transfer rates become even more important, the fiber-optic interface can become a significant bottleneck. Laser data and large program files require faster and faster data transfer rates. One of the most promising new interfaces is the fiberchannel, which offers data rates up to 1.26 Gbps. The complexity and cost of implementing fiberchannel interfaces has limited them to only extremely crucial applications. TriQuint semiconductor has pushed the envelope, recently unveiling a three-chip set that forms a complete high-speed fiberchannel interface using the new TQ 9501/02 and TQ9303 encoder/decoder (ENDEC) chips. The new fiberchannel interface can extend fiber cabling distances up to 10 km.

Research continues to widen the bandwidth of new fiber-optic cable and systems. AT&T claims its Truewave optical fiber has a full bandwidth of up to 80 Gbps. Conventional optical fibers of the past have were nonlinear, which limited the number of channels and data that could be sent through the fiber. These nonlinearities created wavelength mixing in multiplexed systems. The multiplexed channels interacted with each other to create new unwanted wavelengths. AT&T has found that by adding a dopant (impurity) to the fiber, they can create a controlled amount of chromatic dispersion. Thus they can minimize the nonlinearities, allowing multiple channels to be spaced more closely together and a greater amount of data.

Researchers at Corning Glass have recently developed a new fiber using a dispersion-shifting technology that allows transmission of multiple 10-Gbps data channels. Both AT&T and Corning have recently developed fibers that have an even wider bandwidth capability. New wavelength multiplexing systems allow designers to run several independently modulated channels on the same fiber-optic cable, with each channel operating at a different distinct wavelength.

Many of the main long-distance telephone trunks have been replaced with high-speed, low-noise fiber-optic cables. The remaining bottlenecks exist between the telephone central office and your home. This problem has been temporarily postponed with the 128-Kbps ISDN service. AT&T is experimenting with higher-speed ADSL transmissions, while the cable companies are preparing high-speed cable modems.

All of these improvements in fiber-optic interfaces and transmission systems are preparing the way for two-way video conferencing, multimedia, and advanced image and data manipulation. The future will be exciting indeed, thanks to the new fiber-optic technologies.

10

CHAPTER

Reflective light

Alexander Graham Bell was the first person to speak over a beam of light. In February 1880, Bell and his assistant directed a beam of reflected light through a pair of comb-like grids, made by scratching parallel lines in the silver coating of two mirrors. One grid was mounted in a fixed position, and the second grid was attached to the diaphragm of a modified telephone microphone. When he spoke against the diaphragm, the grid on which it was attached moved back and forth in response to the audio signal. This caused a fluctuation in the sunlight passing through the two grids. The fluctuating light was detected by a homemade selenium light detector designed by Bell. Bell's new communicator was dubbed the Photophone and soon he and his assistant were able to talk over light beams hundreds of feet through the air. Bell believed that the Photophone was fundamentally a greater invention than the telephone. In actuality, the Photophone did not enjoy great success during Bell's lifetime, but it did lead to a number of other inventions discussed in this chapter.

Two-way Photophone transceiver

A modern day Photophone transceiver is shown in Figure 10-1. You can use it to communicate over long distances using sunlight or light from a small laser pointer. The key to the Photophone transceiver is a speaker with aluminized Mylar supported in front of it. The mylar is stretched between two supports directly in front of a speaker, as shown.

A silicon solar cell is mounted at the end of a six-inch cardboard tube to eliminate stray light from reaching the detector. In the transmit mode, the electret microphone is fed to the op-amp at U1 via switch S1. The output of U1 is passed to U2 through potentiometer R5. The audio amplifier at U2 amplifies the microphone output and causes the speaker and Mylar to vibrate, which in turn modulates the sunlight or laser carrier beam. The sunlight or laser beam is then directed toward a duplicate Photophone transceiver located some distance away.

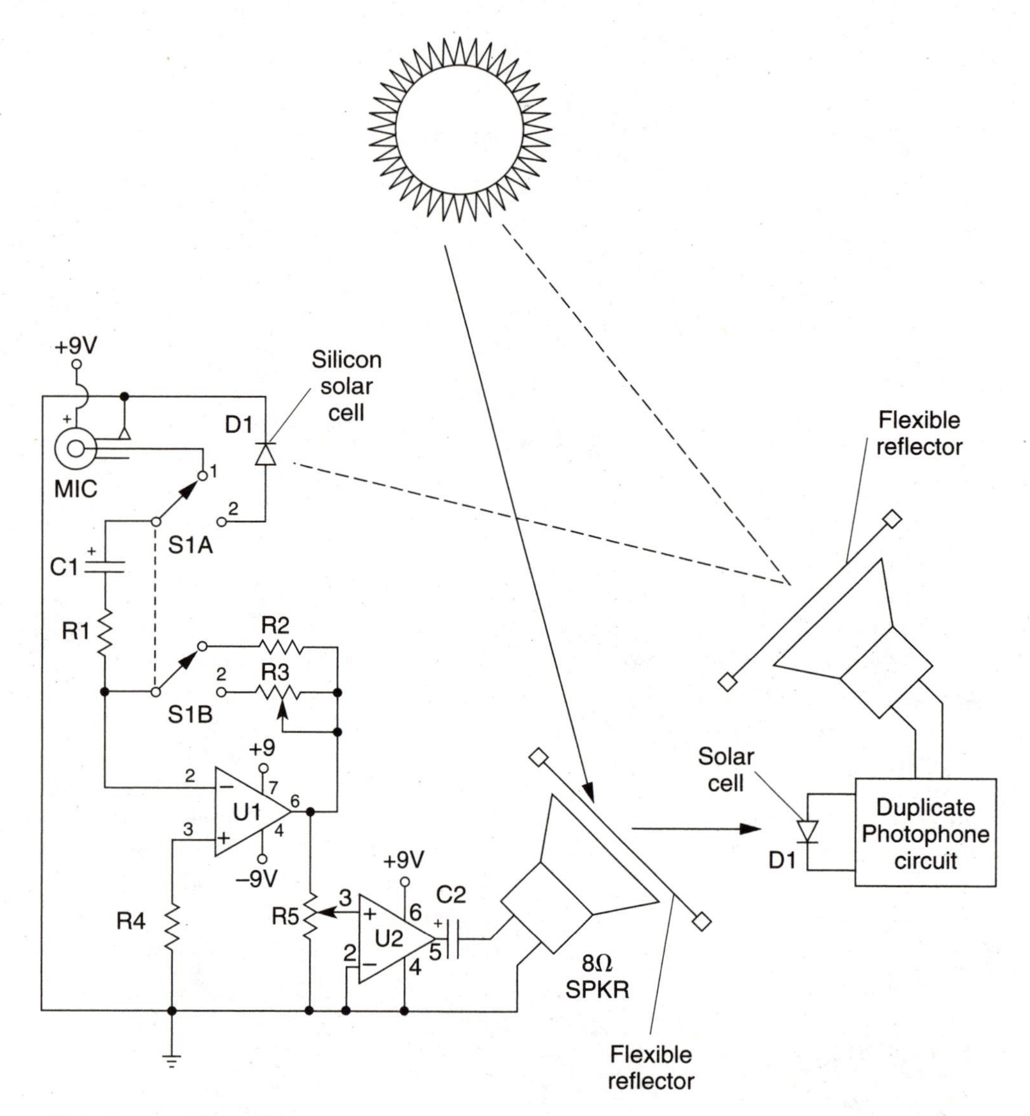

10-1 2-way Photophone communicator. From *Engineer's Mini-Notebook: Communications Circuits* (Radio Shack, 1986). Copyright by Forrest Mimms III. Used with permission.

In the receive mode, modulated light from the second Photophone is directed to the solar cell, D1. The signal from D1 is switched by S1 into the amplifying system at U1, which passes along its signal to be amplified by the audio amplifier at U2. The output from U2 is then amplified to allow you to listen to the remote Photophone.

You can substitute a laser for sunlight, but be extra careful when aligning the two Photophones to prevent eye damage. Both Photophone operators should wear sunglasses and avoid staring at reflected light. Also, be extremely careful when aligning the lasers. Both Photophone transceivers should be mounted on camera tripods for best results.

Two-way Photophone transceiver parts list

D1	Silicon solar cell
R1,R4	1-kilohm, ¼-watt resistor
R2	10-kilohm, ¼-watt resistor
R3	1-megohm potentiometer
R5	10-kilohm, ¼-watt resistor
C1	.1-µF, 25-volt disc capacitor
C2	100-mF, 25-volt electrolytic capacitor
U1	LM741 op-amp
U2	LM386 audio amplifier
SPKR	8-ohm speaker
MIC	Electret microphone
S1	Two-position rotary switch
B1,B2	9-volt batteries
Misc	Chassis, aluminized Mylar, and rubber band

As an aside, a concept similar to the Photophone was used by the Russians in the 1960s. Since very high frequency radio waves behave similarly to optical or light waves, the Russians devised a microwave listener to spy on the American Embassy. The Russians presented the American Embassy with a wall plaque, which was graciously accepted and unknowingly hung inside the Embassy conference room. The Russians had cleverly concealed their microwave listener inside plaque. An unmodulated microwave signal from a van outside the Embassy was directed toward the plaque. Voices from the conference room modulated a diaphragm placed over a microwave cavity inside the plaque. The modulated microwave cavity returned the modulated radio signal to a second van parked outside the Embassy. The Russians could then listen in to all the meetings that took place in the conference room, and the Russians enjoyed this setup until it was later discovered.

Surveillance experts and private investigators have one-upped Alexander Graham Bell, devising a laser listening device for eavesdropping on distant conversations inside a building using reflected laser light. Rather than breaking and entering the premises, conversations can be remotely monitored. I will discuss this device in more detail later in the chapter.

There are many applications for reflected light, from entertainment to measurement, and I will describe a number of reflected-light projects.

The diagram in Figure 10-2 illustrates a fun and educational project using reflected laser light. A helium-neon or semiconductor laser is aimed at a medium-sized loudspeaker. A small front-surface mirror is glued to the speaker midway between the outer edge and the center cone area. A radio or audio amplifier with a microphone attached is connected to the speaker. A projector screen, sheet, or light-colored wall can display the reflected light from the laser, which when modulated is reflected and displayed on the viewing screen. Rather than modulating the reflected laser beam, you can feed an audio amplifier with the output from an audio signal generator for some interesting results. This demonstration project is sure to please.

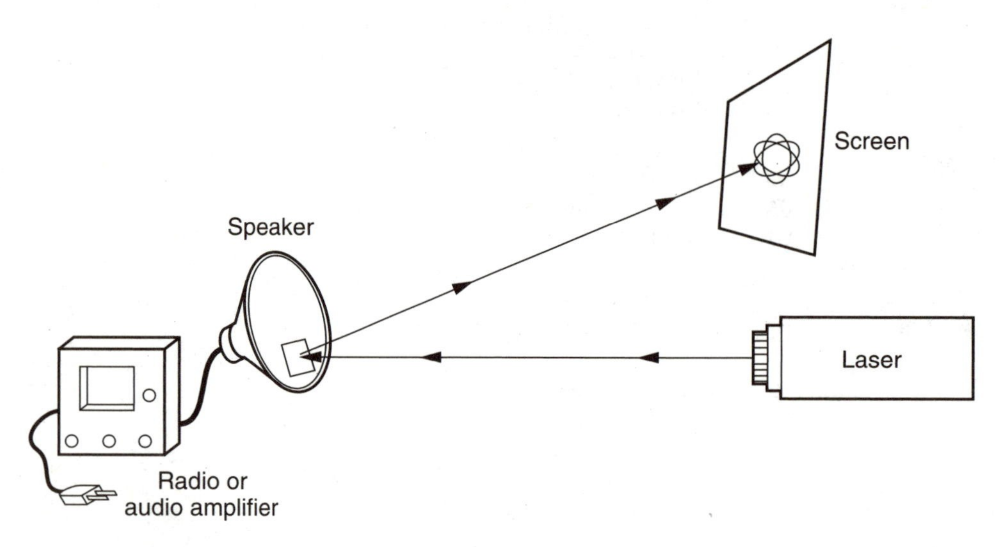

10-2 Reflective laser audio design maker.

Reflective laser audio design maker

The diagram in Figure 10-3 is an educational and fun demonstration project using two mirrors, two motors, and a laser beam. You can create your own spectacular laser light show using this relatively simple setup. First locate two 1.5- to 6-volt dc hobby motors. Measure the diameter of the motor shaft and then drill a hole in the exact center of a penny, using a drill bit slightly smaller than the motor shaft. Strive for a press fit between the shaft and the hole in the center of the penny. Now apply epoxy-glue to secure the penny to the mirror shaft. Obtain a one-inch-square front-surface mirror and glue the mirror to the penny. Repeat this procedure for a second motor. Now you will need to secure the motors to a wood or metal base plate with one-inch plumbing pipe hangers, hose clamps, or whatever you can find. The next item you will need is two motor-speed controllers to animate your laser light show.

The schematic in Figure 10-4 illustrates a dc motor-speed controller, the heart of which is a CMOS CD4011 quad Nand gate. The first section of the CD4011 is set up as an oscillator whose frequency is determined by C1 and R2. The second Nand gate then inverts and drives the FET at Q1, an IFR511. The output from the drain pin on the FET is coupled to a DPDT switch at S1, which reverses the 6-volt motor. The motor-speed controller can be powered with a 6-volt lantern battery or any 6-volt power supply. Now you need to build a second motor-speed controller for the second motor. Once both motors are secured, aligned, and wired, you can shine a laser beam onto the first rotating mirror, which will bounce the beam to the second mirror, which in turn will bounce the beam to a viewing screen or wall. By varying the speeds of both motors independently, you can obtain many interesting light patterns on your viewing screen. You can perform great light shows for very little cost.

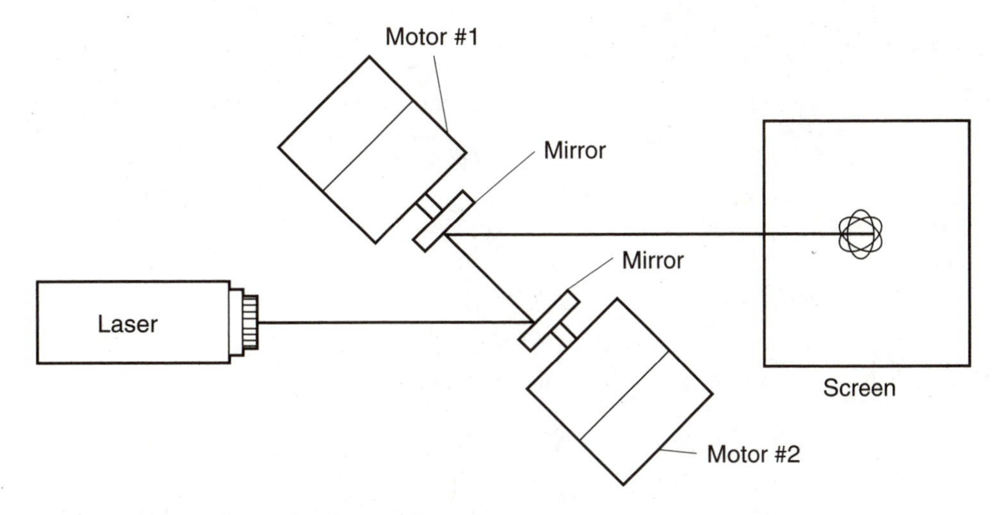

10-3 Reflective laser display maker.

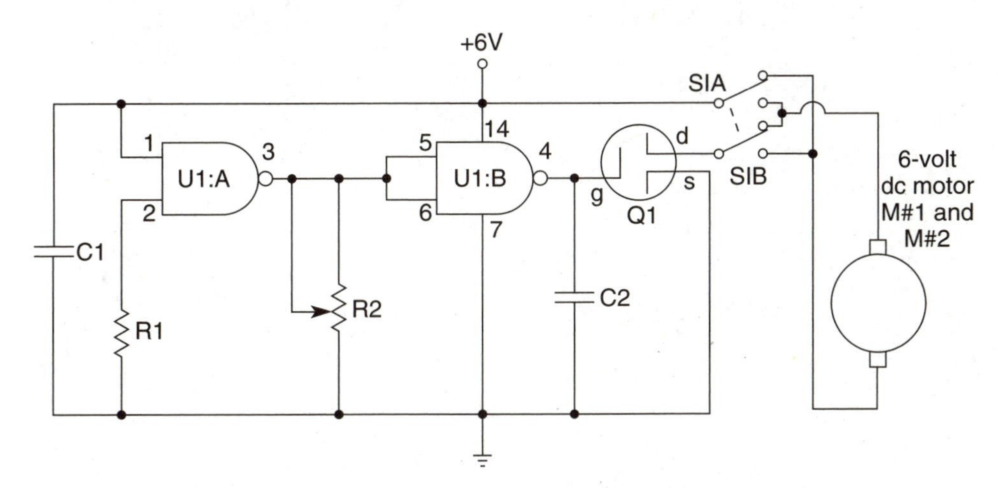

10-4 Motor-speed controller for laser display maker.

Motor-speed controller parts list

R1	1-megohm, ¼-watt resistor
R2	100-kilohm potentiometer
C1	.1-µF, 25-volt disc capacitor
C2	.01-µF, 25-volt disc capacitor
U1	CD4011 quad Nand gate
Q1	IFR511 FET
S1	DPDT toggle switch
M1,M2	6-volt dc motor
Laser	Semiconductor laser or laser pointer
Mirrors	Two front-surface mirrors

Laser pulse viewer

If you want to amaze yourself and your friends just a little bit more, use a piece of double-sided adhesive to fasten a small front surface mirror to your wrist directly over your pulse point, as shown in Figure 10-5. Aim your laser at the mirror and observe its reflection on a nearby wall. Even when you are holding your wrist steady on a table, small movements in the mirror due to your pulse can be easily observed as large movements on the nearby wall (refer to the section *Bar-code scanner principles*, later in this chapter, for a discussion of the optical lever).

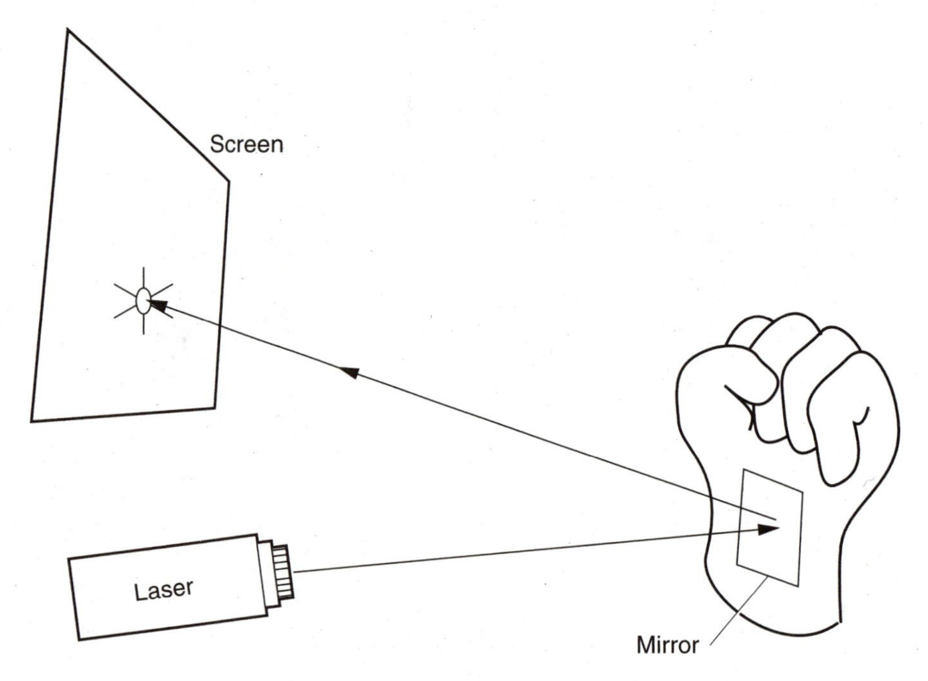

10-5 Laser pulse viewer.

Reflective-light wheel-alignment device

You can use the power of reflected light in another, more serious way, as shown in Figure 10-6. Mount a small front-surface mirror in the center of a wheel, so its surface is perpendicular to the laser beam, as shown. Rotate the wheel and observe the reflected light on a distant wall or screen. If the wheel or shaft is out of alignment, the reflected light will form a circle on the wall; if the wheel is properly aligned, the reflected light will form only a spot of light. This is a simple yet effective means of aligning shafts or wheels.

Bar-code scanner principles

A demonstration of how bar-code scanners work is shown in Figure 10-7. In this measurement-by-scanning application, you mount a mirror or a many-sided prism on

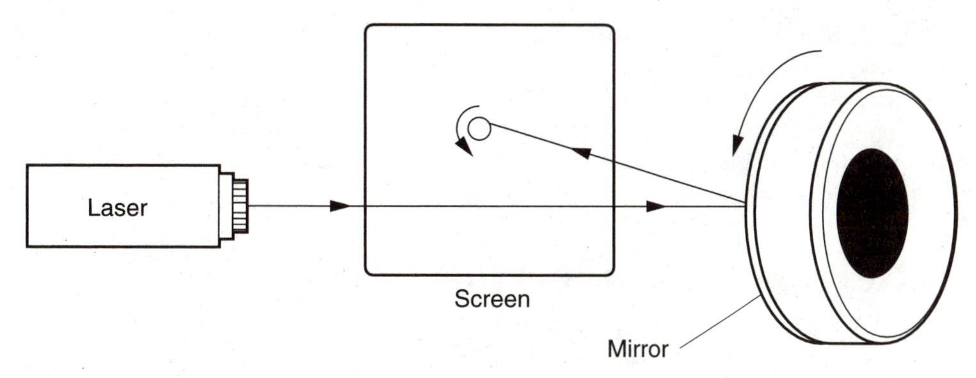

10-6 Reflective-light wheel-alignment device.

a motor shaft that rotates at a constant speed. Them direct a laser beam at the mirror or prism and place a photodetector in the reflected beam path. As the motor rotates the mirror at a known rate, the reflected beam scans the target and its width can be measured by the time the detector is activated. Laser bar-code scanners measure the width of lines or bars on bar-code symbols to thousandths of an inch with this method.

The next few applications of reflective light are based on the optical lever, which is often used for displacement readouts and measurement displays. First you attach a small front-surface mirror to an object whose movement you would like to

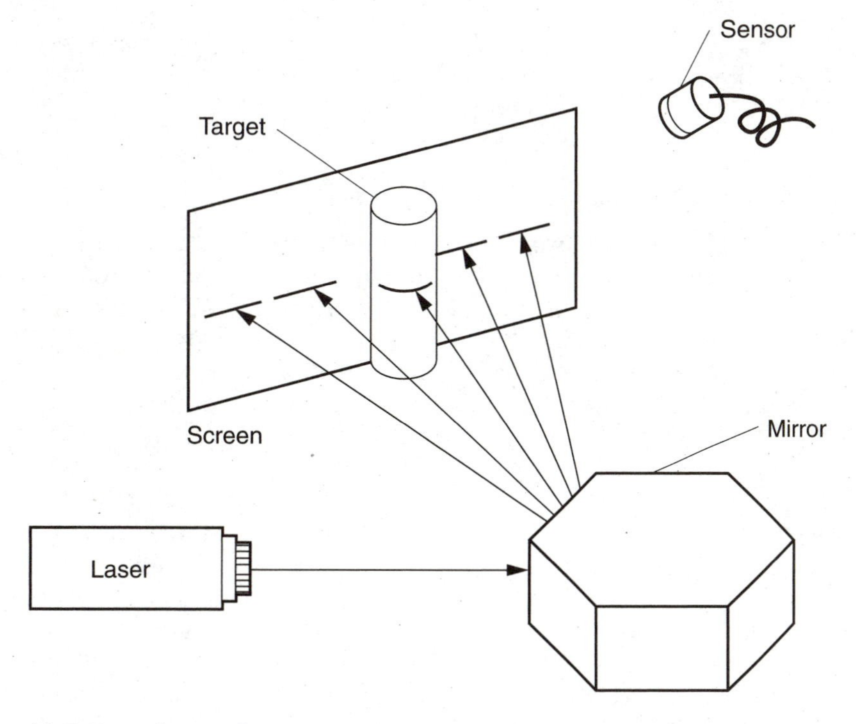

10-7 Laser bar-code scanner.

measure. Then aim a laser beam at the mirror and allow the reflected light beam to fall on a distant screen. Any small movements or changes in the object's position will cause a large displacement of the laser's dot on the screen or nearby wall. The key to successful measurement is a well-calibrated measuring rule on the viewing area and a long distance between the mirror and the viewing screen. As you will see, this type of optical lever can be put to many useful applications.

Reflective-light electroscope

You can use reflective light to produce a reflective-light electroscope, shown in Figure 10-8, which can detect minute differences in electrostatic charges. The reflective electroscope consists of a glass bottle with a piece of insulated copper wire forced through a cork stopper. The copper wire is bent into a circle above the cork and also bent at the opposite end to allow you to hang a thin piece of aluminum foil, which acts as a charge indicator. As an electrostatic charge is brought closer to the electroscope's probe wire, the charge causes the foil to move upward in proportion to the charge. A high electrostatic charge will move the foil upward a great deal, while a small charge will deflect the foil only slightly.

The reflective-light electroscope is simply a basic electroscope with a twist. A semiconductor laser or laser pointer shines a beam onto the electroscope's leaves or foil. A calibrated scale is placed opposite the laser beam to catch the reflected laser beam. When an electrostatic charge is brought near the electroscope, the foil leaves change position, thus deflecting the laser beam onto the scale, as shown. This simple reflective-light electroscope can resolve very small differences in electrostatic charges. Remember that the longer the distance from the electroscope's leaves to the viewing screen, the higher the resolution of the readings, allowing small differences or changes to be easily viewed.

Reflective-light galvanometer

The next reflected-light project is the reflected-light galvanometer, shown in Figure 10-9. A laser galvanometer allows you to measure small differences or variations in electrical current. The laser galvanometer is extremely sensitive and can be constructed for almost nothing. First you need to create the galvanometer coil. The prototype coil consists of 100 turns of #30 magnet wire, which is first wrapped around any nonmetallic form six to eight inches in diameter. The more turns on the coil, the stronger the magnetic field created and hence the more sensitive the galvanometer. Make a wood support for your galvanometer coil by taking two 6 × 8-inch pieces of wood, one inch thick, and nail or screw them together. Then you need to secure the coil on top of the wood support with small plastic cable clamps.

Locate a small, square, front-surface mirror and epoxy a small, strong magnet to the back of the mirror. Then attach a thread to the magnet/mirror assembly and hang it from the center of the galvanometer coil. Shine your laser or laser pointer onto the galvanometer mirror and place a white card or ruled paper opposite the laser spot.

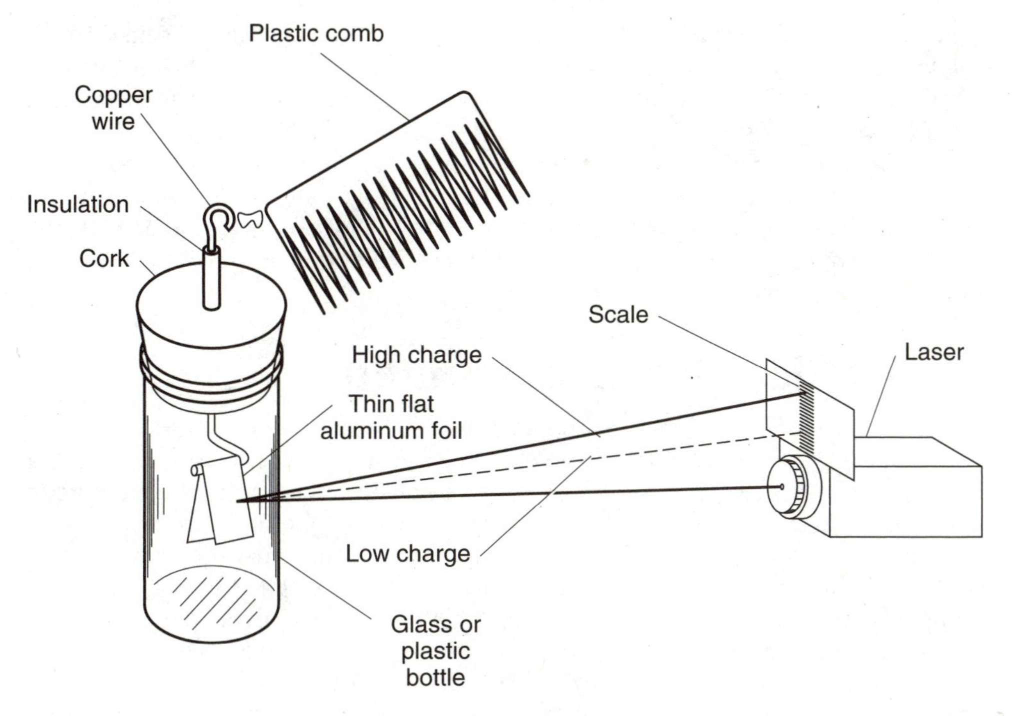

10-8 Reflective-light electroscope.

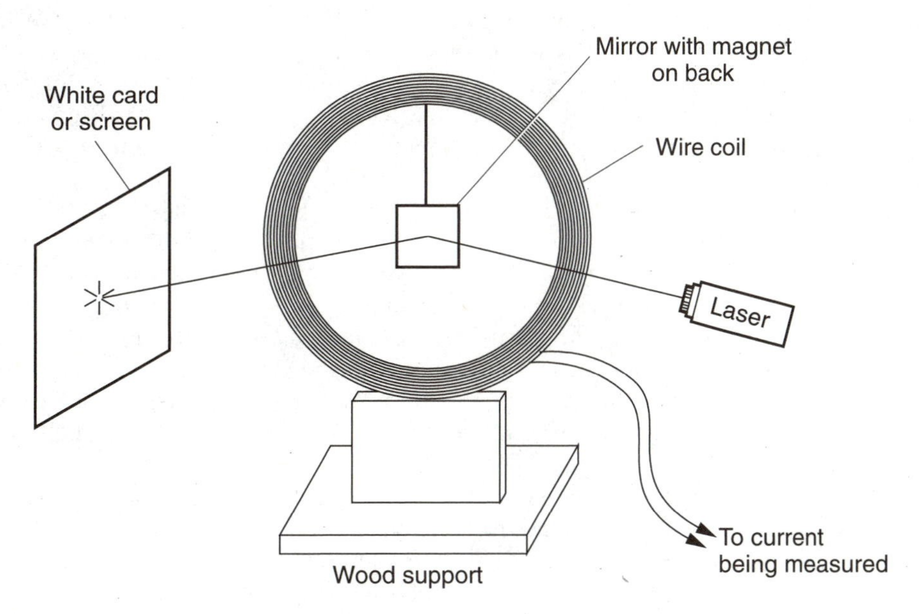

10-9 Reflective-light galvanometer.

For normal operation, the laser is placed about six to eight inches from the mirror, and the distance from the mirror to the view screen is about four feet. Remember the optical lever principle? Now apply a small current to the galvanometer coil, say a 1.5-volt AA battery in series with a 100-ohm resistor. The magnet/mirror assembly should move when current is applied and the mirror should move, deflecting the laser beam. Reverse the leads and the assembly to rotate in the opposite direction. Try to halve or double the current flow through the laser galvanometer. By using the appropriate shunts or multipliers, you can calibrate your new galvanometer to read current or voltage.

Optical jam-jar magnetometer

As mentioned earlier, the Photophone inspired a number of other interesting projects using the reflected-light concept. The drawing in Figure 10-10 illustrates a project in which you can monitor the earth's magnetic field from the comfort of your home. The novel magnetic field detector shown is affectionately known as an optical jam-jar magnetometer. It is a low-cost project that provides excellent results. A small bar magnet with a front-surface mirror attached to it is enclosed and suspended inside a tall jam jar or glass bottle. A front-silvered mirror about 2 cm square is first glued or attached on top of a bar magnet. The suspended mirror/magnet is blackened on either side of a center vertical line and the whole assembly rotates freely on a flexible, small-diameter, glass optical fiber. The original jam-jar magnetometer had a light bulb placed in a metal tube with two razor blades forming an optical slit to allow a sliver of light to project onto the moving mirror. An incandescent bulb, of course, can be replaced by a laser diode or laser pointer mounted firmly at a distance of 20 cm from the mirror assembly. A fixed measuring scale or meter stick is then mounted about two meters away from the jam-jar magnetometer's moving mirror, so reflected light from the moving mirror assembly can be projected onto a nearby wall or viewing screen. The 2-cm to two-meter distance relationship amplifies the small movements of the mirror/magnet assembly.

A simple damping scheme will work to prevent random magnet swing due to eddy currents. Place a copper or aluminum ring at the bottom of the jam-jar magnetometer just beneath the magnet, to act as an electronic brake. The brake is formed by the eddy currents between the magnet and the copper ring.

The key to successfully operating the jam-jar magnetometer is to use a room relatively free of vibration and magnetic interference. The direction of magnetic north relative to the room layout is also important in this project. It is also very important to have ample clear space between the magnetometer and the ruled screen.

So does this simple magnetometer really work? Amateur aurora watchers around the world on all continents have used this type of simple instrument to observe changes in the earth's magnetic field. Suspended magnet-type magnetometers can measure only changes in the direction of the horizontal component of the earth's magnetic field, but this is usually enough to obtain observable readings tied to actual magnetic storms.

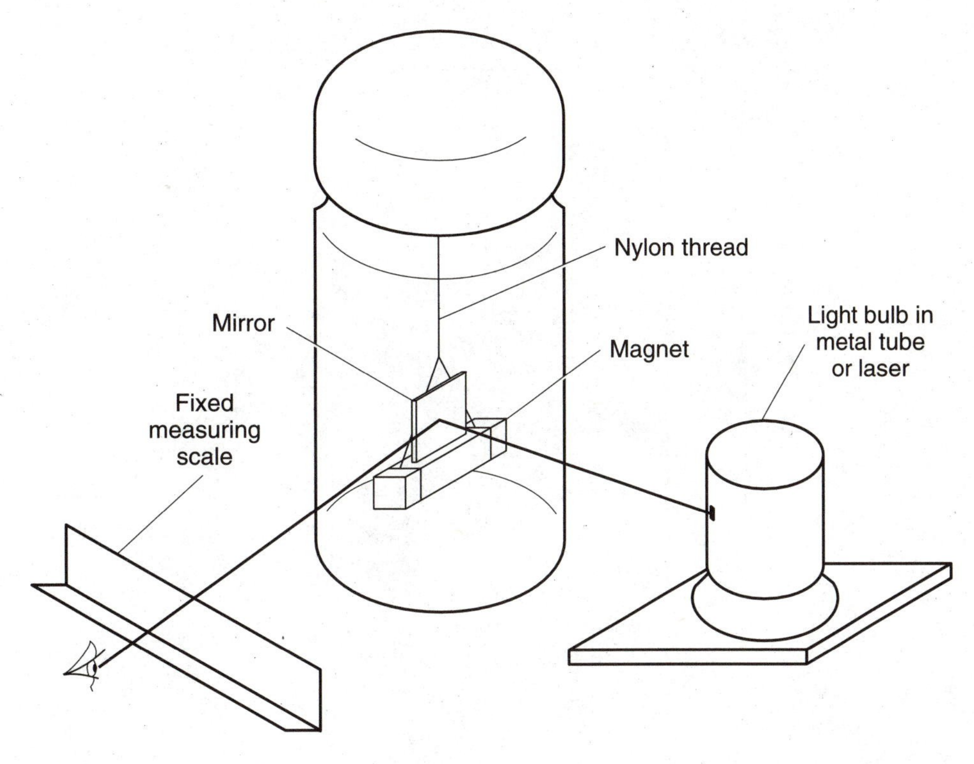

10-10 Optical jam-jar magnetometer.

The geomagnetic field of the earth is closely tied to the condition of the solar wind, to the interplanetary magnetic fields, and ultimately to solar activity, so think of magnetic observations as an extension of sunspot activity. Amateur observers located in temperate or tropical regions can take part in monitoring the magnetospheric dynamo that drives the aurora found on long winter and early spring nights.

Reflective-light laser listener

Surveillance experts and private investigators have discovered the Photophone in yet another configuration. Alexander Graham Bell would never have suspected that his humble Photophone idea could have been turned into a laser listening device capable of eavesdropping on distant conversations inside buildings. Rather than breaking and entering into a building to plant a "bug" or listening device, it is now possible to listen to a conversation remotely through a window.

Sound waves from conversations within a room cause the glass in window panes to vibrate slightly. A laser beam can be bounced off a window of a targeted room, the vibrating window panes modulating the laser beam, which returns to a distant receiver (shown in Figure 10-11). The laser receiver then demodulates the laser beam and the distant room sounds can by heard remotely at another location. It is difficult to align and set up the laser beam and laser receiver, but the rewards are often well worth the effort to private investigators or law enforcement officers.

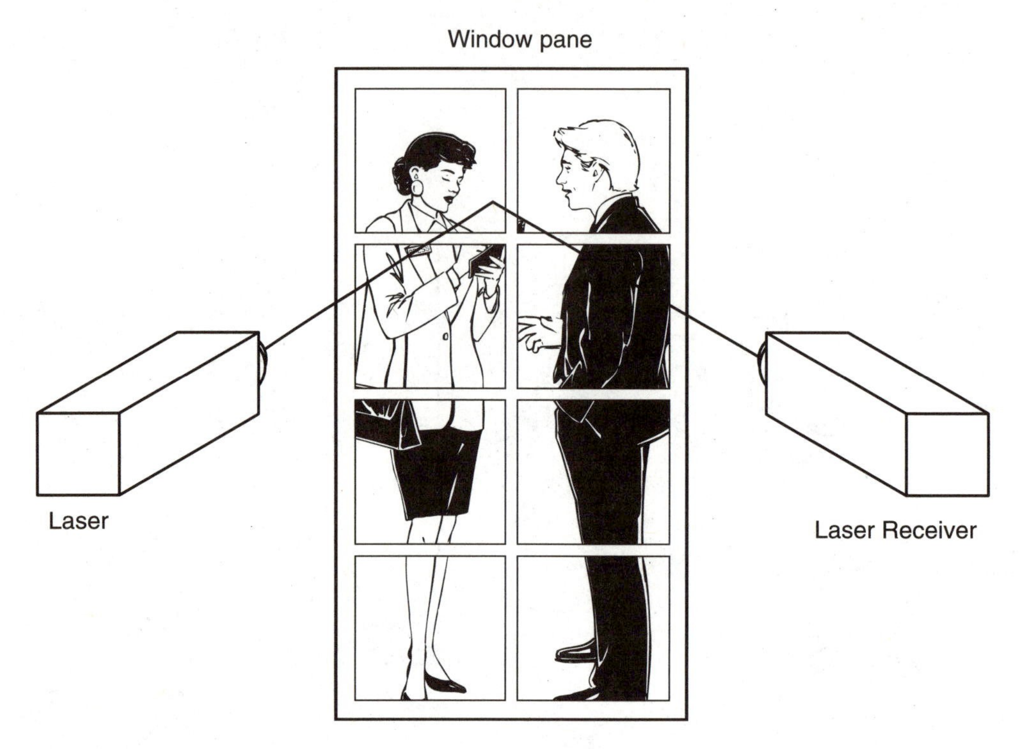

10-11 Reflective-light laser listener.

Commercially available laser listeners use infrared laser beams with an output of up to 35 milliwatts for long-range listening. These power levels can cover very long distances but can also cause serious eye damage, so be cautious.

Short- to medium-range laser listeners can be constructed with visible or infrared laser diodes, while longer-range listeners can be made with helium-neon lasers. Basically, any laser can achieve the desired results, the range of course depending on the power of the laser.

A medium-range laser listener is shown in Figures 10-12 through 10-16. This laser listener uses a visible laser diode for demonstration purposes, a Toshiba 9200 or Hitachi 6700. You can substitute infrared laser diodes such as the CQL60A or equivalent with minimal current adjustment.

I will begin this discussion with the laser diode and driver assembly shown in Figure 10-12, which can be installed in a small-diameter PVC tube, as shown in Figure 10-13. The circuit for the laser driver begins with the batteries at the left of the diagram, two 3.6-volt lithium batteries to power the laser diode driver. You can select other laser drivers and laser diodes for this project, from the many suppliers listed in the Appendix. You must, however, select the proper laser driver to match the laser diode you want to use.

In the laser driver circuit shown, the system is activated when Q2 begins to conduct. Touch switch S1 consists of two small pieces of metallic tape. When bridged, a small amount of current flows into Q1. The collector current of Q1 flows

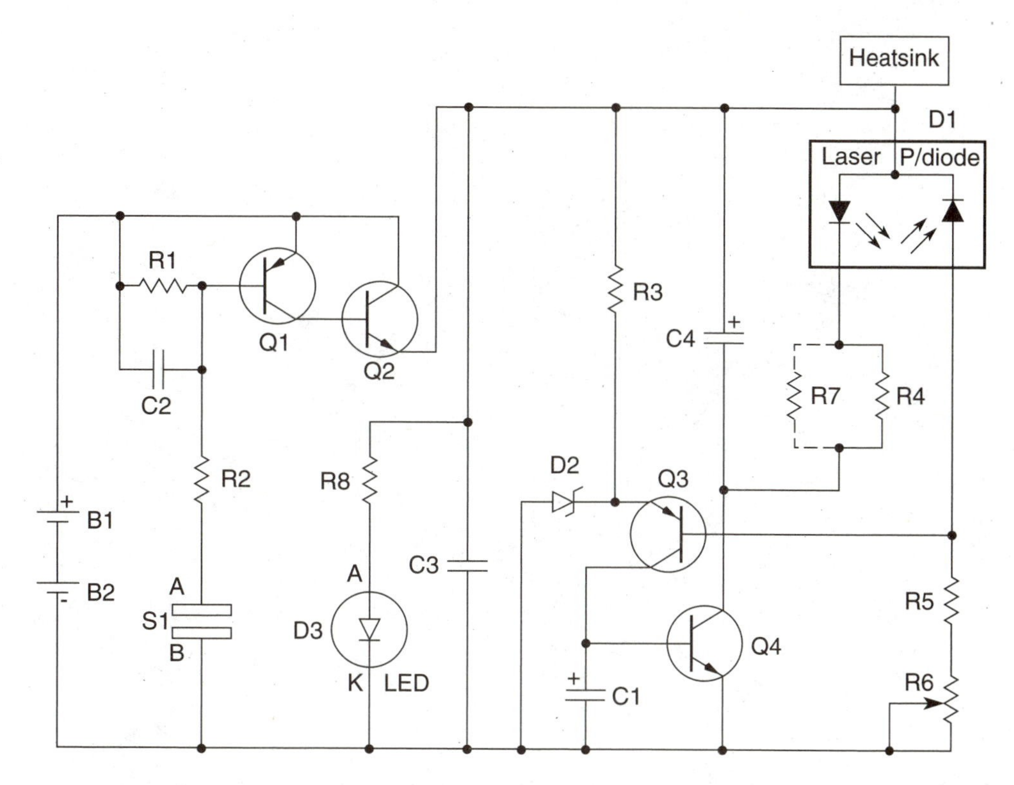

10-12 Laser listener transmitter unit.

into the base of Q2, causing it to saturate and supply current to the laser diode. Base current to transistor Q1 is limited by resistor R2, while resistor R1 and capacitor C2 reduce the circuit's sensitivity to stray ac fields, which might cause the laser to turn on prematurely. Zener diode D2 maintains the voltage across transistor Q3, and R3 limits the zener current. The current through Q4 is controlled by Q3. The collector current of Q3 is the same current supplied to Q4 and is controlled by its base, which is connected across R5 and R6. Current from the photodiode develops a voltage across these resistors that is proportional to the output energy of the laser, which constitutes the feedback required for laser diode stabilization. Increased output causes Q3 to conduct less base current to Q4, resulting in less laser diode current. Potentiometer R6 presents the value of quiescent current, while capacitor C1 limits transients at the base of Q4 and C4 limits them from the Vcc line. The laser diode is actually an assembly that contains a laser diode and a photodiode. The photodiode allows the circuit to monitor the laser diode's output and produce feedback, as mentioned, which is necessary to protect the laser diode from excessive drive currents that could destroy it. The laser diode is connected in series with current-limiting resistor R4 and the collector of Q4. Resistor R7 is used parallel to R4 if another laser diode is selected, since current requirements for other laser diodes might be slightly different. Light-emitting diode LED1 in series with R8 is used as an "on" indicator when S1 is touched.

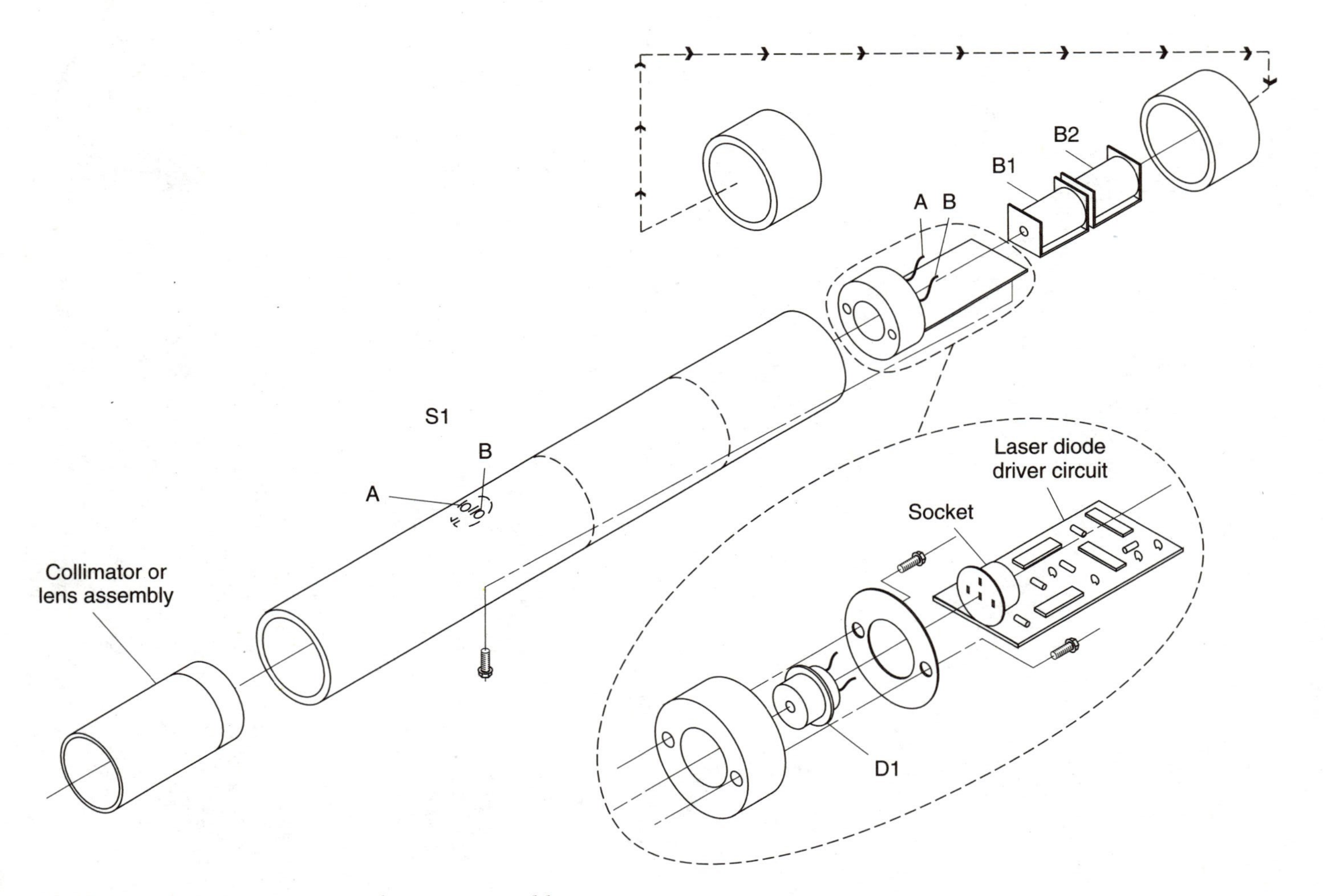

10-13 Laser listener transmitter housing assembly.

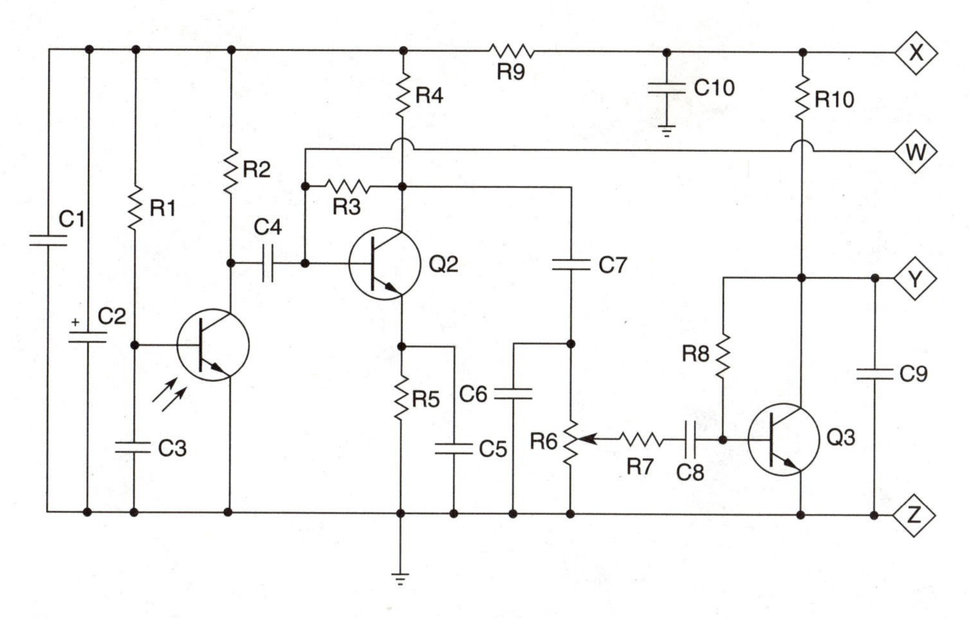

10-14 Laser listener receiver I.

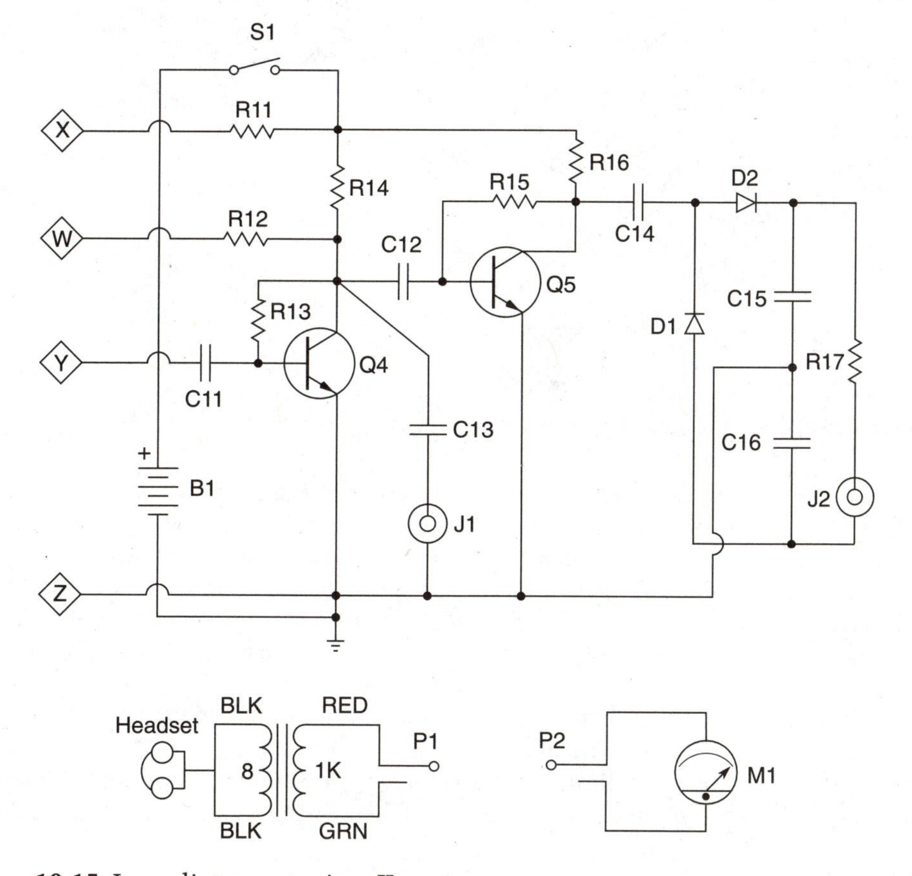

10-15 Laser listener receiver II.

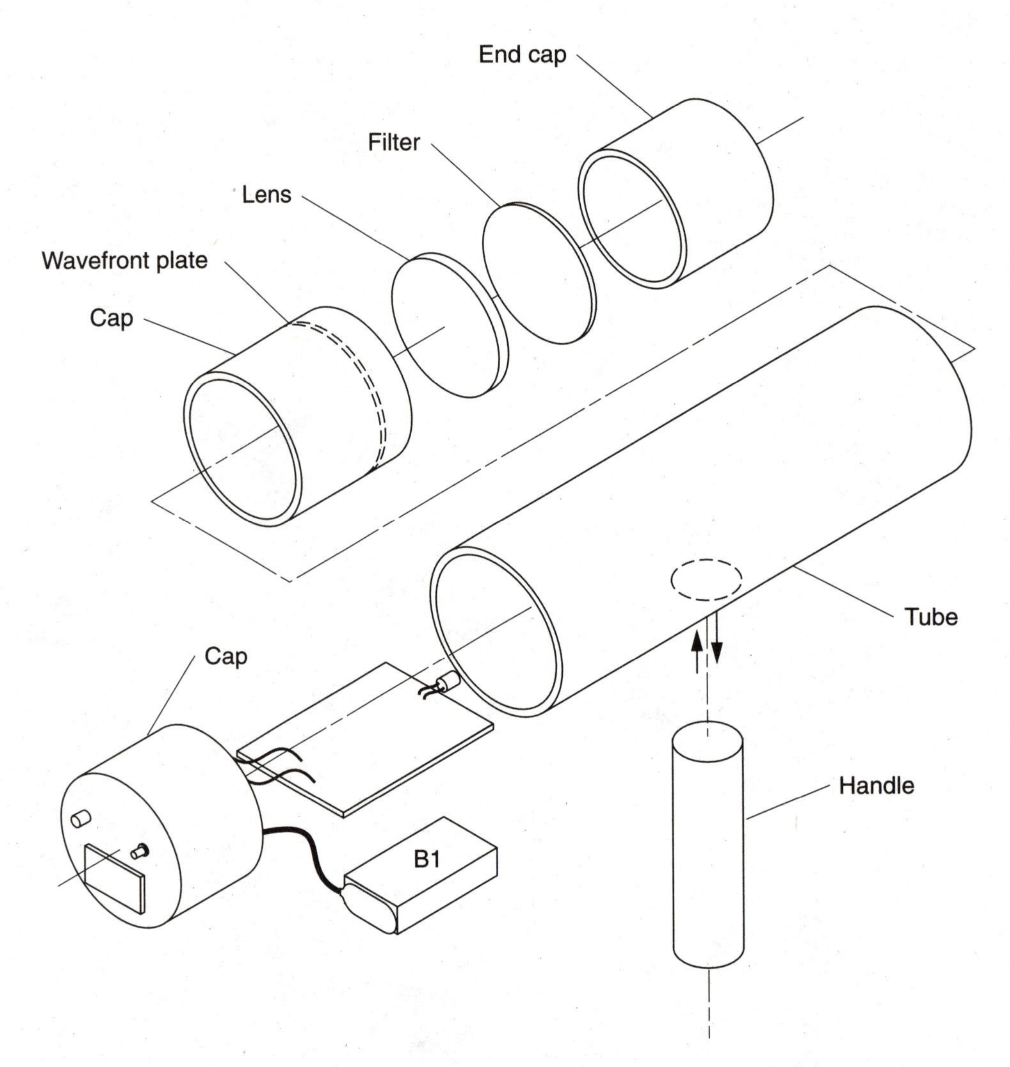

10-16 Laser listener receiver housing assembly.

The assembly of the laser diode driver is shown in Figure 10-13. The enclosure for the laser driver uses PVC plastic pipe. Note that switch S1 is two pieces of aluminum foil or copper sheeting cut into two small squares. You could use a toggle switch instead. The circuit board for the laser driver was fabricated on a narrow $1\frac{1}{2} \times 6$-inch glass epoxy circuit board to which the laser socket was soldered. The two batteries were then mounted behind the circuit card assembly. The collimator assembly must be a bit larger than the mounting tube so the collimator can telescope. Use glue or epoxy to secure a 1 × 6-mm double convex lens inside the collimator tube. The lens should be able to focus near the laser diode. Note that the value of R6 should be 50 kilohms for the Hitachi 6700 series red laser diode, while the value of R6 is 5 kilohms for the Toshiba 9200 series laser diode. If you are using other laser diodes, you might have to change the value of R6 to compensate for the required monitor current. The values of operating and monitor currents vary from laser to laser. You might want to parallel resistor R7 for lasers that require more drive current.

To ensure a long life for your laser diode, measure both the operational current and the photodiode monitor current to make sure you are within the specifications for the particular laser diode. A laser power meter could help you verify that the laser is not overdriven. Table 10-1 illustrates monitor versus operational current for a few typical laser diodes.

If you already own a semiconductor laser or helium-neon laser, skip this part of the project. Note that a helium-neon laser should be used for long-range applications.

The laser listener receiving unit is shown in Figures 10-14, 10-15, and 10-16. The receiving unit uses discrete semiconductors rather than ICs since they provide a simpler approach to learning the function of each component and all the parts are readily available. The laser receiver circuit diagrams are shown in Figures 10-14 and 10-15. The heart of the laser listener receiver is the sensitive phototransistor Q1. Varying light levels across R2 produce a changing voltage level at the collector of Q1. The output of Q1 is capacitively coupled via C4 to the base of the preamplifier transistor Q2. Resistor R3 biases the base of Q2 and sets the gain of Q2. Emitter bias is obtained via R5, with signal current by C5. This combination provides a gain of about 40 volts for the first stage. The amplified output of Q2 at R4 is capacitively coupled to potentiometer R6. From R7, the signal is again coupled capacitively by C8 to transistor Q3. Capacitors C6 and C9 stabilize the circuit by bypassing any unwanted oscillations. The gain of this second amplifier at Q3 is again approximately 40 volts, which is set by R8 and R10. The output of transistor Q3 is then capacitively coupled to Q4 by capacitor C11 (shown in Figure 10-15). The gain of this stage is set, yet again, to 40 volts by resistors R13 and R14. Resistor R12 provides a small amount of degenerative feedback for the system. The output of Q4 is capacitively coupled by C13 to the output jack at J1, which drives a set of headphones via the audio transformer at T1. The 1-kilohm output at J1 is converted to 8 ohms to drive the headphones. The output of Q4 is also coupled capacitively by C12 to Q5.

This final stage has a gain of 10 volts, which is set by R15 and R16. The output of Q5 is now rectified and integrated onto capacitors C15 and C16. The dc

Table 10-1 Laser Diode Comparison.

Laser diode	Output power	Operational current	Montior current
Hitachi 6711	4-5mW	60mA	.1mA
Hitachi 9200	4-5mW	100mA	.5mA
CQL60A (IR)	10mW	100mA	500mA
Toshiba 9211	4-5mW	60mA	1mA

level drives an external meter at jack J2. Resistor R7 limits the current output to .5 mA. You can connect a 100-μA meter to J2, if desired, to assist in aligning the system. Note that capacitor C4 causes the frequency response of the system to "roll off" below 100 Hz. This helps to reduce the 60-Hz signal component from ac power lines.

The completed laser listener receiver unit can be assembled in a PVC tube measuring 9 inches long, with a 2⅜-inch diameter, as shown back in Figure 10-16. One end cap houses a small-meter headphone jack and volume control. The circuit board containing the amplifier/demodulator also carries the phototransistor at one edge of the board, as shown. The phototransistor should come directly up to the lens and filter at the opposite end of the PVC enclosure. The cap on the light-receiving end of the laser listener holds both the lens and filter assembly. This end cap must be open to accept the incoming laser beam. The filter helps reduce interference in high ambient light conditions. The passband of the filter is approximately 1000 nm, which should match the output of the laser, either visible or IR light. The filter prevents Q1 from saturating and becoming inoperable. A 45 × 90-mm lens was used, which is made for looking at distant sources. The length of the PVC enclosure must accommodate the almost four-inch focal length of this lens. The 9-volt battery that powers the laser listener receiver can be stored in the PVC handle, as shown.

The reflected laser light from a distant window pane consists of interference bands, which are best observed by directing the returned light to a white surface. This white surface can be a wavefront plate or a simple white card, placed just behind the working surface of the phototransistor.

The reflected light from the laser beam bouncing off a distant window pane contains interference bands, as mentioned. This is primarily due to the phase interference that occurs on relatively flat surfaces. A slight motion or distortion from this window surface will cause these interference bands to vary in position, which is precisely what is needed to modulate the phototransistor. It is very difficult to properly position the phototransistor relative to the correct view of these interference bands. A slight change can cause a tremendous difference in reception. You will notice that it is not absolutely necessary to view the laser beam at an exact right angle to the reflection surface. Any reasonable acute angle will produce the interference bands needed to receive the distant laser beam.

A good sturdy camera tripod is a must for both the laser listener transmitter and receiver. Ideally, both the sender and the receiver would be positioned so the reflected beam would have nearly an equal axis. This might be difficult to achieve since you have two planes, azimuth and elevation, to deal with. Try to position the test reflecting or vibrating surface to achieve the coincidental axis alignment.

You might want to experiment with a few other methods of receiving the reflected light beam. An effective remote reflector for experimental use is a small front-surface mirror attached to a speaker cone inside the target room you want to monitor. The movement of the speaker cone from the remote room sounds will cause the remote mirror to modulate the laser beam, assuming of course that you can place a small speaker in the target room. In applications where double-pane glass is used, it is possible to aim the laser beam through the double glass to the front surface of a glass china cabinet in the target room. Another approach is to aim your laser at a hanging picture frame within the target room. A small reflective dot placed on the china cabinet glass or picture frame glass would greatly enhance the reflective beam.

The key to effectively using the laser listener is patience and perseverance in adjusting the reflected laser beam for optimum results. You can learn much from the various experimental approaches to using reflective laser light in this project. I, however, do not condone or recommend using a laser listener for other than educational or experimental uses.

Laser listener sender unit parts list

R1	5.6-megohm, ¼-watt resistor
R2	1-kilohm, ¼-watt resistor
R3,R5	470-ohm, ¼-watt resistor
R4	15-ohm, ½-watt resistor
R6	5- or 50-kilohm trimmer (see text)
R7,R8	100-ohm, ¼-watt resistor
R9	27-ohm, ¼-watt resistor
C1	10-µF, 25-volt electrolytic capacitor
C2	.1-µF, 25-volt disc capacitor
C3	.01-µF, 25-volt disc capacitor
C4	1-µF, 25-volt electrolytic capacitor
D1	Laser diode (see text)
D2	1N5221 zener diode
D3	LED
Q1,Q3	PN2907 NPN transistor
Q2,Q4	PN2222 NPN transistor
S1	Aluminum/copper foil switch
B1,B2	3-volt lithium batteries and 6-volt Duracell DL123H
Misc	Laser socket, circuit board, 1 × 6-mm double convex lens, battery holder, PVC tube, and focus collimator tube

Laser listener receiver/demodulator parts list

R1	100-megohm, ½-watt resistor
R2,R4,R10,R15	10-kilohm, ¼-watt resistor
R3,R8	390-kilohm, ¼-watt resistor
R5,R14,R16	1-kilohm, ¼-watt resistor
R6/S1	10-kilohm potentiometer and switch
R7	2.2-kilohm, ¼-watt resistor
R12	5.6-megohm, ¼-watt resistor
R13	39-kilohm, ¼-watt resistor
R17	22-kilohm, ¼-watt resistor
R9,R11	220-ohm, ¼-watt resistor

Laser listener receiver/demodulator parts list continued

C1	470-pF, 25-volt disc capacitor
C2,C10	100-µF, 25-volt electrolytic capacitor
C3,C9	1000-µF, 25-volt electrolytic capacitor
C4	.05-µF, 25-volt disc capacitor
C5	10-µF, 25-volt electrolytic capacitor
C6	.01-µF, 25-volt disc capacitor
C7,C8,C11	2.2-µF, 25-volt nonpolar electrolytic capacitor
C12,C13,C14	2.2-µF, 25-volt nonpolar electrolytic capacitor
C15,C16	1-µF, 25-volt electrolytic capacitor
Q1	L14G3 ultra-sensitive phototransistor
Q2,Q3,Q4,Q5	PN2222 NPN transistor
D1,D2	1N914 silicon diode
J1	RCA phono jack
J2	3.5-mm phono jack
P1	RCA phono plug
M	100-µA micro-ammeter
T1	1-kilohm to 8-ohm matching transformer
Lens	54/89 double convex lens, 45/89
Filter	2-inch IR filter, cut from a 2 × 2-inch gel filter
B	9-volt transistor radio battery
Misc	9-inch × 2⅜-inch-diameter PVC tube, PVC end caps, PVC handle, and battery clip

11
CHAPTER
Lasers

The laser has changed our society in a great many ways since its development in July of 1960. Albert Einstein first proposed the idea of a laser in 1916, but the spark lay dormant until the first practical ruby laser was developed by Ted Maiman, a researcher at Hughes Laboratories. Einstein reasoned that when a photon hit an atom that was excited at a high energy state, a proton of light identical to the first proton would be released. If enough atoms could be excited, the photons hitting them would be greatly increased. This would then lead to a chain reaction where photons would hit atoms to make new photons, and the process would continue until the original source was removed. Raising atoms to a high energy state is now often referred to as *pumping*. Atoms can be pumped in a variety of ways, include electrically, thermally, and now chemically.

Early lasers consisted of a synthetic ruby rod with mirrored ends. A spiral strobe lamp was placed around the ruby rod and the flash tube was connected to a high-voltage power supply. The power supply sent short pulses of high-energy electricity to the flash lamp. The lamp would flash on and off, rapidly bathing the ruby rod in white light. Chromium atoms give the ruby its red color, absorbing just blue and green light. The absorption of these other colors, in fact, raises the energy of the chromium atoms. As the flash tube went on and off, photons would go into an excited state in just milliseconds. Since the ends of the ruby rod were mirrored, some of the photons would bounce back and forth. After being bounced to and fro and amplified via the two mirrors, photons were emitted from the half-silvered mirror at one end of the ruby rod laser. As the atoms were pumped back and forth in a high energy state, a "population inversion" occurred, which is when the number of high-energy atoms exceeds the number of low-energy atoms. This inversion process must occur for a laser to operate.

I will limit the discussion of lasers in this chapter to the helium-neon gas laser and the semiconductor laser. These two types of lasers are relatively inexpensive and easy to obtain, either new or from the surplus market. Helium-neon lasers emit a bright, deep red glow at 632.8 nm, which is perfectly suited to most laser experiments. The helium-neon laser possesses four main properties. Its light is both monochromatic and spatially coherent, which means the crests and troughs of the light waves are in step. It also exhibits temporal coherency, meaning that the laser light waves are emitted in even, precise, and accurately spaced intervals. The light from a helium-neon laser is also collimated. Since its light is coherent and monochromatic, the resulting laser beam does not diverge as much as ordinary light waves.

A helium-neon laser is generally made of a glass vessel filled with 10 parts helium to 1 part neon. The tube is then pressurized to 1 mm/Hg. Electrodes are placed at both ends of the tube to provide a means of ionizing the neon gas, which in turn excites the helium and neon atoms. Mirrors mounted at both ends of the laser tube form an optical resonator. The mirror at the rear of the laser tube is fully silvered, while the one at the output end of the laser tube is partially silvered. Actually, most laser tubes are comprised of two glass tubes, an outer vacuum or plasma tube that contains the helium-neon gas, and a shorter smaller-bore tube where the "lasing" action is performed. The bore is attached to only one end of the outer laser tube, and the loose end of the inner bore tube is held in place by a metal element called a "spider" at the output end of the laser tube (see Figure 11-1).

The internal facing mirrors in the laser tube are commonly called Fabry-Perot resonators and they are aligned parallel to each other. In actual practice, a confocal resonator is used, which is formed from two concave spherical mirrors instead of the flat mirrors.

Helium-neon or HeNe lasers are available in three basic varieties: bore, cylindrical, and self-contained. The bore type is ideal for fitting into confined places. It is somewhat dangerous, however, since all high-voltage leads are exposed. Cylindrical-head lasers are generally more protected, but tend to be larger and not easily mounted into a hand-held device. They are ideal for optical bench and holography experiments. Self-contained lasers are protected against tube breakage and electrocution, but they also tend to be larger than bore types and are generally used in school labs for experimentation.

Another major component of the helium-neon laser is its high-voltage power supply. Laser power supplies usually take 117 volts ac or 12 volts dc and produce from 1200 to 3000 volts dc at 3.5 to 7.5 mA. A helium-neon laser kit is shown in Photo 11-1.

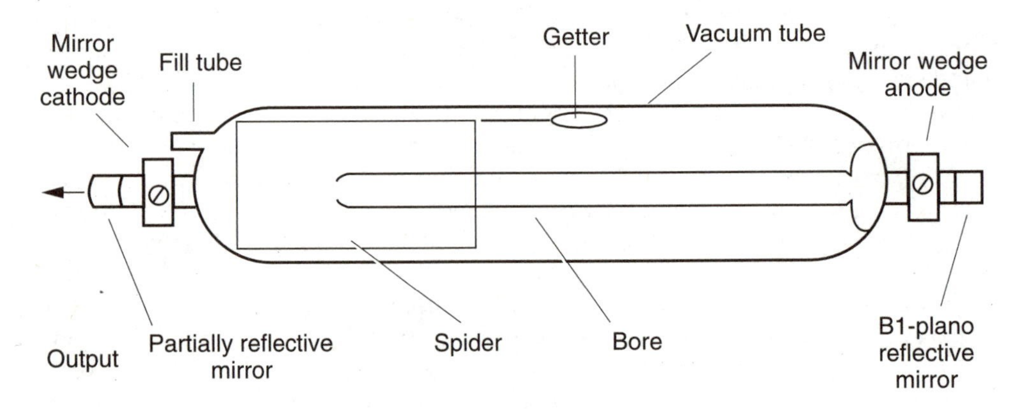

11-1 Helium-neon laser tube.

You can obtain laser power supplies from a number of different sources. You can purchase commercially made laser power supplies, either ac-to-dc or dc-to-dc type. You can also obtain a variety of surplus high-voltage laser power supplies, available at reasonable prices, from the sources listed in the Appendix. There are also a number of high-voltage power supply kits in the $30 to $120 range. Or you can build your own power supply from scratch for about $20 to $40. Salvaged high-voltage TV flyback transformers and HV modules can often be obtained quite inexpensively, and can be used to create a laser power supply. A simple guide to laser power-supply requirements dictates that a .5-milliwatt laser tube generally requires 900 volts dc at 3.5 mA, and a 2-millivolt laser tube requires 1400 volts dc at 5.6 mA. At the higher end of the spectrum, a 7-milliwatt helium-neon laser requires a 2400-volt power supply at 7 mA of current.

Pulse-width-modulated laser power supply

The pulse-width-modulated (PWM) laser power supply, shown in Figure 11-2, can power most 1- to 5-milliwatt helium-neon laser tubes. The PWM laser supply is built around an LM555 IC timer, which is configured as a pulse-width modulator. Potentiometers R12 and R13 are used to adjust the width of the LM555's output pulses. Resistors R8, R9, R12, and R13 as well as C5 determine the pulse width of the LM555. Operationally, R12 and R13 are set to their center position and relay RL is deenergized, thus removing R8 and R13 from the circuit. As 12 volts dc is applied, the LM555 triggers and Q1 is activated. This, in turn, drives transformer T1, which steps the 12 volts up to 1000 volts dc. Capacitors C7 through C10 and diodes D4 through D19 form a four-stage voltage multiplier that increases the 1000 volts to about 3500 volts. Adjust potentiometer R12 to increase the duty cycle of the LM555

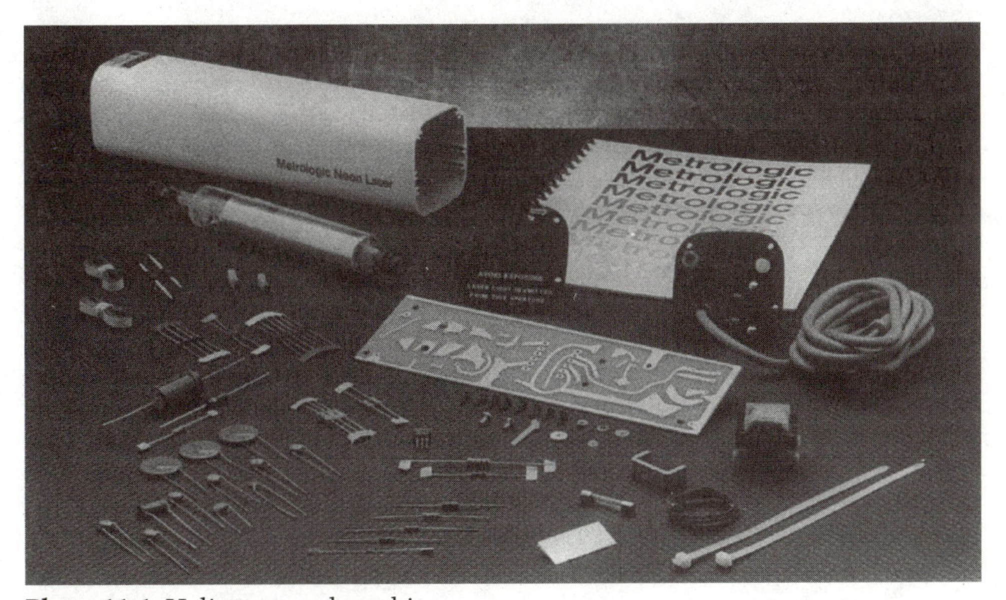

Photo 11-1 Helium-neon laser kit.

if the laser tube does not light. When the tube begins to "lase," sensing resistors R3 through R6 trigger Q2, which energizes the relay. Note that a shorter duty cycle decreases current to the tube, and lengthening the duty cycle increases current to the laser tube. You need the ballast resistor at R11 to determine the amount of current applied to the laser tube. The ballast resistor shown is a 75-kilohm, 3- to 5-watt resistor, but you can experiment with other resistor values to best suit your particular laser tube. Connect a milliampere meter in series with the ballast resistor to ensure that the current to the tube does not exceed 6 to 10 mA. The completed laser supply can be powered with a 12-volt lantern battery or a D-cell battery pack.

Semiconductor lasers have become ubiquitous in our modern world, and there are two basic types. Single-heterostructure (SH) laser diodes are thought of as older, and can operate only in the pulse mode if cryogenically cooled. SH laser diodes are capable of multiwatt outputs, but pulses must be kept to 200 nanoseconds or shorter. Double-heterostructure (DH) laser diode can operate in both continuous-wave (CW) and pulse modes. DH lasers are also capable of multiwatt output if operated in the pulse mode. Most DH laser diodes are designed for CW output in the 1- to 10-milliwatt range. It is interesting to note that the first laser diodes were developed concurrently with the ruby laser discussed earlier.

PWM laser power supply parts list

R1,R2	100-ohm, ¼-watt resistor
R3,R4,R5,R6	22-megohm, ¼-watt resistor
R7	3.9-kilohm, ¼-watt resistor
R8	1-kilohm, ¼-watt resistor
R9	220-ohm, ¼-watt resistor
R10	10- kilohm, ¼-watt resistor
R11	75-kilohm, 3- to 5-watt ballast resistor
R12,R13	2-kilohm potentiometer
C1	.1-µF, 50-volt disc capacitor
C2	4.7-µF, 50-volt electrolytic capacitor
C3,C4	10-µF, 50-volt electrolytic capacitor
C5	.01-µF, 50-volt disc capacitor
C6	.06-µF, 50-volt disc capacitor
C7,C8,C9,C10	.15-µF, 3-kilovolt capacitor
C11	.47-µF, 50-volt disc capacitor
D1,D2,D3	1N914 diode
D4-D19	3-kilovolt HV diode
Q1	TIP-146 (on heatsink)
Q2	2N2222
U1	LM-555 timer IC
RL	12-volt SPST relay
T1	Step-up transformer, 9-volt primary and 375-volt secondary
HS	Heatsink for Q1
MISC	Chassis box and switch

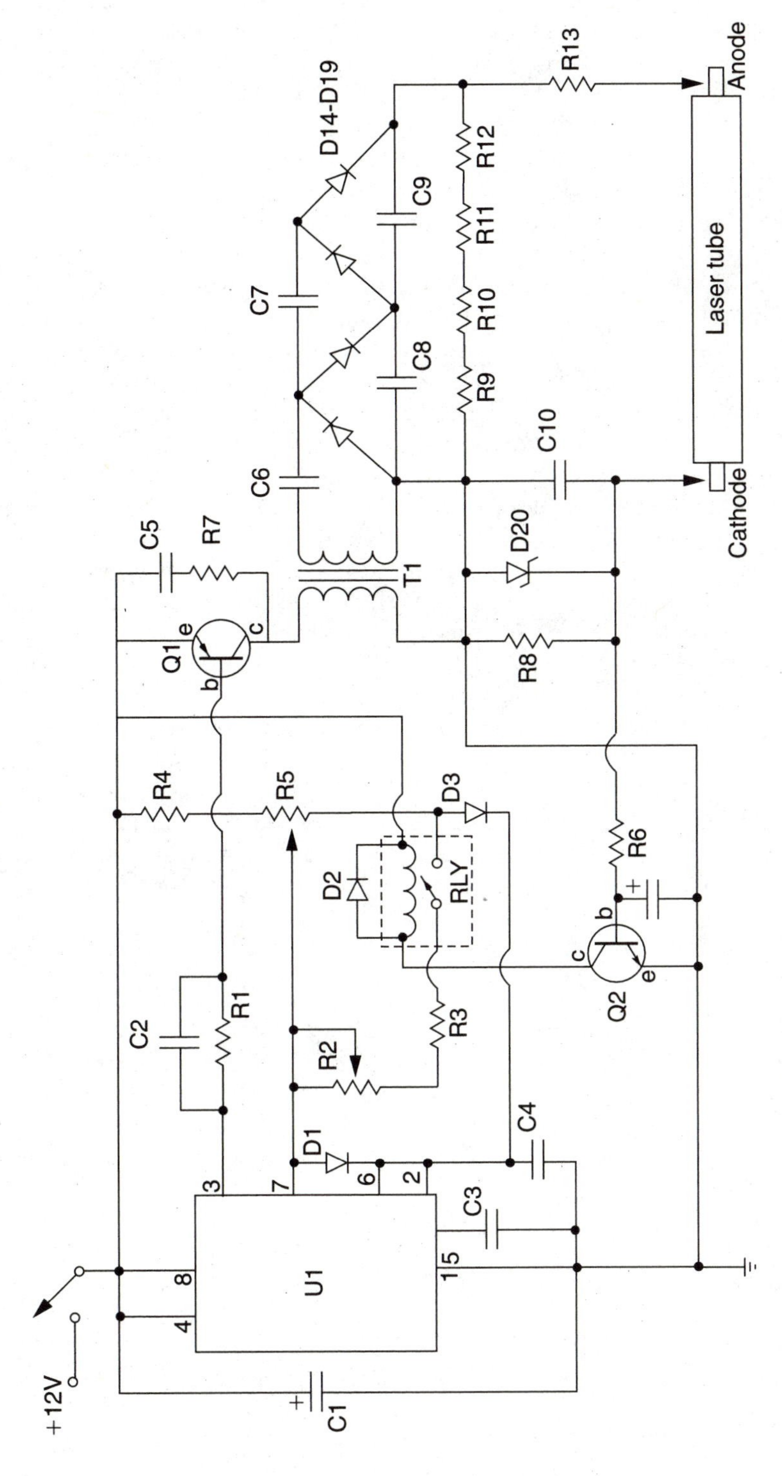

11-2 Helium-neon pulse-width-modulated power supply.

A basic laser diode is shown in Figure 11-3. The semiconductor or injection laser is composed of a PN junction similar to the junctions found in transistors and LEDs. The double-heterostructure laser diode is made from sandwiching a GaAs junction between two AlGaAs layers. The sandwich helps to confine light generated within the chip and allows the laser diode to operate in the CW mode. You can alter the wavelength of light by varying the amount of aluminum in the AlGaAs material. The output of these laser diodes varies between 680 and 900 nm.

When current is applied to the laser diode chip, light is produced at its junction, but at this point the light output is not coherent. The cleaved faces shown in Figure 11-3 are partially reflective mirrors that bounce the emitted light back and forth within the junction area. Once amplified, the light exits the laser diode chip and is temporally and spatially coherent, with a 10- to 30-degree elliptical beam. All laser diodes are susceptible to temperature changes. As the temperature of a semiconductor laser increases, the chip becomes less efficient and the light output decreases. Most CW laser diodes incorporate a monitor or feedback loop to monitor the temperature or output power of the laser diode and adjust the current correspondingly. Temperature sensing is an elaborate affair, and most laser diodes incorporate a built-in photodiode as a monitoring device. The photodiode is placed at the output side of the laser diode to sample the output power. Most often the photodiode is connected to a comparator circuit to monitor the current changes of the laser diode. This feedback circuit tracks changes and can adjust the current applied to the laser diode. Laser diodes generally have three leads, as shown in Figure 11-4. Either the anode of the laser diode is connected to the cathode of the photodiode, or the cathodes are grouped together. This allows the laser diode to be connected to a control or monitoring circuit. Follow the connection diagrams for your particular laser to avoid damage.

Laser diodes generally require a heatsink to keep them cool. You can construct your own heatsink in a number of ways. One method is to use a clamp-type fuse holder. Simply place the laser diode into the fuse holder and press firmly. You can

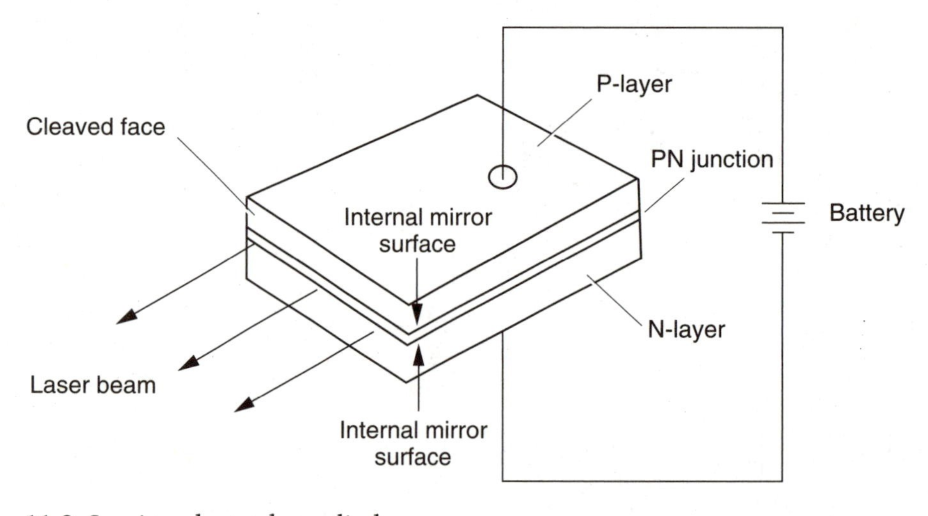

11-3 Semiconductor laser diode.

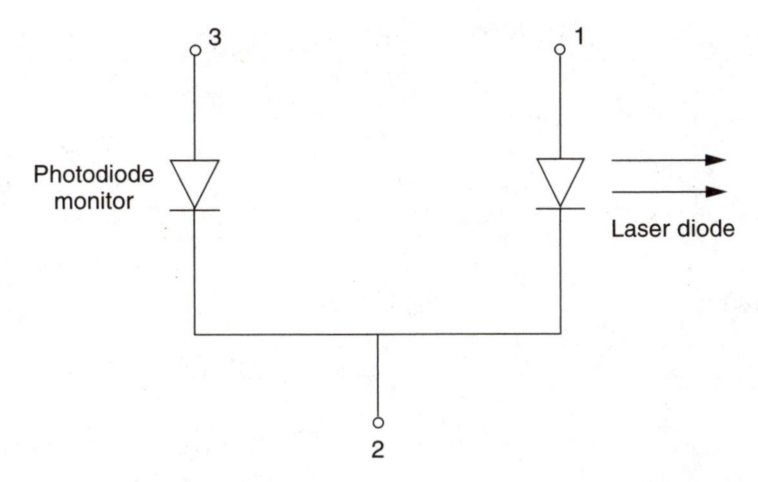

11-4 Semiconductor laser diode with photodiode monitor.

then use the screw holes at the base of the fuse holder to secure the assembly. A second approach is to use a flexible copper retaining ring to hold the laser diode to a TO-220 heatsink. Use heatsink compound between the laser diode and the heatsink.

Virtually all the light radiated from the front face of a laser diode can be collected and collimated into a thin beam with a simple lens arrangement. The newest-generation CW laser diodes emit beams with a coherence that rivals that of helium-neon lasers. The Sharp LT026MD CW laser, for example, is inexpensive and only about $25 (see the Appendix). Consumer applications such as compact disk players and bar-code scanners have contributed to a plentiful supply of surplus double-heterostructure laser diodes. Laser pointers are another tiny version of the semiconductor laser that have become very popular in the last few years due to falling prices. The laser pointer shown in Photo 11-2 could be used in a number of short-range experiments and demonstrations.

Semiconductor laser power supply

The circuit shown in Figure 11-5 illustrates the Sharp IR3CO2 laser driver chip. This low-cost (under $1.50), eight-pin DIP driver can readily drive and monitor DH lasers. The current output terminal of the IR3CO2 on pin 1 is connected to the anode of a Sharp LTO2OM/LTO22MC laser diode through a series resistor, as shown. The series resistor limits the current through the laser diode. Resistor R1 controls the current delivered to the laser at pin 1. When the current is operated for the first time with a new laser, make sure to adjust R1 for the highest resistance. As the resistance is decreased gradually, the current to pin 1 is increased slowly to "break in" the new diode. Pin 5 of the laser driver chip allows the laser to be gated on and off. When pin 5 is connected to 5 volts, the laser is on, while a zero or ground on pin 5 turns the laser off. The chief drawback to this circuit is the –5-volt supply. You can construct a portable laser drive circuit with a plus and

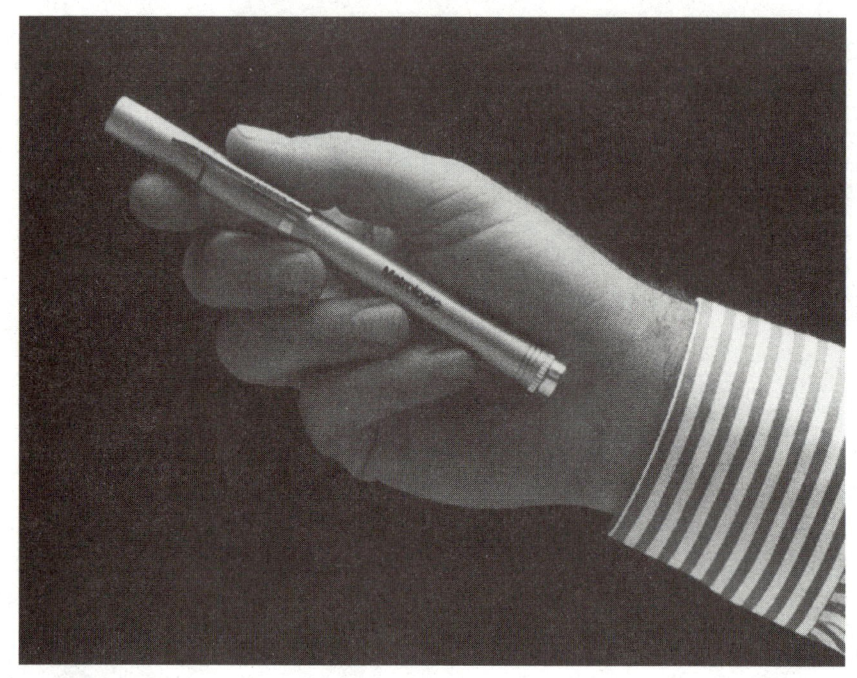

Photo 11-2 Laser pointer.

minus battery supply connected to a 5-volt regulator or two 6-volt lantern batteries with a series resistor, as shown.

A second laser power supply is shown in Figure 11-6. In this discrete component supply, an op-amp functions as a high-gain comparator. It checks the current from the photodiode monitor; as the current increases, the output of the op-amp decreases and the output of the laser drops. Resistors R1, R4, and R5 determine the gain of the circuit. In this circuit, an RCA C86002E laser diode is used. The output of the op-amp is fed to two transistors that are cascaded together with R6, a 30-ohm, 10-watt resistor, in the emitter path to –12 volts dc. A single power transistor such as a TIP-120 could be substituted for both transistors shown, and you can readily modify the circuit for use with another laser diode.

So far I have discussed only visible-light laser diodes, but there is another realm of laser diodes. Infrared laser diodes are used for many applications, including security systems and communication links. IR laser diodes operate exactly the same as optical lasers, but the issue of laser safety becomes much more important when using IR lasers because the beam cannot be readily seen.

Lasers are divided into six classifications, depending on their power output, emission duration, and wavelength. Class I lasers are considered harmless. A class IIa laser is confined to wavelengths between 400 and 710 nm, with a power output of up to 1 milliwatt. The beam cannot be on or be motionless for a period of greater than 1000 ns. A class II laser also operates between 400 and 710 nm, but it cannot be on or motionless for a period of greater than $\frac{1}{4}$ second and the power can be as high as 1 milliwatt. Most helium-neon lasers whose outputs are less than 5 milliwatts are

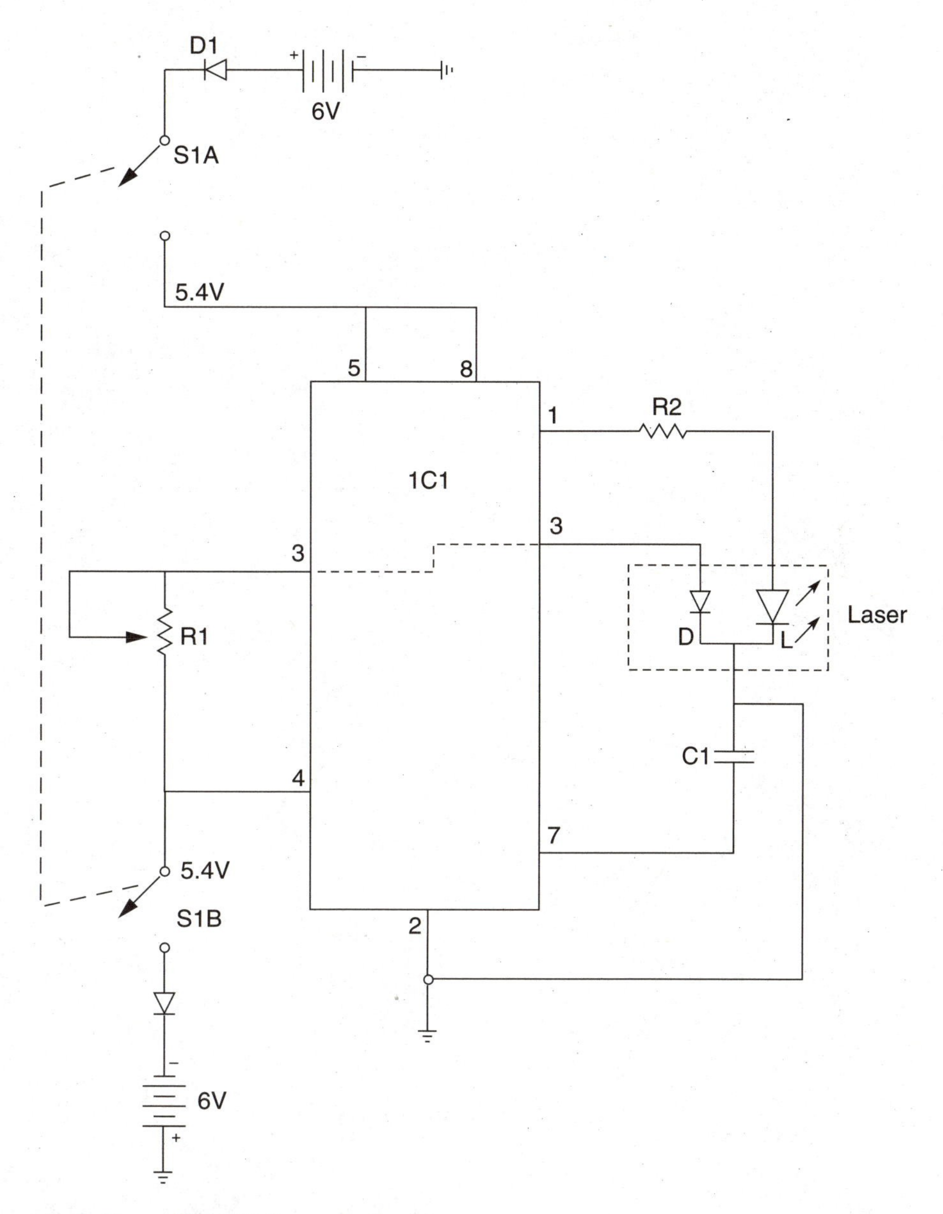

11-5 Sharp IR3C02 laser diode driver.

classified as class 111a lasers. Class IIIb lasers have a power output from 5 to 500 milliwatts in the visible spectrum and less power for an invisible beam. A class IV laser is usually dangerous to the eye and is reserved for multiwatt lasers, which are usually used for cutting. It is important to observe laser safety precautions since a 20- to 50-milliwatt laser beam can cause temporary and in some instances permanent eye damage. Eye damage from 5- to 10-milliwatt lasers is rare, but this should not give you a sense of false security. Wear laser goggles for any class IV laser beam, without

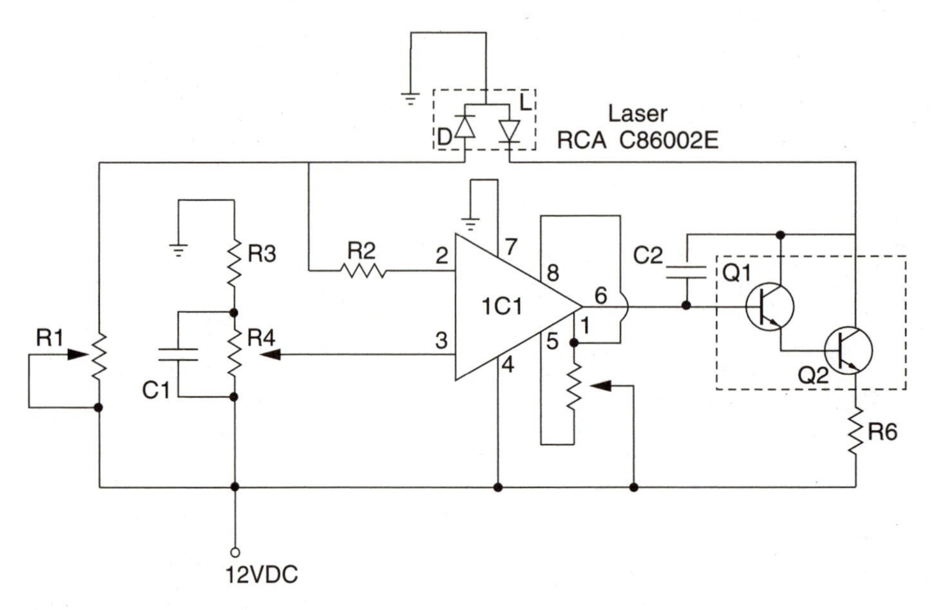

11-6 Op-amp laser diode driver.

Op-amp laser driver parts list

R1,R5	100-kilohm potentiometers
R2	10-kilohm, $\frac{1}{4}$-watt resistor
R3	3.3-kilohm, $\frac{1}{4}$-watt resistor
R4	10-kilohm potentiometer
R6	30-ohm, 10-watt resistor
C1	100-µF, 25-volt capacitor
C2	.1-µF, 25-volt disc capacitor
Q1	2N2101 transistor
Q2	2N3585 transistor
1C1	RCA 3130 op-amp
Laser	RCA C86002E
Misc	Chassis box and power switch

question. Also, follow simple safety precautions to avoid eye damage. Never look directly into a laser beam and be careful when laser beams are bounced from one location to another or when a beam is scattered.

High-voltage power supplies also pose a risk of electrocution. While most hobby lasers cannot cause death because the currents involved are generally quite small, the real danger exists with high-power lasers. High-voltage and high-current supplies pack quite a "jolt" and should be enclosed to prevent the potential hazards of shock and death.

When purchasing a laser, consider that the intensity of a laser beam is a function of the output power, which generally ranges from .3 to 8 milliwatts. Those with the lowest power are usually reserved for general-purpose classroom use, such as pointers, holograms, and experimentation. The brighter lasers are of higher power, so they are better for large lecture-hall demonstrations and research that requires detailed observation and measurements of diffraction patterns and interference fringes. Brighter laser beams also mean better holograms because of the shorter exposure intervals. You can use a laser power meter like the one shown in Photo 11-3 to determine the output of a laser.

The next consideration when selecting a laser is the beam diameter. If the laser is going to be used for a research project that involves beam shaping, knowing the beam-diameter specifications will be helpful when you are planning for accessories. The larger the original beam diameter, the less expansion is required to illuminate a given area and the greater is the intensity when the beam is focused to a fine point with an additional lens.

Beam divergence is another concern when selecting a laser. An ideal laser has a fine beam with perfect parallel edges. In reality lasers do come close, but most have a small amount of divergence that makes the beam spread out and lose power with distance. If you are going to use a laser in a laboratory without collimating optics, then it is important to choose a laser with the smallest divergence available.

Polarizing helium-neon lasers with fixed internal mirrors produce a beam that is elliptically polarized. If the laser you select is to be used for research or demonstrating the polarizing plane rotation that occurs when the beam is transmitted through certain liquids or gasses, then choose a laser with the longest beam because they offer the best stability.

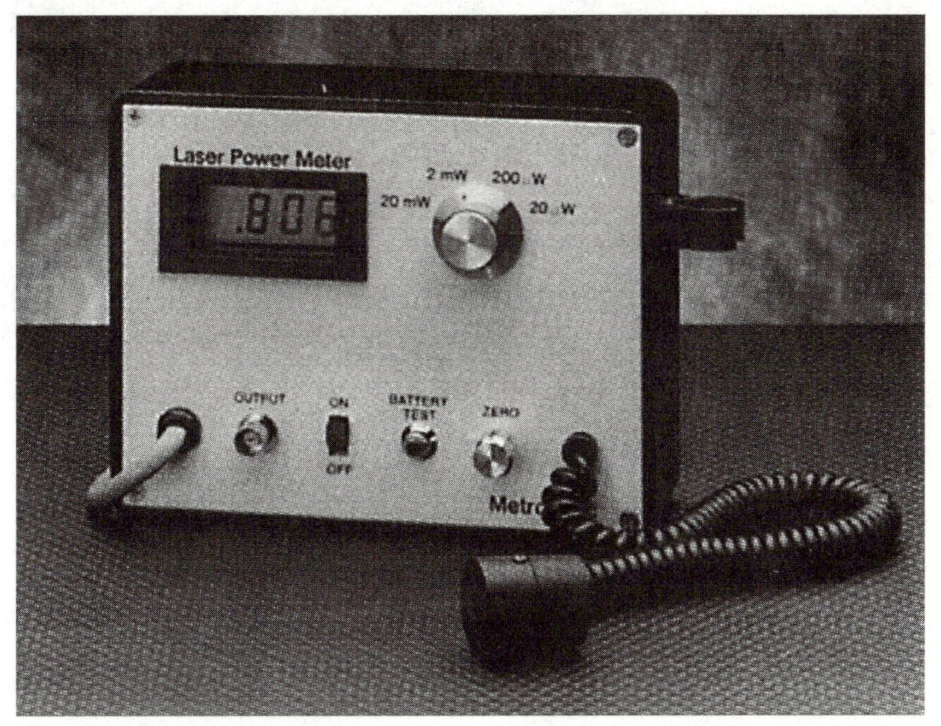

Photo 11-3 Laser power meter.

Ruggedness and longevity are also often a concern, and most helium-neon lasers use hard-sealed tubes and solid-state power supplies for long life and ruggedness. Semiconductor lasers are by nature more rugged than helium-neon lasers or other glass-tube varieties. If you carefully observe current limits, then both laser types should have long, useful lives. If you are going to use your laser for communications or holography, then consider a laser that can be modulated or easily modified.

Once you have purchased or constructed your laser, what happens next? First you should know a little about beam shaping and collimation. Collimating a laser beam is often necessary if you plan to send your laser beam over a distance of tens to hundreds of meters. To minimize the adverse effects of beam spreading, plan to collimate the laser beam. First you will need to spread the beam out with a diverging lens and then make the edges of the beam parallel with a converging lens. By adjusting the distance between diverging and converging lenses, you can collimate the beam so it remains at a constant width of about 2 cm over distances of several kilometers (see Figure 11-7). Generally several lenses are paired together to focus or collimate a laser beam. You can make a compound lens from two simple lenses, and predict the compound lens's focal length if you know or can determine the focal length of each lens and the distance between them. If you know the values, you can use the following lens maker's formula:

$$1/F = 1/F_1 + 1/F_2 - d/F_1F_2$$

If either of the two lenses used is divergent, use a negative focal length for this lens when substituting values into the formula. If the calculated answer for the combined focal length is a negative number, this indicates that the lens system has a virtual rather than real focus. Check out the accuracy of you calculations by actually putting the lens into a laser beam path, as shown in Figure 11-8.

Simple laser experiments

Now let's take a look at some simple laser experiments. First we will try some beam spreading. Locate a glass stirring rod, which will serve as a cylindrical lens. Place the glass rod horizontally in front of the laser beam, as shown in Figure 11-9. This setup will spread the laser beam out so it forms a line of light rather than a round spot. This technique is often used in saw mills to provide guidelines for cut-

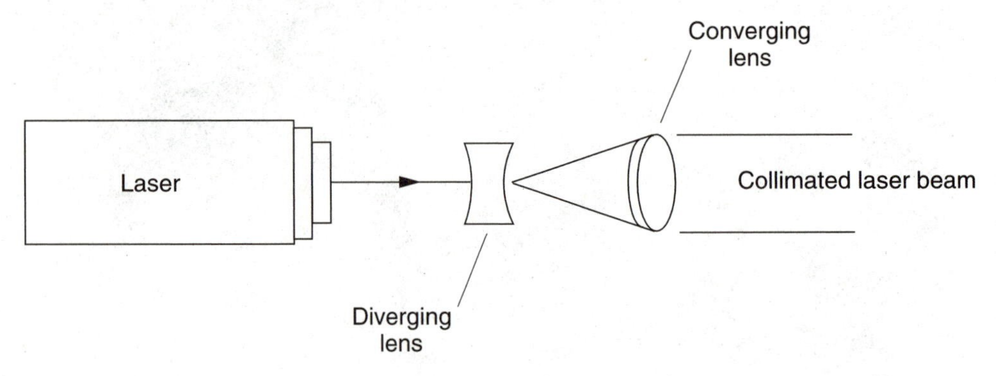

11-7 Laser with diverging and converging lenses.

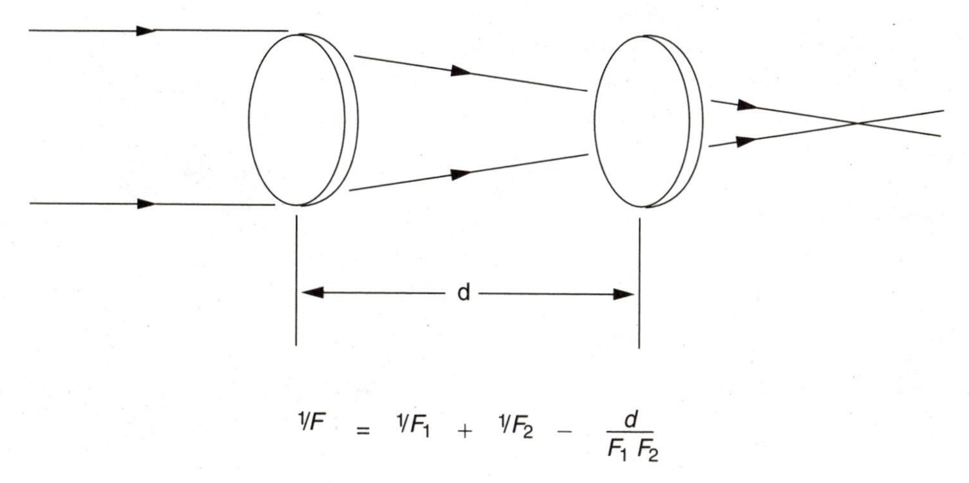

$$1/F = 1/F_1 + 1/F_2 - \frac{d}{F_1 F_2}$$

11-8 Lens calculations.

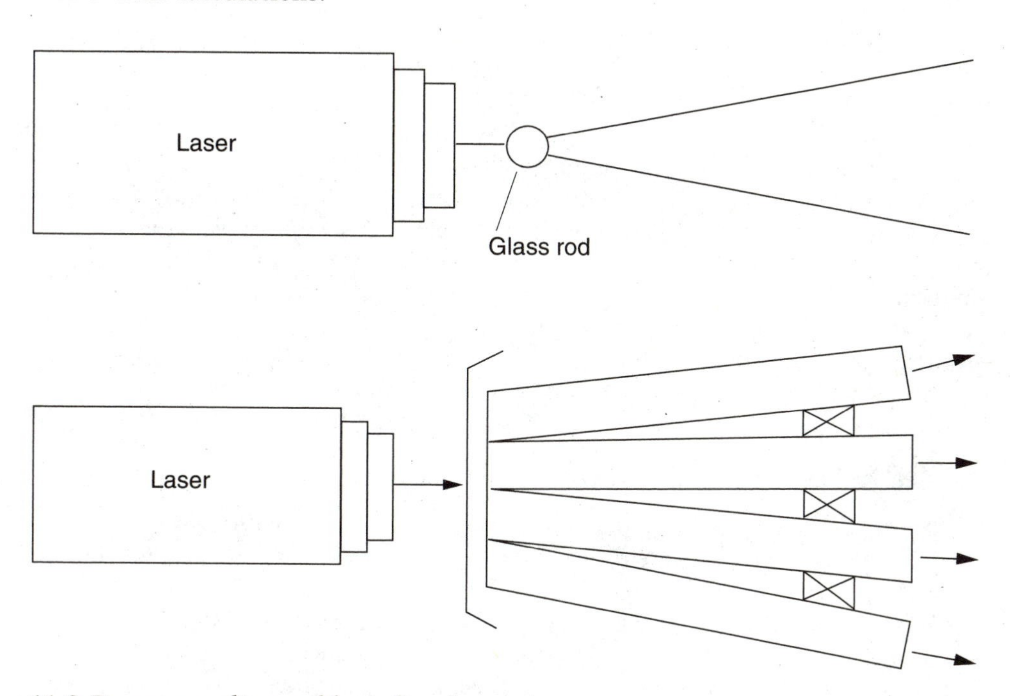

11-9 Beam spreading and laser line formation.

ting lumber. Commercial laser line generators are commonly available through surplus suppliers and can often be located in discarded laser printers.

Another simple experiment you can perform is to split a single laser beam into many different beams. The second diagram in Figure 11-9 depicts multiple microscope slides fastened together. Tape all the microscope slides together at one end. At the opposite end, place small wedges of cardboard between each slide and then tape the bundle together. Now send a laser beam from the unwedged side and multiple laser beams will emerge, one from each of the microscope slides.

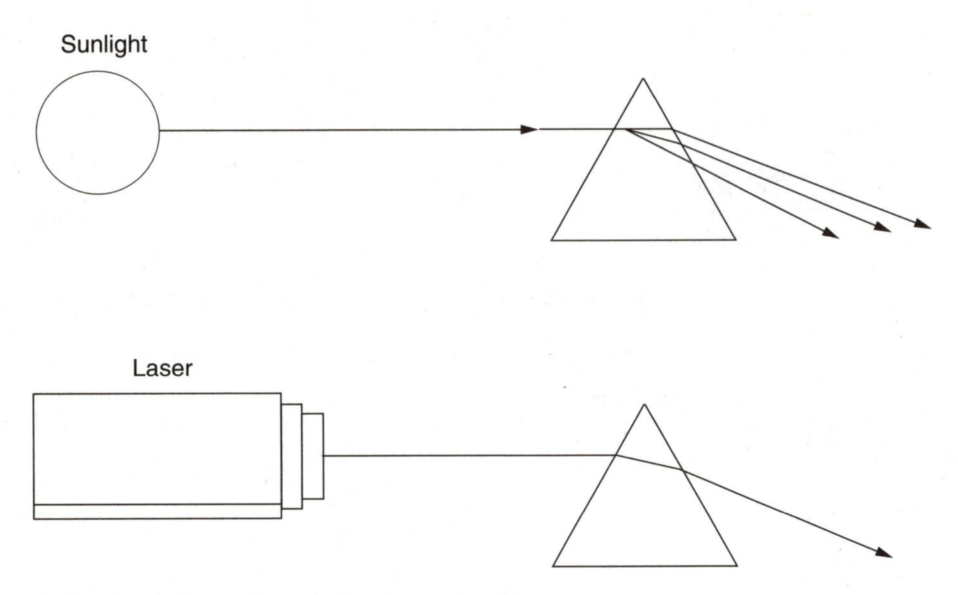

11-10 Sunlight vs. laser light through a prism.

The next experiment demonstrates the monochromaticity of laser light. First form a narrow beam of white light by passing sunlight between two razor blades. Allow this sunlight beam to pass through a prism and a spectrum of colors will appear, as shown in Figure 11-10. Since there are no prominent wavelengths other than red emitted from most lasers, the laser beam cannot be divided into colors. To observe this, pass a laser beam through the prism and you can verify the single color emitting from the prism. Note that while most lasers produce only one wavelength, some lasers can generate several closely related but different wavelengths.

Index of refraction

Next we will observe the index of refraction through a liquid with varying optical density. If the optical density of a liquid or gas varies, then a light beam passing through it will gradually bend as the beam is transmitted through the fluid. You can perform this experiment by partially filling a fish tank with clear water and then adding several large lumps of sugar or a number of sugar cubes. The sugar solution will gradually become less dense at the surface and more dense at the bottom of the tank. Aim your laser beam horizontally into the side of the fish tank and observe how the beam gradually bends as the sugar solution's index of refraction increases (see Figure 11-11).

Color reflection and absorption

The diagrams in Figure 11-12 illustrate color reflection and absorption. Color filters can act as selective absorbers of light; for example, most green filters are strong absorbers of red light. If you were to pass a red laser beam through a green filter, you

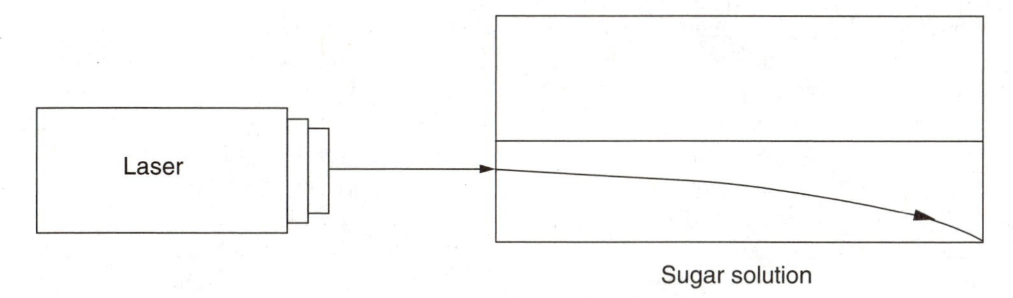

11-11 Laser beam through a sugar solution.

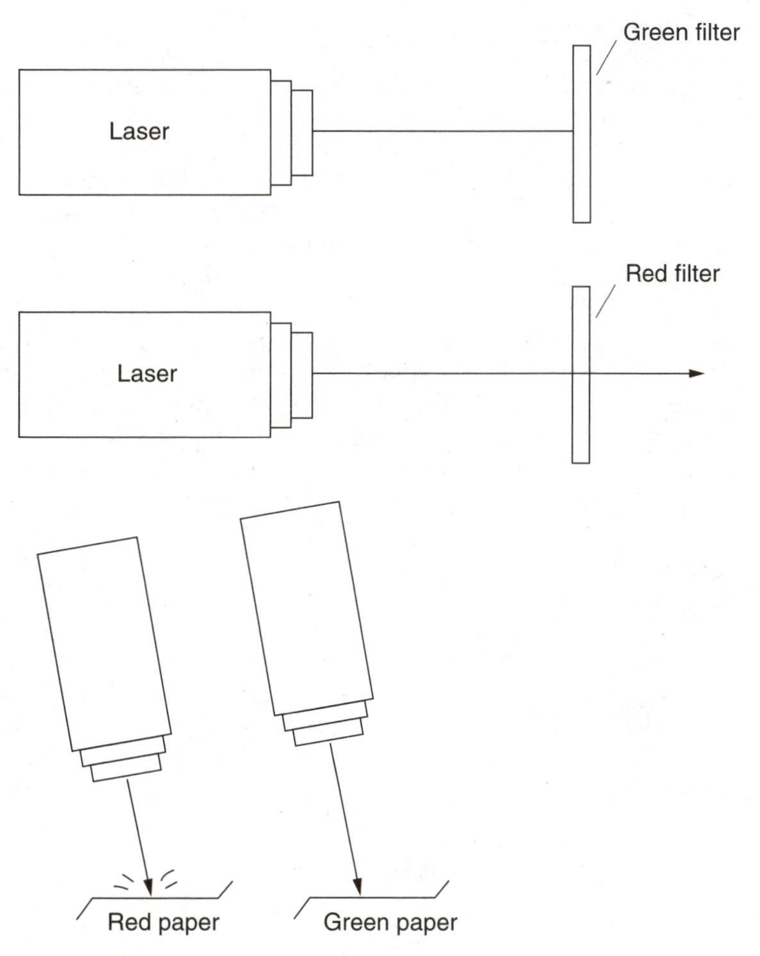

11-12 Color reflection and absorption with a laser.

would notice very little light emerging from the filter. On the other hand, you can pass a laser beam through a red filter quite easily. Some colors are absorbed and others are not. Try experimenting with colored cellophane or colored plastic to observe this effect.

Opaque materials selectively reflect different colors. You can demonstrate this by expanding a laser beam with a diverging lens and then holding various colored papers in front of the laser beam. Observe how some papers reflect light better then others. Red paper will reflect the beam, for example, and green will not. Consider bar-coded store packages. If the bars are printed with black ink, the background of the package must appear bright when illuminated with a red laser light. Only a tiny portion of the original laser light is reflected from the symbol on the package back to the photodetector. This small amount of light is separated from ambient light with a red filter. The red light passes through the filter and is bright to the detector, while the background is dark.

Reflection and refraction

Objects such as mirrors provide specular reflections; that is, they change the direction of the laser beam without scattering or diffusing the light. Rough objects provide diffuse reflections that scatter light. Shine your laser beam at various objects and plot the reflection strength versus the angle. Notice how an aspirin tablet reflects 100 percent of the light over a large angle. The laser offers convincing proof that light actually bends, or is diffracted, around small objects. You can demonstrate knife diffraction by pointing a laser beam at a screen, such as a sheet of glossy white paper, about three meters from your laser. Slide the edge of a new razor blade partway into the laser beam and observe the interference patterns on the white paper screen. A close observation will show that there is now a sharp shadow of the edge of the razor blade on the screen; in fact, there is a diffraction pattern consisting of a series of light/dark fringes parallel to the edge of the razor blade.

Now take two razor blades, as shown in Figure 11-13, arranged so the laser beam can pass between them. Observe the effects when you bring the two razor blades

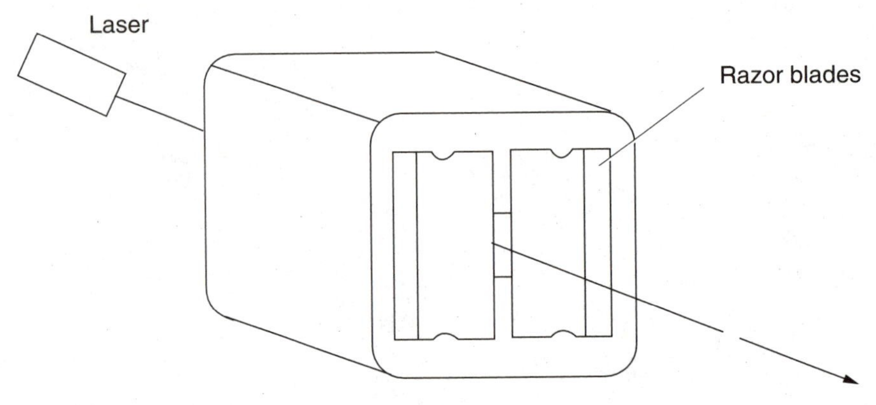

11-13 Laser beam diffraction through a razor slit.

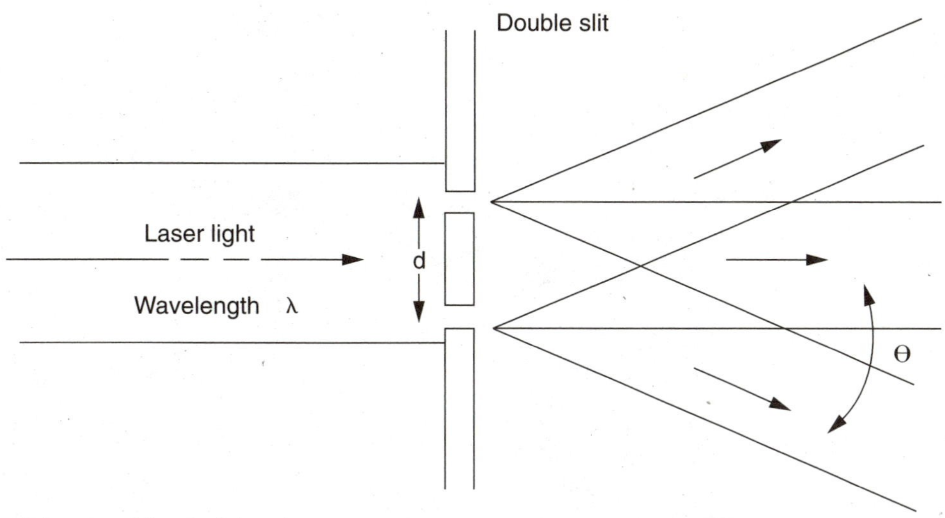

11-14 Double slit/interference viewing.

closer together to form a narrow slit. Observe the variations in intensity of the fringes and the distance between them as the razor blades move closer together. You can also perform this experiment by taping a razor blade to either side of a pair of vernier calipers so you can easily move the blades together.

When a laser beam is sent through two narrow parallel slits, each slit produces an identical diffraction pattern. If the slits are close together, the diffraction patterns overlap and a phenomenon called *double-slit interference* occurs. You can investigate the characteristic diffraction and interference patterns by varying the width and spacing of the two slits (see Figure 11-14). For best results, use a Cornell diffraction slit plate or a diffraction slide mosaic, which can be obtained from Metrologic (see the Appendix). You can sketch a typical diffraction pattern and explore the intensity of illumination at various locations by using a photometer (refer to Chapter 4).

An interesting diversion and good demonstration is the Schlieren effect, or variable index of refraction, which can be seen in Figure 11-15. Place your laser on a flat table and place an expanding or diverging lens in the path of your laser beam. Place the flame of a bunsen or propane burner near the expanded laser beam and observe the image of the flame on a nearby slide projector screen. The heated and rising convection air currents cause interesting variations in the surrounding air's index of refraction. These moving shadows on the screen, known as the Schlieren effect, are very interesting to observe. Schlieren photography would be an interesting follow-up project.

If you are getting really interested in lasers and want the challenge of more complex experiments and projects, you might want to obtain an optics lab such as the one shown in Photo 11-4, which contains many of components needed to conduct laser experiments. You might also consider purchasing an optical bench, shown in Photo 11-5. You use it to align lasers and optics when conducting more complex experiments or holograms.

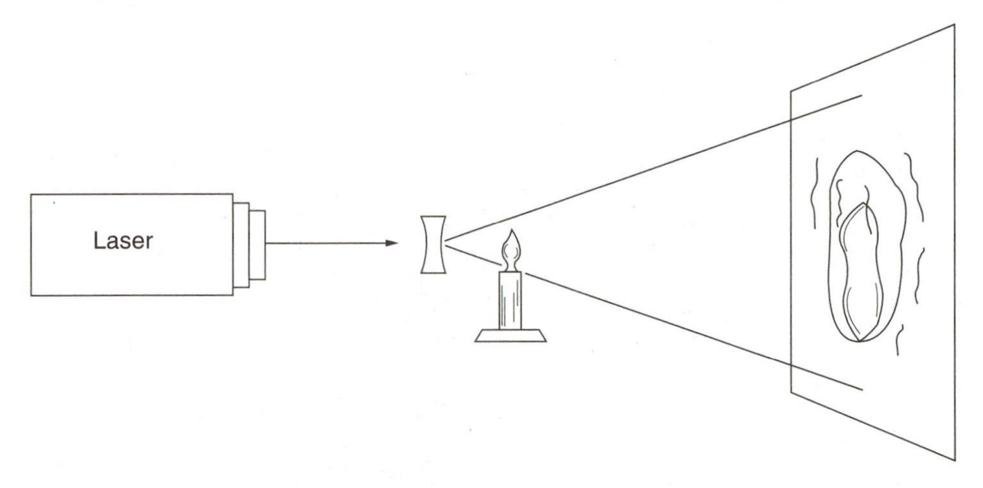

11-15 Schlieren effect demonstration.

Photo 11-4 Optics lab.

Michelson interferometer

The Michelson interferometer, shown in Figure 11-16, opened new vistas in the science of light and optics at the turn of the century. Albert Michelson was able to accurately measure the speed, wavelength, and frequency of light emitted by different sources, as well as the consistency of light speed through any medium long before the introduction of the laser.

The function of the Michelson interferometer was to split nearby monochromatic light from a lamp into two paths. The two beams would then travel at right angles to one another, proceeding the same distance, and recombine at some common converging point. As with all types of waves, fringes appeared as the result of alternating reinforcement and canceling of waves. Where the crests of two waves meet, the light is reinforced and a bright glow can be seen. Where the crests and valleys of two waves meet, the light is canceled, causing a dark spot. Michelson used his new interferometer to test for the existence of the famed "ether," the debate over which has never been fully resolved.

Michelson was plagued with many problems, including weak fringes in his early interferometers, since his light sources were not entirely monochromatic and definitely not coherent. With the aid of today's helium-neon lasers and semiconductor lasers, however, you can construct a good working laser interferometer with the ability to discern dimensional differences of less than ¼ wavelength.

The interferometer consists of an acrylic plastic base with four adjustable bolts to level the base. A single plano-concave or double-concave lens spreads the incoming laser beam to a larger spot. In the center of the base is a glass-plate beam splitter, which is positioned at a 45-degree angle to the incoming laser beam, as shown in Figure 11-16. The beam splitter breaks the light into two components, directing it into two fully silvered mirrors. Mirror 1 is placed directly across from the incoming laser beam, while mirror 2 is mounted on an adjustable aluminum "sled" that can be moved forward and backward by means of an adjustable-screw micrometer. The

Photo 11-5 Optical bench.

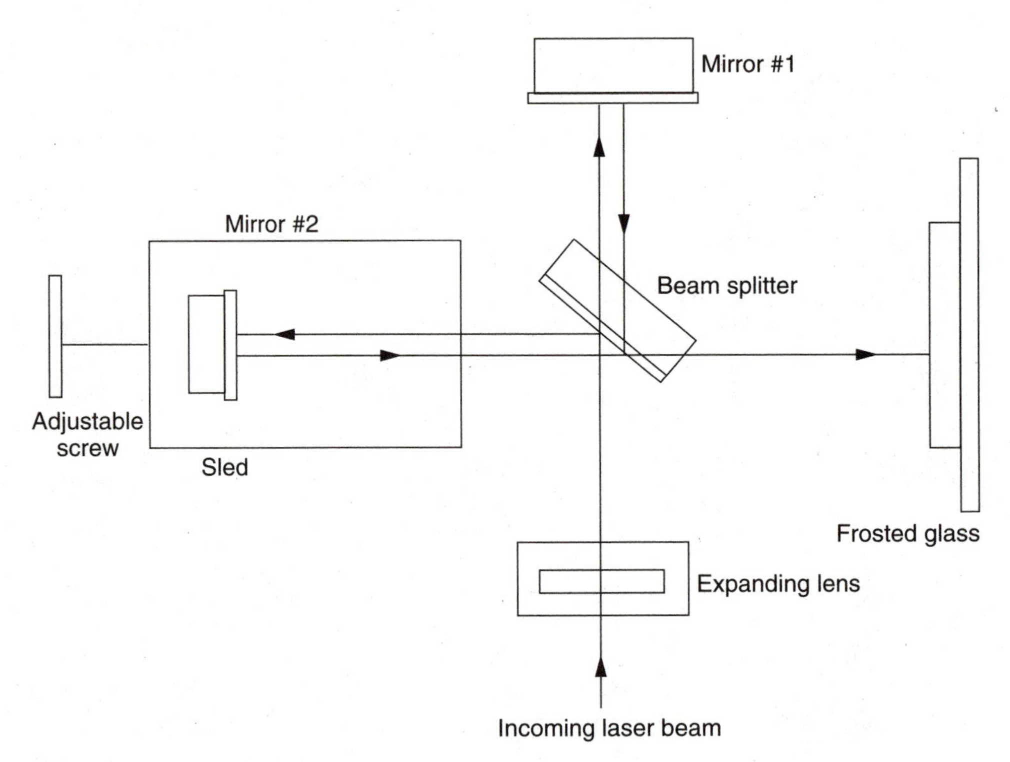

11-16 Michelson laser interferometer.

final component of the interferometer is a ground-glass viewing screen placed opposite mirror 2. After reflecting off the mirror, the two laser beams are redirected by the beam splitter, and a single beam is projected onto the rear of the viewing screen. When properly adjusted, the two laser beams will exactly coincide and fringes will appear on the frosted screen.

The interferometer is highly sensitive and susceptible to vibration, so take care to place the instrument on a smooth, level, hard, vibration-free surface. Ensure that the mirrors are mounted securely to the base to prevent them from vibrating, which would cause the fringes to shift. When adjusting the mirrors for coincidence, allow 20 to 30 seconds before actually viewing to allow for mechanical stability. Constant temperature and an absence of moving air currents will help to provide a stable environment for the interferometer. Once you have a stable working instrument, you can perform a variety of interesting experiments.

With a Michelson interferometer, you can study the effects of refraction, perform linear measurement, determine structural stress, and make a number of other interesting observations. First place an object in front of either mirror 1 or mirror 2. This will cause a shift in the time it takes for light to traverse the two paths. You can study the effects of refraction in the air by gently blowing through a straw or tube. Place the end of the tube in either optical path and watch the fringes move about. You can also use the interferometer to measure the size of small objects with amazing accuracy. Attach a small pointer to the "sled" at mirror 2. Each light-to-dark transition of interfering rings represents ½ wavelength of light, or 316.4 nm. You can then use the pointer on the "sled" to gauge the distance between rings.

To view structural stress on a wall, for example, you can remove mirror 2 from the "sled" and mount it on a wall. Position the interferometer base close to a wall, making sure it doesn't actually touch the wall. Now place the interferometer so the light from and to mirror 2 on the wall can be recombined. Apply pressure to the wall and you should see a shift in the fringes. You can obtain a relatively low-cost Michelson interferometer from the Edmund Scientific Company (see the Appendix).

Electronic fringe counter

By mounting a phototransistor in back of a frosted glass viewing screen and using the circuit shown in Figure 11-17, you can electronically count fringes. Light from the interferometer strikes the phototransistor Q1 and is amplified by op-amp U1. The output of Q1 is coupled through R1 to U1. You can use potentiometer R2 to adjust the sensitivity of the fringe detector.

The network formed by the components ahead of U2 provides clean, square-wave transitions to the counter chip at U3. For resistor R4, you can select from between 10 kilohms and 10 megohms for the desired sensitivity. The switch at S1 is configured as a reset control for the counter. The counter outputs A through D drive transistors Q2 through Q5. These transistors activate the seven-segment common cathode LED displays. Note that resistors R6 through R13 are connected from pins A to G of the counter chip. If the fringe counter is powered with 5 volts, then you can eliminate these resistors. An on-off switch is connected to pin 18 of U3 to supply power to the circuit.

To use the fringe counter, apply power to the circuit and press the reset button to clear the counter. Place the phototransistor behind a simple focusing lens, such as a biconvex lens with a 20- to 4-mm focal length, at least two feet from the interferometer. At this distance, the fringe pattern should be fairly large and the phototransistor/lens arrangement should be able to discriminate separate circular fringes.

Holograms

One fascinating aspect of laser applications is creating and viewing holograms. There are many books and monographs on holographic experiments, so I will only touch the surface of this topic.

A hologram is a photographic plate that contains interference patterns representing the light waves from both a reference source and the photographed object itself. These patterns contain information about the intensity of light and its instantaneous phase and direction. Together, these elements make holograms, or three-dimensional reproduction, possible. These interference patterns constitute a series of diffraction gratings. The orientation of the gratings, along with their size and width, determines how the image is reconstructed when viewed in light. A hologram stores an almost unlimited number of views of a three-dimensional object, and you

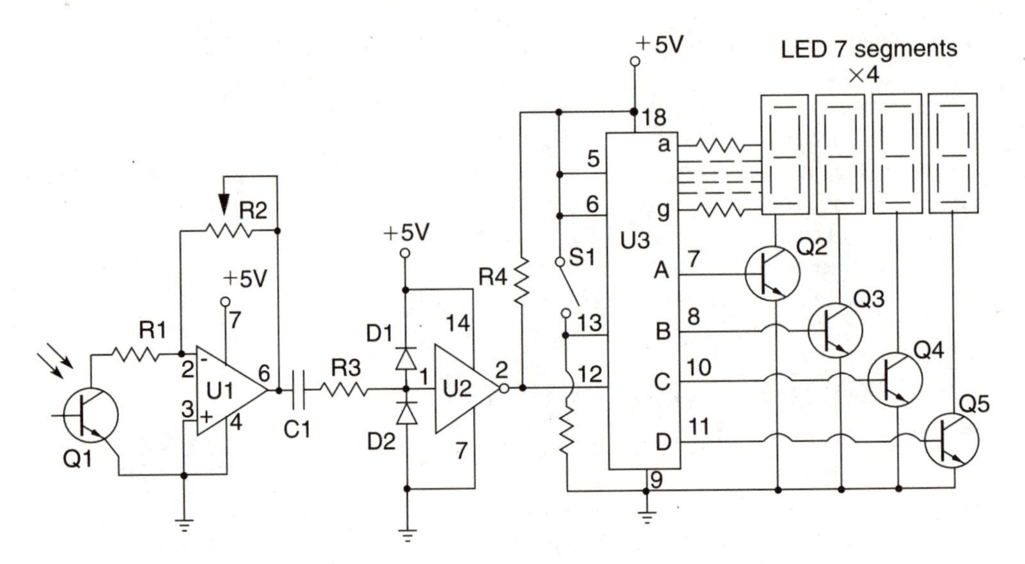

11-17 Electronic fringe counter.

Electronic fringe counter parts list

R1	1-kilohm, ¼-watt resistor
R2	250-kilohm potentiometer.
R3	1-megohm, ¼-watt resistor
R4	10-kilohm to 10-megohm, ¼-watt resistor (see text)
R5	100-kilohm, ¼-watt resistor
R6-R13	330-ohm, ¼-watt resistor
C1	.01-µF, 25-volt disc capacitor
D1,D2	1N4001 silicon diode
Q1	Infrared phototransistor
Q2,Q3,Q4,Q5	2N2222 transistor
U1	LM741CN op-amp
U2	74C14 or CD40106 Schmitt trigger
U3	74C926 counter
LED	Four seven-segment common cathode displays
S1	SPST switch
Misc	Chassis box and hardware

can see these various views by moving your head up and down and from side to side. There are basically two types of holograms: reflected-light holograms, which can be viewed in ordinary light, and transmission holograms, which must be viewed with a laser light.

Reflective-light hologram

Creating a single-beam reflection hologram is really quite straightforward. Locate a solid, vibration-free table and arrange your laser and a diverging lens, film, and small object to be captured on film all in a straight line (shown in Figure 11-18). The best results will be achieved if you use a relatively small object. Reflective-light holograms require an antihalo type of holographic film, which needs to match the wavelength of your laser.

When you are ready to make your laser exposure, turn off the room lights, turn on your laser, and place a thick card or *shutter* in front of the laser, allowing it to warm up for about 10 minutes. Next, remove the holographic film from its container and place a piece of the film between two pieces of clean glass. Press the glass plates together with firm constant pressure for 10 to 20 seconds to remove any trapped air bubbles. Clip the glass plates together with two strong metal paper clasps and place them in a vertical mount. Align your diverging lens so the outer ⅓ of the diameter falls off the edge of the film (only ⅔ of the film is normally active). Reflective holograms require a ratio between reference and object of from 1:1 to 2:1 for best results. Use a light-meter to determine the proper beam ratios. Exposure time for the holographic film depends on the laser power output. Typical exposures are usually for one to two seconds for a 3- to 6-milliwatt laser, and three to five seconds for a 1-milliwatt laser.

When everything is set and ready, you can remove the shutter from the laser and expose the holographic film. Place your exposed film in a light-tight container and proceed to the darkroom to process the film. Dissolve two grams of potassium dichromate with 30 grams of potassium bromide in a liter of water. After mixing thoroughly, add 2 cm^3 of sulfuric acid and mix again. Now pour this mixture into a shallow tray, and pour some Kodak D-19 developer in a second shallow tray. Dip the exposed hologram in the developer for two to five minutes, and then wash the film in running water for five minutes. Dip the film in the sulfuric bath for two minutes or until the film turns clear. Then wash the film again for five to ten minutes in running water, and dry for 30 minutes.

It's easy to view a reflective hologram; you can use sunlight or an incandescent lamp. Place the film in front of the light source and move your head up and down or from side to side to observe the different views of the object. Avoid diffused light such as fluorescent lamps or point-source lighting.

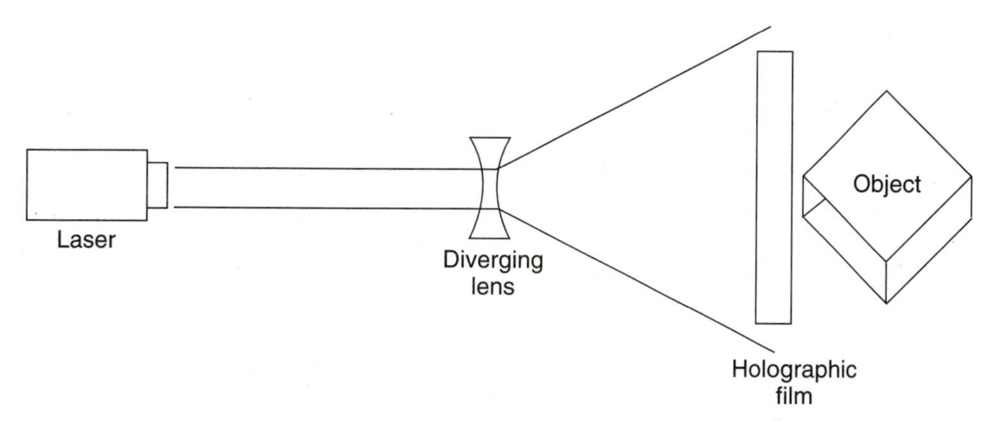

11-18 Viewing reflective holograms.

Transmission hologram

Constructing a single-beam transmission hologram takes a somewhat different approach to exposure than making a reflective hologram. The single-beam transmission setup is shown in Figure 11-19. Note that the object is placed in front of the film in this configuration. To create your first transmission hologram, choose a small object that is smooth but without highly reflective surfaces. Arrange the laser and optics as shown. The distance between the diverging lens and the object should be about 1.5 to 2 feet.

Block the laser-beam path with a thick card or shutter. Turn off the room lighting and turn on the laser. Load your holographic film into a holder with the emulsion side towards the laser, making sure the wavelength of the laser matches the film. Allow the laser to stabilize for about ten minutes; then with the setup in place remove the shutter and expose the holographic film. The exposure time depends on the power output of the laser. The rule of thumb is three to four seconds for a 1-milliwatt laser, and one to two seconds for a 3- to 5-milliwatt laser. Replace the shutter in front of the laser and head for the darkroom to develop the film. Be sure to keep the film in a light-tight container while transporting the film to the darkroom.

To develop your transmission holographic film, you need to purchase some Kodak D-19 developer, which can be poured full-strength into a shallow tray. Then you need to mix a batch of bleach solution. Mix one tablespoon of potassium ferrocyanide with one tablespoon of potassium bromide in 16 ounces of water, and pour the solution into a shallow tray. Be sure both the developer and the acid bath are both kept at between 68 and 76 degrees, for best results.

In the darkroom, remove the holographic film from the light-tight container and place the film in the developer tray. Swish the film around and agitate slowly for about two to five minutes. Plastic tongs work best for handling the film. The image should gradually begin to appear. Use the tongs to pick up the film, and rinse it in running water for about 20 to 30 seconds. Place the film in the bleach solution for two to three minutes. Rinse the film again for ten seconds in running water. Then dip the developed film briefly in a solution of Kodak Photoflo, squeegee the film, and dry for about 20 to 30 minutes.

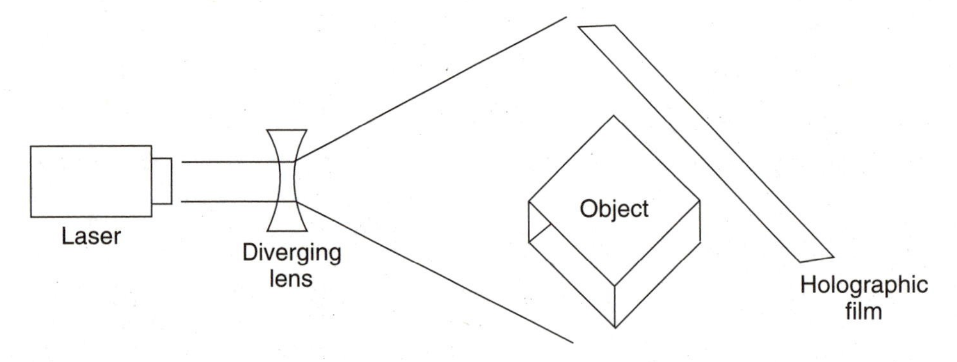

11-19 Transmission hologram viewing.

Viewing a transmission hologram is easy, but you must use your laser to see the image. If you haven't disturbed the laser setup, just replace the film holder with your newly developed hologram. With the emulsion side facing the laser, you can view the image by looking through the film. You will have to move your head up and down or side to side to observe the different views of the hologram.

So far I have discussed only single-beam holograms. The most realistic holograms with the best depth of field and sense of perspective, however, are obtained with dual- or split-beam holograms. A beam splitter and a more complex arrangement of optics are required for this type of holography. If your interest is peaked, you might want to get a copy of one of the many books available on the subject, such as *The Laser Cookbook*. You might also want to obtain a complete hologram kit, like the one shown in Photo 11-6, to begin your own experiments. Lasers and supplies are available from Edmund Scientific and Metrologic Corporation, listed in the Appendix.

Laser seismometer

The next project is a solid-state optical seismometer. Earthquakes are among the most dramatic and disastrous of all natural phenomena. They are feared most because they occur with little or no warning, and predicting earthquakes is far from an exact science. Even though geologists and seismologists have been measuring earthquakes for decades, predicting them a still a long way off. Fortunately, many of the largest earthquakes occur at sea. Earthquakes on the West Coast of the United States are plentiful and generally quite small. Earthquakes can happen anywhere, at any time, but most quakes occur along or adjacent to known fault lines or crustal plates. Tremors can often be felt and recorded, hundreds and often thousands of miles away. Most seismographs use massive weighted and pivoted arms that carry electromagnetic coils.

You can construct your own laser seismograph without winding any coils, by using the fiber-optic seismograph shown in Figure 11-20. The laser fiber-optic seismograph can be easily constructed with low-cost components. Basically, the laser fiber-optic seismograph detects changes in coherency through a length of fiber-optic cable.

When a laser beam is transmitted through a stepped-index optical fiber, some of the light waves arrive at the opposite end of the fiber before others. This varying arrival of waves reduces the coherency of the laser beam in proportion to the design of the fiber, its length, and the amount of curvature or bending of the optical fiber. The object is not to remove the coherency of the laser beam, but to alter it slightly through the length of the fiber. Movement or vibration of the optical fiber causes displacement of the coherency, and that displacement can be detected with a phototransistor or PIN photodiode. You can even listen to the change in coherency by connecting the photodetector to an audio amplifier.

The heart of the laser seismograph is the 6½-foot-long piece of PVC pipe. A stepped-index plastic optical fiber is essentially strung between the two end caps at either end of the PVC pipe. Both ends of the optical fiber must be polished to ensure

Photo 11-6 Hologram kit.

the maximum light transfer from the laser to the fiber and from the fiber to the photodetector. Once cut, carefully polish the fiber by rubbing the flat edge in a circular motion on wet 400-grit fine sandpaper until the face of the fiber end is smooth. At the top end of the PVC pipe, make a fixture to secure the fiber, as shown in Figure 11-21. Drill a small-diameter hole through the end cap in order to pass the optical fiber through to the fiber clamp assembly. The free end of the top fiber is coupled to a semiconductor laser, shown in Figure 11-22, which is attached via a transistor socket to the top end cap. The top end cap is drilled to accept the evacuation tube, which controls the vacuum inside the seismograph tube.

At the opposite or bottom end of the 6½-foot-long PVC pipe, the optical fiber is secured to a second end cap, as shown in Figure 11-21. As shown in the bottom of the diagram, use a knurled screw to fasten or secure the fiber in place, adjusting the fiber so it can be pulled taught but not stretched (more on this later). Then couple the fiber end at the bottom of the PVC tube to the photodetector chip, as shown in Figure 11-22. The laser seismograph tube, once completed, can be supported via a substantial metal bracket attached to the floor.

The top end of the optical fiber is coupled to a laser diode (a Sharp LT020MC/LT022MC was used in the prototype). Either of these two low-cost diodes can be mounted on top of the PVC end cap. The three laser diode leads are then directed to the laser diode driver circuit, shown in Figure 11-23, which can be mounted in a small box at the side of the seismograph sensing tube.

The laser diode driver chip is a Sharp IR3C02 configured in a simple circuit that consists of only four components. The 100-kilohm potentiometer at R1 controls the cur-

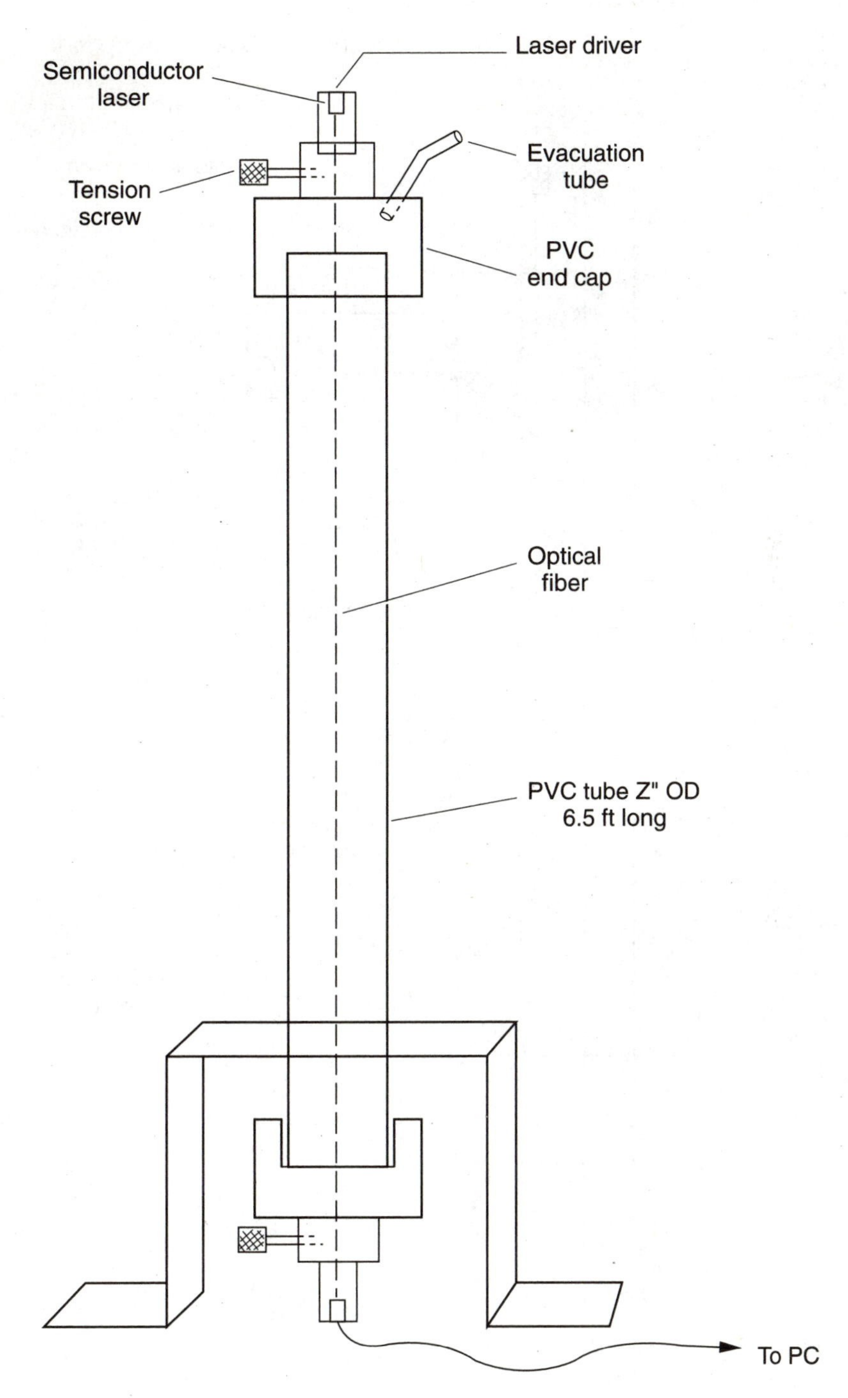

11-20 Fiber-optic laser seismometer.

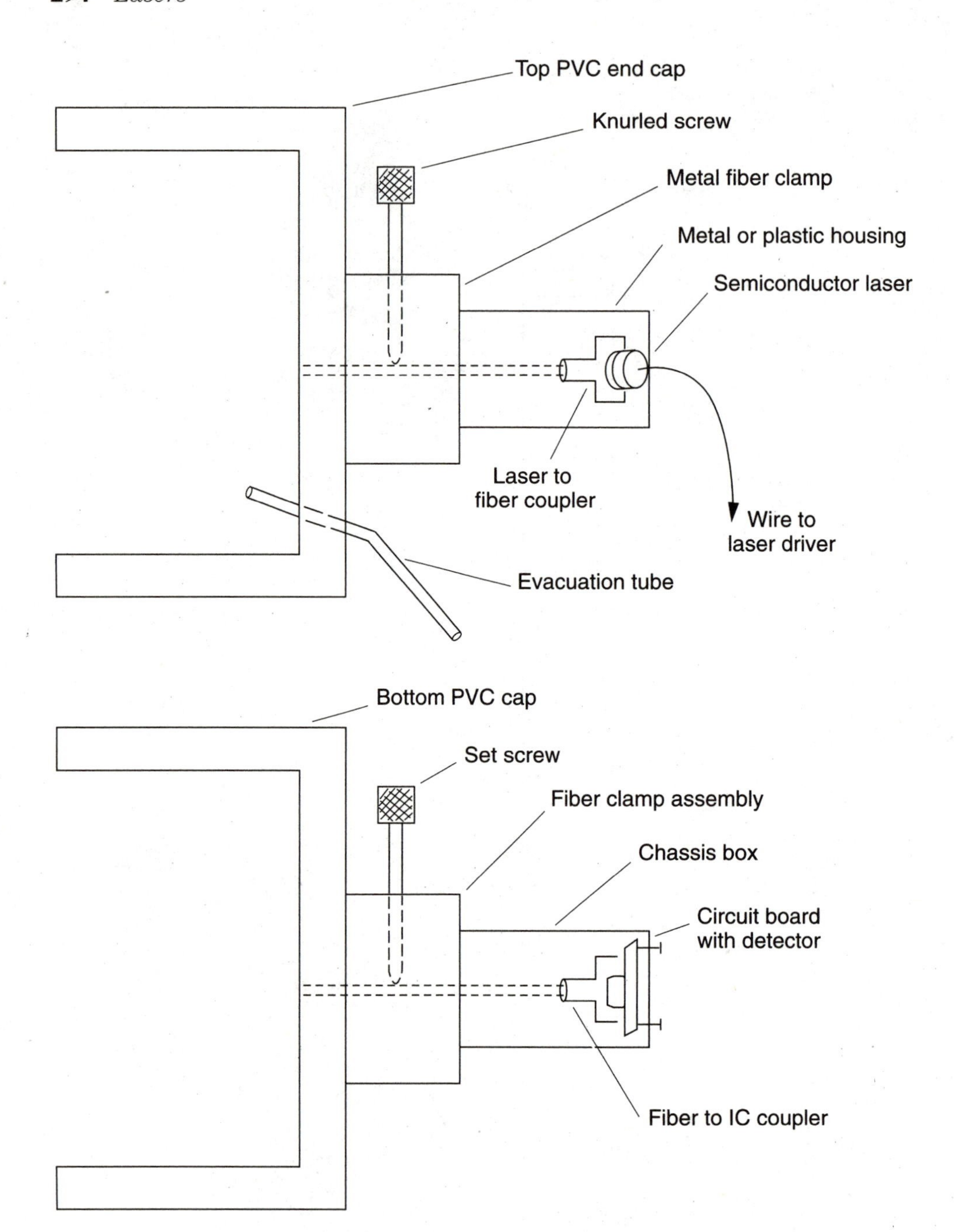

11-21 Fiber-optic tension assembly.

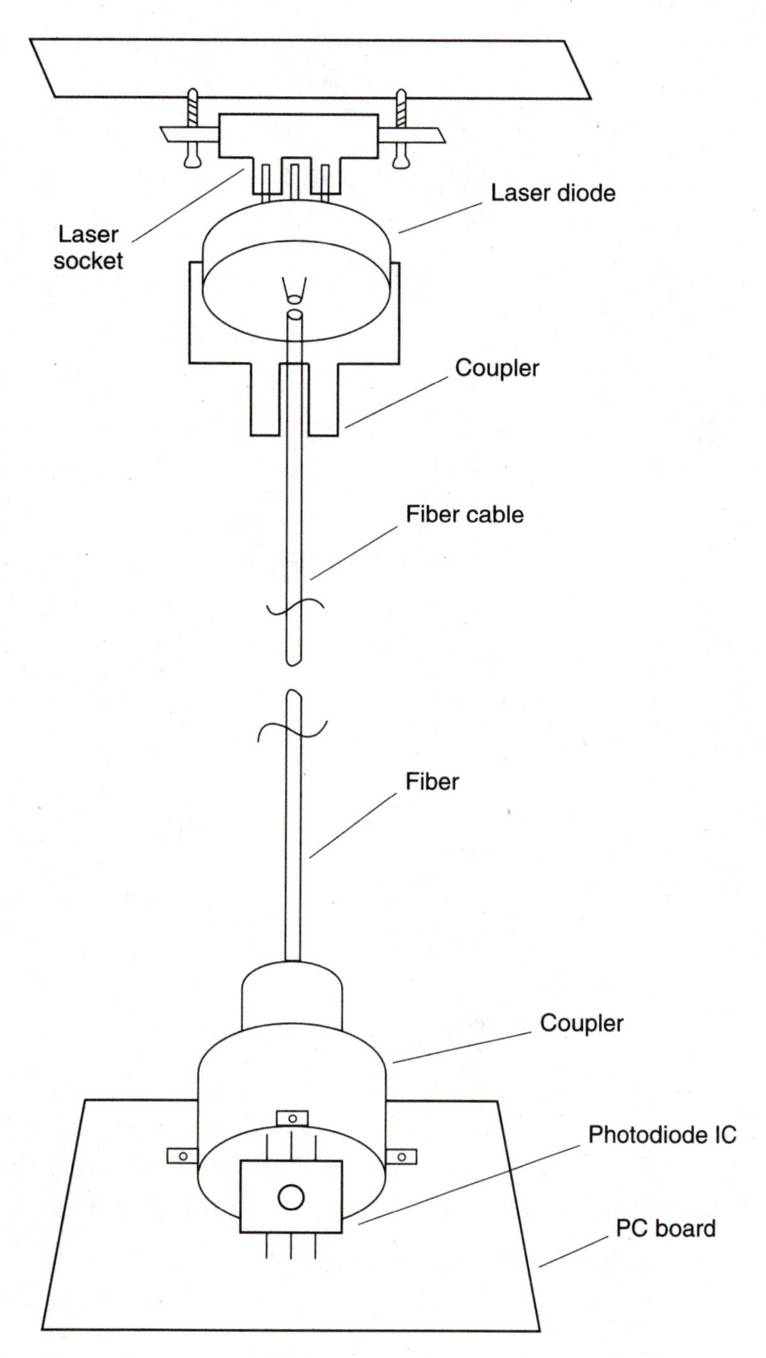

11-22 Laser diode and sensor diagram.

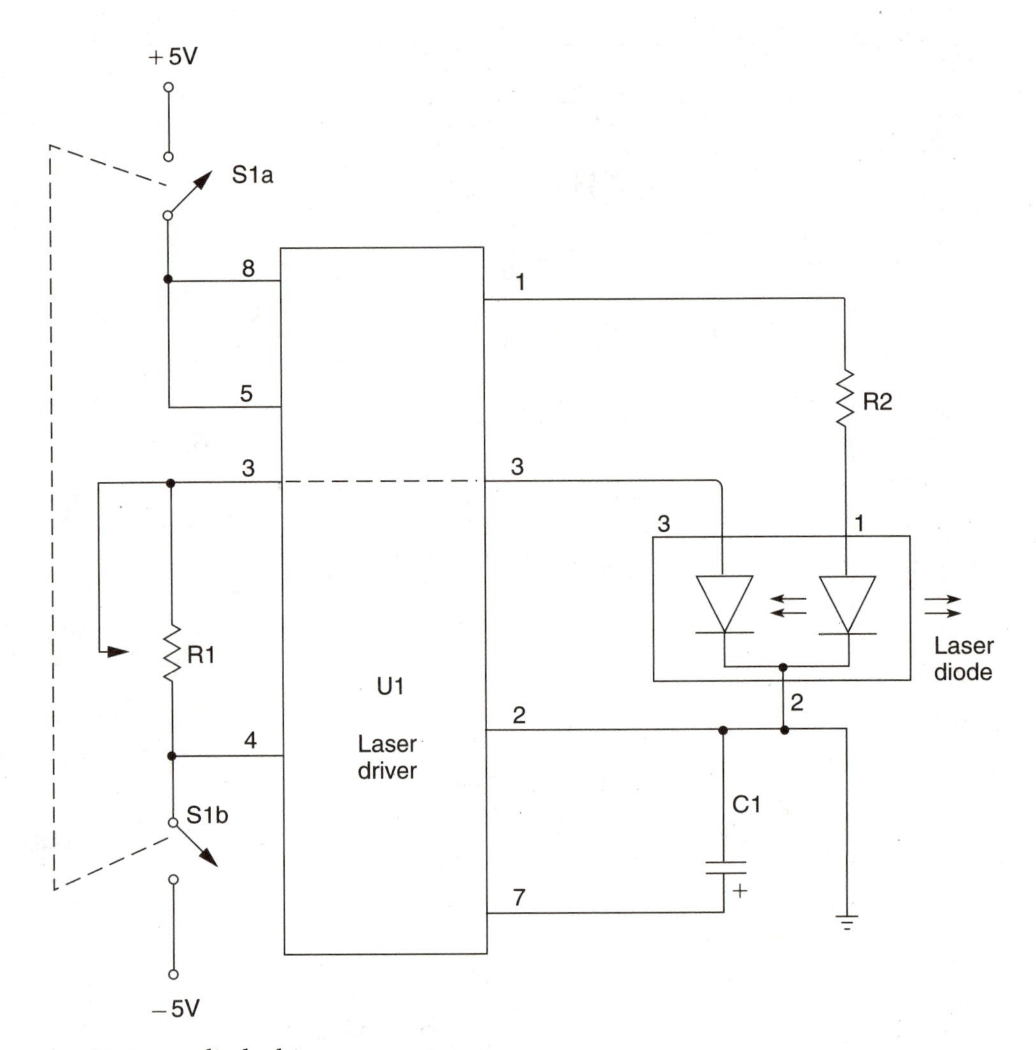

11-23 Laser diode driver.

rent to the laser diode. The laser diode is driven from pin 1 through a 22-ohm resistor at R2. A laser monitor diode is connected from ground to pin 3 of the driver chip. Note that this circuit is powered from a dual 5-volt supply. Two 6-volt lantern batteries could be used with a series diode at each battery to reduce the voltage/current to the 5 volts that is needed. Or you could purchase a laser and driver from one of the vendors listed in the Appendix. Couple the optical fiber at the bottom of the seismograph tube to a photodetector chip, as shown. The coupler links the fiber to the photodiode. The receiver portion of the laser seismograph is centered around a Burr-Brown OPT-211 integrated photodetector amplifier chip. The OPT-211 is a combination photodiode and transimpedance amplifier on a single chip, which is ideal for the laser seismograph.

Set up a gain path between pins 2 and 5 of the OPT-211 chip. A 100-megohm resistor is connected across a .5-pF capacitor, which provides up to a 5-kHz bandwidth. The resistor at R1 and capacitor C3 drive a length of coaxial cable at the output. The output of the OPT-211 chip can be interfaced to either a chart recoder or an analog-to-digital card in a personal computer (see Figure 11-24). An A/D card with 12- to 16-bit resolution is recommended for this project. Analog-to-digital cards can be obtained for about $80 to $100 from various vendors listed in the Appendix.

Aligning the fibers in the PVC is also quite important. In operation, the fiber is secured at one end of the tube and the opposite end of the fiber is tensioned. The fiber should be tensioned to remove any slack, but not pulled to stretch it too tightly. Remove air from the seismograph with a vacuum pump, and then secure a short length of plastic tubing to the evacuation tube with a hose clamp. At the free end of the hose, clamp the evacuated tube. You can then use the free end of the hose clamp to adjust the air entering into the tube. You can allow some air back into the tube in order to damp the vibrations of the optical fiber.Once your laser seismograph is constructed, you need to find a quiet place for the instrument. A laser seismograph, like any seismograph, is susceptible to the effects of local noise and vibrations caused by passing cars, people, etc. In order to be effective, the seismograph needs to be placed in an area that is not be affected by local noise. People who live in the country will find this problem a bit easier to solve than city folks. The seismograph should be placed on firm ground. Try to avoid placing it indoors on a wooden floor. Most buildings are flexible and do not accurately transmit seismic vibrations. For best results, attach the seismograph to a big rock or boulder, away from any buildings if possible. If no large rocks are available, you could attach the seismograph to a cement piling. You could also partially bury 6 to 8 cement blocks and fill them with sand to support your seismograph.

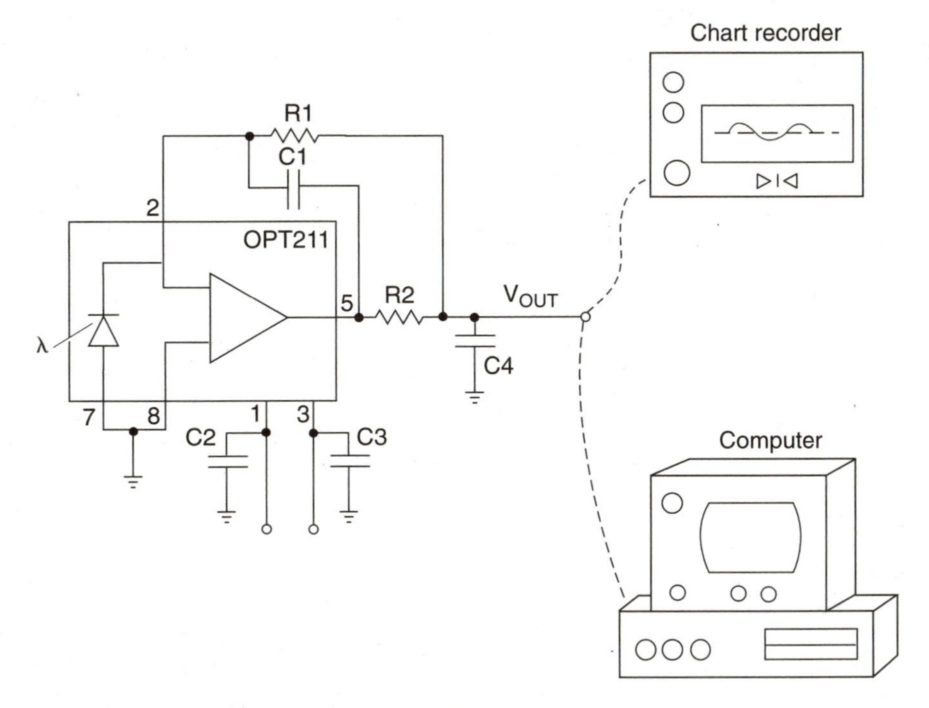

11-24 Laser seismograph receiver/display.

The best-known scale of earthquake monitoring is the Richter scale, a logarithmic measuring system ranging from 1 to 10. Each increase of 1 represents a ten-fold increase in earthquake magnitude. Most earthquakes occur at fault lines or fissures in the earth's crust. The majority of seismic events occur at the boundaries of crustal plates, many of which are located in the world's oceans. These crustal plates slip and slide upon each other, thus releasing stress and strain built up over time.

Seismic waves are generally composed of two major components. P waves are longitudinal compression waves, which are basically sound waves that travel at about five miles a second. S waves are shear waves, which influence the seismograph at the surface after moving through the Earth. Shear waves travel about one half as fast as P waves, and the vibrations are at right angles to the wave direction. This action causes the shaking of rocks and wall structures.

Laser seismometer mechanical parts list

Amount	Item
1	10-foot stepped-index plastic fiber
1	6½-foot length of 2-inch OD PVC tube
2	PVC end caps
2	Tension screw fastener assemblies
2	Fiber-optic couplers (laser and detector)
1	Evacuation tube
4	Support brackets

Laser diode transmitter unit parts list

R1	100-kilohm potentiometer
R2	22-ohm, 2-watt resistor
C1	22-µF, 50-volt electrolytic capacitor
S1	DPDT switch
U1	Laser driver chip (Sharp IR3C02)
Laser	LT020MC/LT022MC

Photodetector receiver unit parts list

R1	50- to 100-megohm, ¼-watt resistor
R2	175-ohm, ¼-watt resistor
C1	.5- to 1-pF mica capacitor
C2,C3	.1-µF, 25-volt disc capacitor
U1	OPT-211
Misc	Coaxial RG174, batteries, chassis box, etc.

Other laser applications

Lasers can be used for many practical applications, including counting objects, as shown in Figure 11-25. A moving conveyor belt moves objects from one point to another. A laser beam can be directed across the belt to a photodetector at the opposite side of the belt in order to count the moving objects. Every time the beam is interrupted, the photodetector senses the absence of light, and the loss of signal advances a counter circuit (refer back to Chapter 5).

A laser is often used in construction and surveying to align walls and ensure accurate borders and boundaries. A laser can be placed on a level tripod, and plastic pipe laid on top of a wall under construction. The laser beam is then directed down the center of the pipe, as shown in Figure 11-26. Lasers are used in many everyday practical applications, including wheel alignment and bar-code scanning (as described in Chapter 10), as well as flaw detection, blood-cell-diameter measurement, medical surgery, burglar alarms, and distance measurement. Table 11-1 lists a number of laser applications in use today, and many new laser applications will be discovered in the near future.

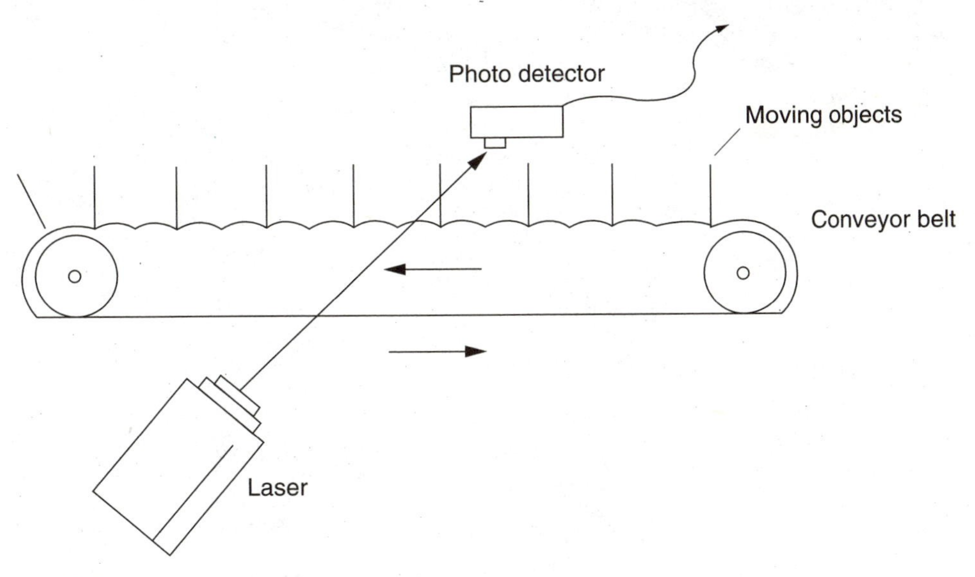

11-25 Laser conveyor belt counter.

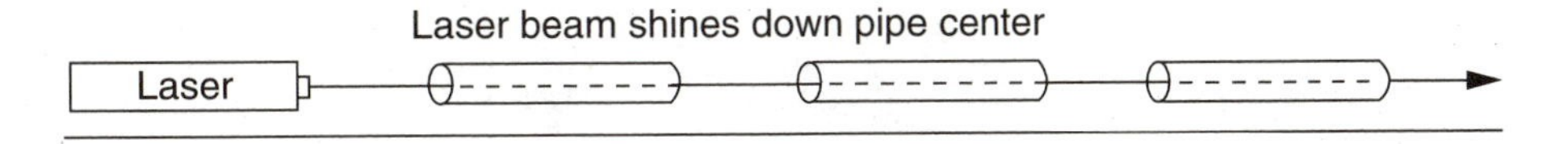

11-26 Construction alignment using a laser.

Table 11-1. Laser applications

Holography	Surgical lasers
Interferometry	Internal organ view systems
Gyroscopes	Dental lasers
Laser wheel alignment	Laser listeners
Bar-code scanners	Signaling systems
Laser gauging and sorting	Laser rangefinders
Pollutant detection systems	Laser seismographs
Laser velocimeter	Laser security systems
Laser spectral analysis	Laser timber-cutting guides
Materials processing	Laser flaw finders
Laser-induced chemical reactions	Communications systems
Laser leveling for construction	Laser anemometer
Conveyor belt counting	Laser cutting and welding

12 CHAPTER

Distance measuring

Measuring distance by laser is a fascinating subject. In this final chapter, I will describe a few different methods of distance measurement using IR LEDs and a laser beam.

The first method of laser measurement uses a tape measure or optical range finder, a protractor or inclinometer, a laser, and a few trigonometric calculations. The next method is based on a triangulation sensor. In this method, light from an IR LED is projected across a distance over a mirror, then reflected back through a lens to a position-sensitive detector. The final distance-measuring method uses a helium-neon laser in an all-in-one electronic range-finder system. A laser beam is first sent out and shown on the distant object of interest. The return laser beam is filtered and referenced against the original laser beam, and the difference in time between the original and the return or echo beam is the basis for electronically calculating the distance from the observer and the target object.

The optical lever shown in Figure 12-1 is a simple means of amplifying or changing small movements into larger displacements that can be shown on a distant screen and used for measurement. You can measure remote objects with an optical lever and a protractor or inclinometer, which resolves the angular movement of a small front-surface mirror. A front-surface mirror is attached to the protractor or inclinometer so the mirror and angle-measuring device can move together, as shown in Figure 12- 2. You can use this mirror and angle-measurement assembly to measure the height of buildings or trees by using it with vertical travel. If the mirror assembly is placed on its side so the mirror travels side to side, you can measure horizontal distances, as you will see.

Note in Figure 12-1 that the laser beam (A) forms one side of a right triangle. The angle of mirror movement is θ. Side C forms the triangle's hypotenuse and side B is the distance to be measured or calculated. If side A is known and angle I can be read off the protractor, then side B or C can be computed.

Armed with an angle-measuring device, a front-surface mirror, and a laser, you now need only some trigonometry to complete the measurements. Figure 12-3 depicts a right triangle. The capital letters A, B, and C represent the triangle sides,

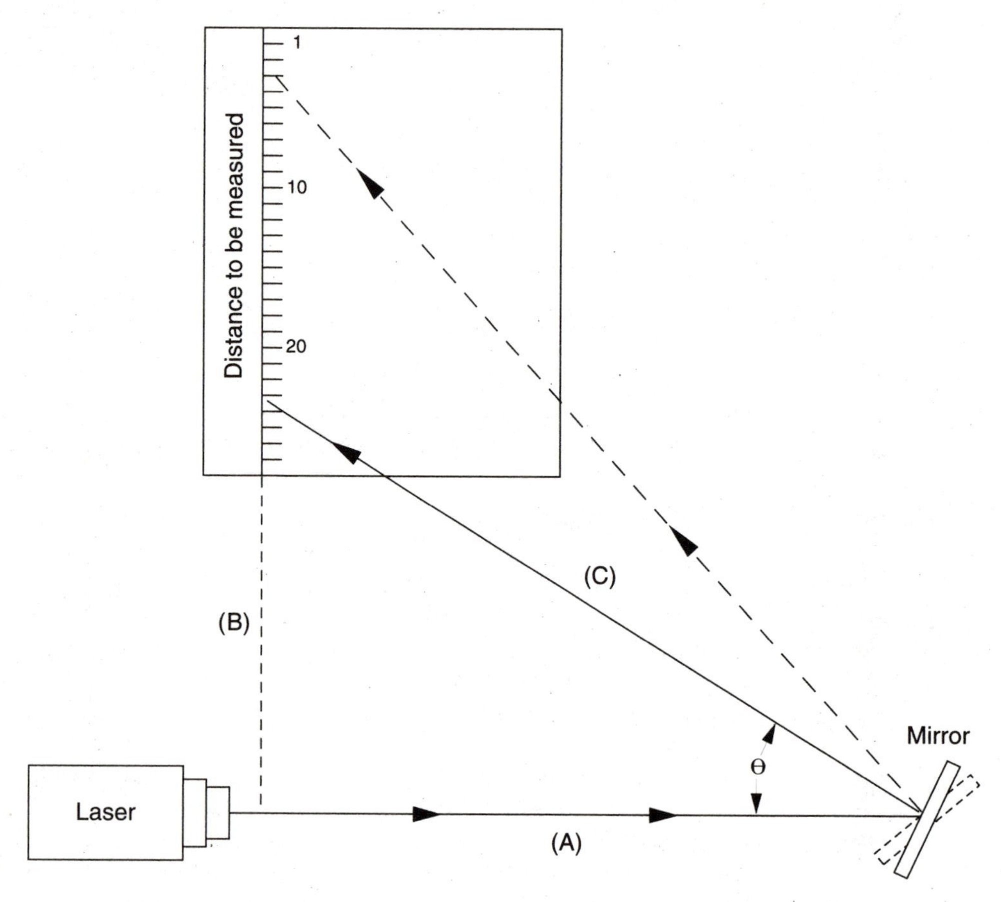

12-1 Optical lever.

while small letters a, b, and c represent the angles formed by the triangle sides. Refer to Table 12-1 for the trigonometric formulas. Rows 1 through 9 represent known values of sides and/or angles, and columns A, B, and C, and ∠a and ∠b represent the unknowns or values to solve for. Using Table 12-1 greatly simplifies the calculations required to perform distance measurement. All you need is a calculator that can derive sine, cosine, and tangent functions from your angle-measuring device.

Figure 12-4 shows another method of range finding, using two lasers or one laser beam with a beam-splitter and a mirror to generate two laser beams separated by at least 50 cm. Leave one of the lasers fixed and rotate the other laser beam or beam-splitter so the beams intersect on a distant object. You can use triangulation and trigonometry or a tape measure to calibrate the rotation angles that are required to target objects in terms of their distance from the laser.

The diagram in Figure 12-5 illustrates a practical means of measuring the height of tall buildings, trees, or flying objects. Mount a laser on a hinge assembly and secure it to a base plate that houses a bubble level and a protractor or inclinometer, as shown. Assume side A to be 100 feet, which you can quickly measure with a tape measure or optical range finder. Now turn on your laser and point it to the top of a

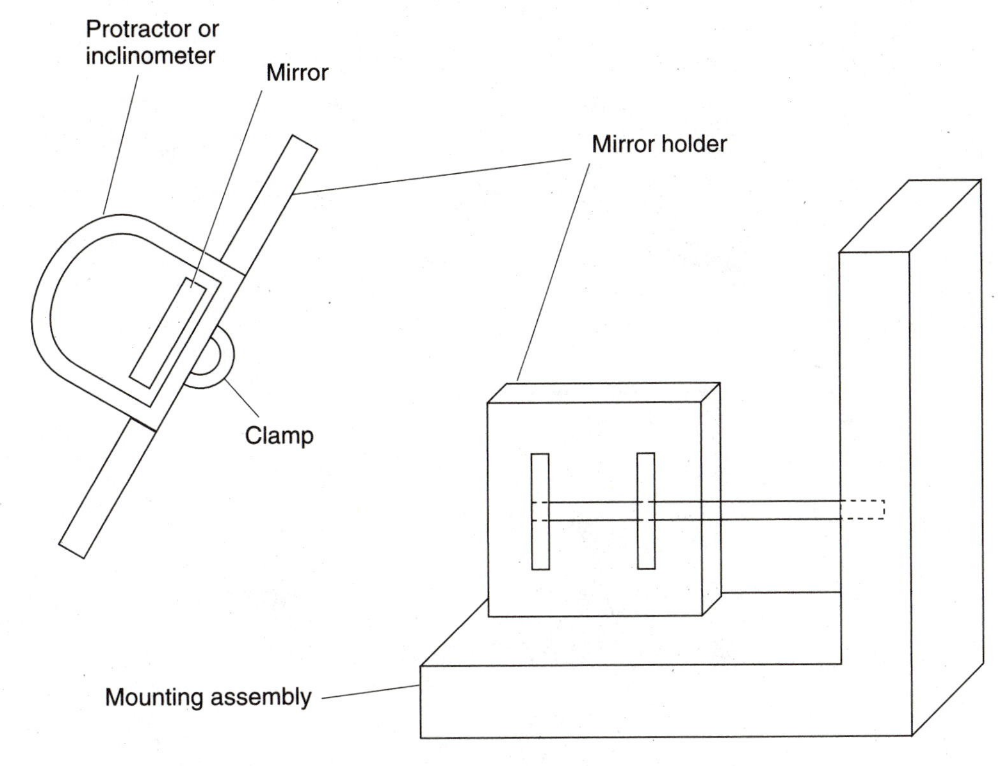

12-2 Optical lever mirror and protractor mount.

tall building. The angle the laser subtends is shown against the protractor or inclinometer, and you can quickly read and use it to calculate the height of the building. Assuming a distance from you to the base of the building, or side A, of 100 feet, and an angle of 60 degrees, you can look at Table 12-1 to find the answer. Look at line 3 of the known values and across to the second column, A tan ∠b. Enter tangent 60 degrees into your scientific calculator, and multiply 1.732 by side A or 100 feet. There you have it: the building height is 173 feet!

Once, while looking out a large picture window on the shore of Lake Superior, I wanted to know how to determine the speed of one of those thousand-foot ships passing in front of me. Actually, there are a number of ways you can determine a ship's speed.

First, let's start by rotating the mirror protractor assembly on its side (refer back to Figure 12-2). Now, instead of measuring a building height with a vertical mirror, flip the assembly so the mirror can travel horizontally, from side to side. To determine the distance to the ship, you could use an optical range finder (see Figure 12-6). Take a sighting when you first spot a ship straight out in front of you. Shine the laser out to the ship while recording the optical range and starting your stop watch. As the ship moves across the field of view, rotate the mirror or laser assembly to the final viewing position and record the time again. Now take the optical range finder's reading out to the ship (A) and multiply the tangent of the angle formed by the mirror. Now you know the distance the ship traveled from the initial

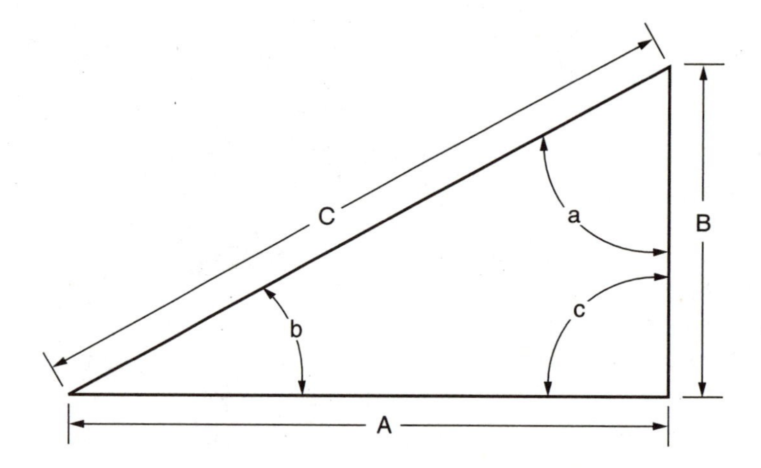

In any right triangle,
the values are vaild if we let:

a equal the acute angle formed by the hypotenuse and the altitude leg,
b equal the acute angle formed by the hypotenuse and the base leg,
A equal the side adjacent to ∠ b and opposite to ∠ a,
B equal the side opposite ∠ b and adjacent to ∠ a,
C equal the hypotenuse.

12-3 Plane trigonometry—right triangle.

Table 12-1 Trigonometric Formulas.

Known	**Formulas for unknown values**				
values	**A**	**B**	**C**	**∠a**	**∠b**
A & B	——	——	$\sqrt{A^2 - B^2}$	arc tan B/A	arc tan A/B
A & C	——	$\sqrt{C^2 - A^2}$	——	arc cos A/C	arc sin A/C
A & ∠b	——	A tan ∠b	A/ cos ∠b	——	90° − ∠b
A & ∠a	——	A/ tan ∠a	A/ sin ∠a	90° − ∠a	——
B & C	$\sqrt{C^2 - B^2}$	——	——	arc sin B/C	arc cos B/C
B & ∠b	B/ tan ∠b	——	B/ sin ∠b	——	90° − ∠b
B & ∠a	B tan ∠a	——	B/ cos ∠a	90° − ∠a	——
C & ∠b	C cos ∠b	C sin ∠b	——	——	90° − ∠b
C & ∠a	C sin ∠a	C cos ∠a	——	90° − ∠a	——

recorded starting time to the final stop time. Since you now know the distance traveled and the time it took, you can easily determine the speed of the ship, with the following formula:

$$\text{Time } (T) = \text{Distance } (D) \,/\, \text{Speed } (S)$$

If a ship travels 1000 feet in 10 seconds, therefore, its speed would be 100 feet per second.

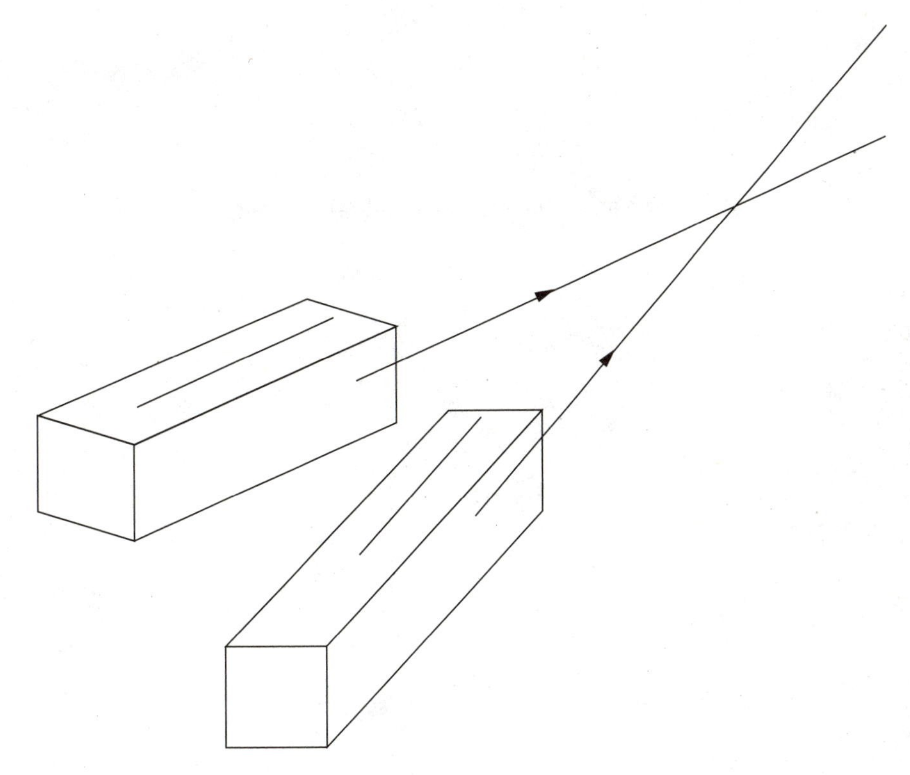

12-4 Dual laser triangulation.

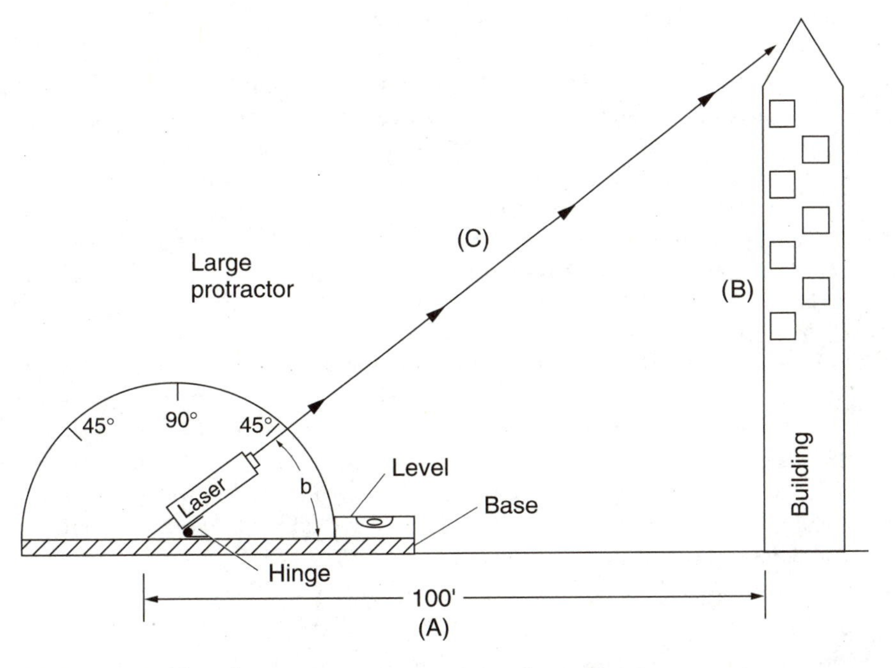

12-5 Vertical height measurement using triangulation.

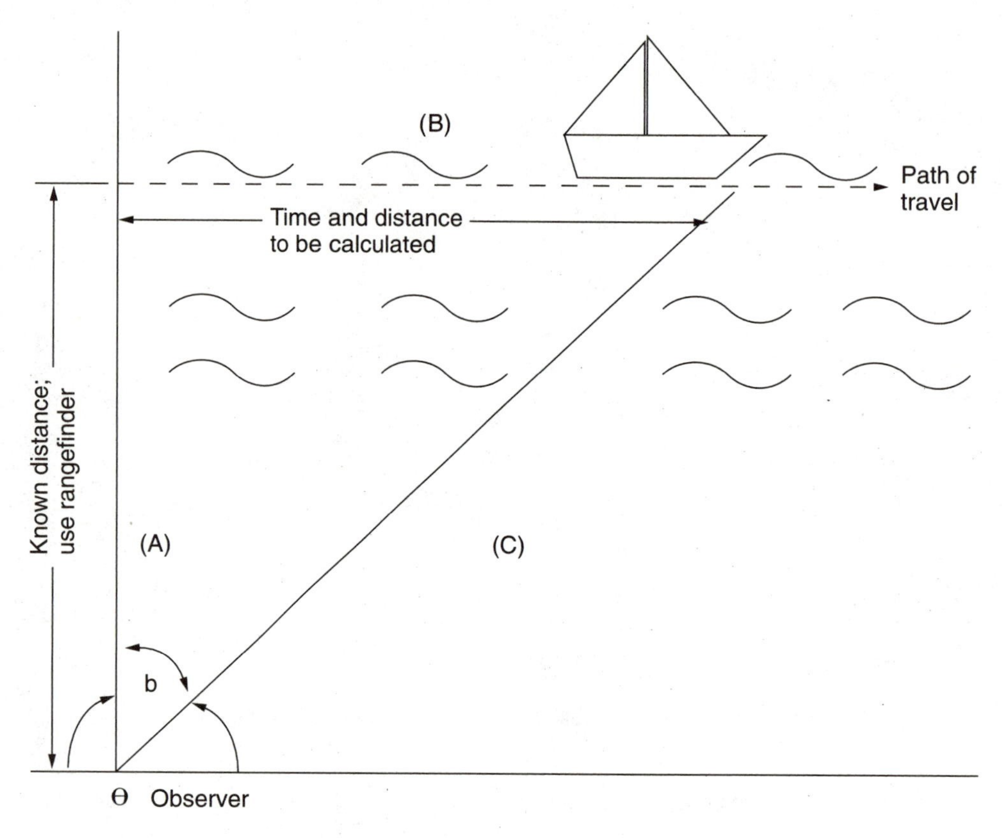

12-6 Horizontal distance and time measurement.

The diagram in Figure 12-7 illustrates an experiment to measure the curvature of the Earth. First, set a helium-neon laser on a tripod a short distance above the ice on a large frozen lake. Collimate and aim the laser beam horizontally over the ice with the aid of an accurate bubble level. Several kilometers away, set up a telescope to intercept the laser beam. Because of the curvature of the earth, the height of the telescope above the ice will be greater than that of the laser. By measuring the difference in height between the laser and the telescope and by knowing the distance from the laser to the telescope, you can calculate the curvature of the Earth. In the summer, try using two boats on a calm lake.

The diagram in Figure 12-8 illustrates the Earth's curvature, or 1/R. First, I need to define some terms. The distance from the laser to the observer is D, the angle formed by the distance between the laser and the observer is α, the radius of the earth is R, and the difference in height between the laser and the distant observer is g. In order to determine R (the radius), you need to solve the formula $R = D^2 / 2g$. The rule of thumb states that for every mile traveled across the earth, the difference in height is 8 inches. So:

$$1 \text{ mile} / 2(8) = 1 / 16 \, (1.578 \times 10^{-5})$$

For example:

$$1 / .00025 = 4000 \text{ miles} = R$$

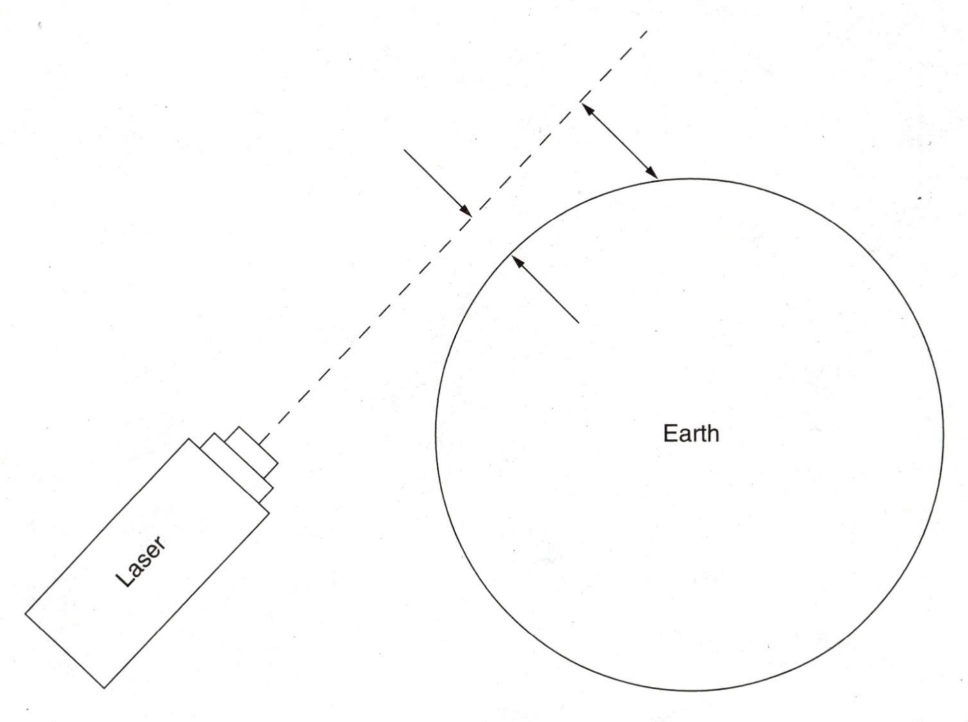

12-7 Curvature of the Earth experiment.

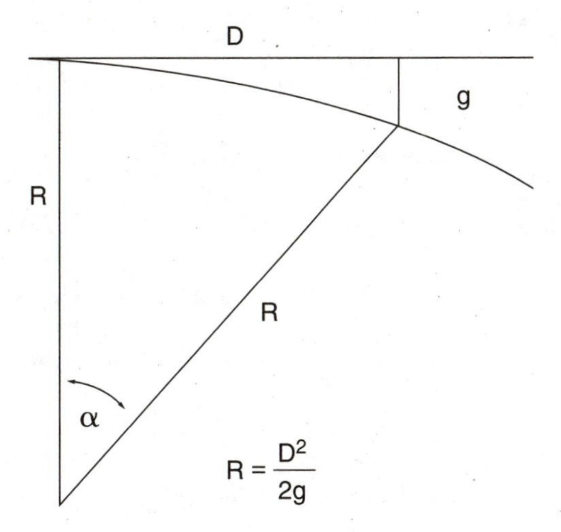

12-8 Earth section measurement.

Optical triangulation sensor

Optical triangulation techniques are widely used for distance measurement in geodetic surveys as well as industrial assembly and inspection applications. Optical

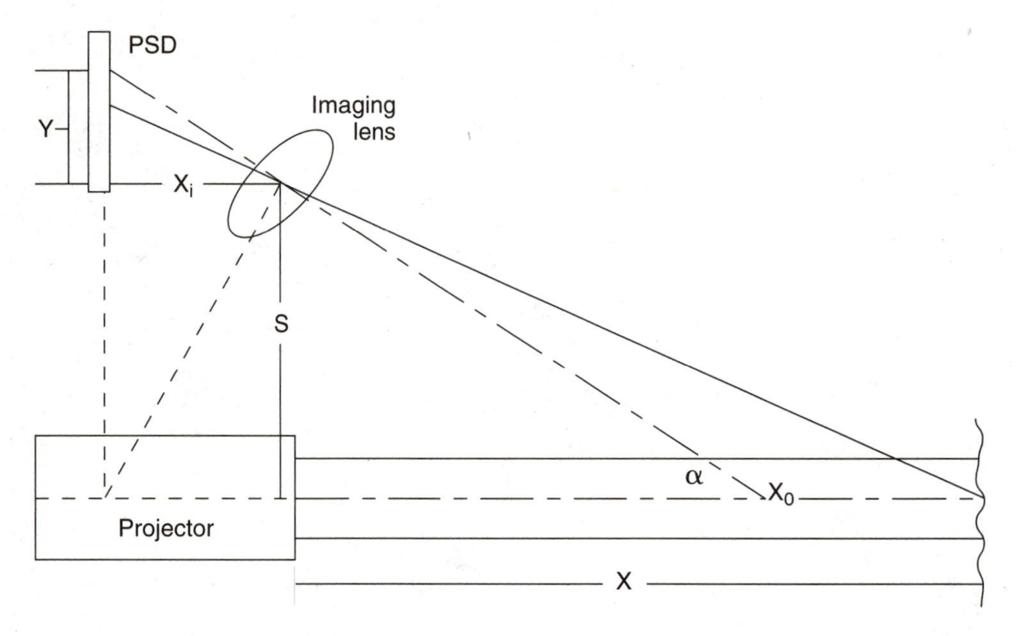

12-9 Triangulation sensor.

triangulation using IR LEDs and lasers combined with position-sensitive detectors or PSDs have become available only in the last few years.

An active triangulation sensor is shown in Figure 12-9. This system measures the distance to a reflecting surface along a projected IR beam by imaging the illuminated area on a position-sensitive detector (refer to Chapter 5). Note that the dimension y indicates the position of the focused spot and is inversely proportional to the object distance x. This system can resolve distances from near zero to about 100 cm. Long-range triangulation sensors used for geodetic surveys have lasers instead of IR LEDs, but they use the same principles. The system can detect surfaces that vary in reflectivity from bright white to absorptive black. The projector consists of an IR LED modulated at 10 kHz. The IR LED produces an 830-nm beam and is placed behind a 25-mm-diameter collimating lens. The diameter of the IR beam varies from 25 to 40 mm at the maximum range. The light reflected from the projector is directed to a 9-mm-aperture imaging lens centered on the linear PSD sensor.

Active triangulation sensors measure the position of the reflecting surface along the centerline of the projected beam, as shown. Consider the similar triangles formed by a ray passing through the principal point of the lens. The basic equation for the spot's position on the PSD is as follows:

$$Y = \frac{SX_i}{X}$$

Where:

X = distance to object
S = triangle base
X_i = image distance behind the principal point

Nominally:

$$X_i = f\,(M_o + l)\cos\alpha$$

Then:

f = focal length of lens

M_o = magnification from the point X_o on the lens axis to its image on the PSD

α = the angle between the lens axis and the beam

The PSD's output signal indicates the center of brightness of the light reaching its surface, so the image does not have to be perfectly focused as long as the lens aperture is fully illuminated. To make this scheme practical over a large range of object distances, position the point X_0 to provide equal field angles at the minimum and maximum ranges, and maintain approximate focus over the range. Select the lens focal length and the axis angle α by orienting the lens for a plane normal to the lens axis and passing through the intersection of the object, and the image planes shown by the dotted lines in Figure 12-9.

Laser range finder

A laser range finder is an electro-optical device that can measure the distance from itself to a distant target and display the distance on a readout device—all of which is done electronically with a semiconductor laser.

The laser range finder in this discussion consists of a number of building blocks, shown in Figure 12-10. The building blocks consists of a semiconductor laser and optic collimator, a reference detector/preamplifier (REF), a target detector/preamplifier (TAR), a beam-splitter and telescope assembly, a digital voltmeter, and the heart of the range finder: the phase comparator.

Your tour inside the laser range finder begins with the laser diode assembly. The laser diode used in this project is a 5-milliwatt, 780-nm Sharp LTO22MD. The prototype range finder used a laser diode assembly and collimating optics purchased from MWK Industries (a second supplier, Meredith Industries, also sells similar laser diode assemblies; see the Appendix). To achieve a long-distance range finder (1000 meters or a 2000-meter round trip), you can substitute a 30-milliwatt laser. Mount the laser diode in a heatsink housing to remove excess heat from the laser diode. Then place laser diode collimating optics in front of the laser diode so the beam can diverge at a rate equivalent to F/4. This means that the beam expands by one inch every four inches of distance away from the laser diode. You can use a divergence up to F/8 for this project.

The laser diode assembly with housing and optics should be about 1.5 × .75 inches in diameter. Attach the assembly to an adjustable mount that provides tilt, translation, and horizontal angle adjustments. The height of the laser beam emitted from the range finder should be about 49 mm above the base of the mount. The mount might have been designed for a rectangular laser diode housing, requiring you to modify it, depending on the laser diode you choose. The optical axis of the laser diode is parallel to the length of the adjustable mount, so the angle entering the beam-splitter assembly is 90 degrees, with respect to the telescope's optical axis (see Figure 12-11).

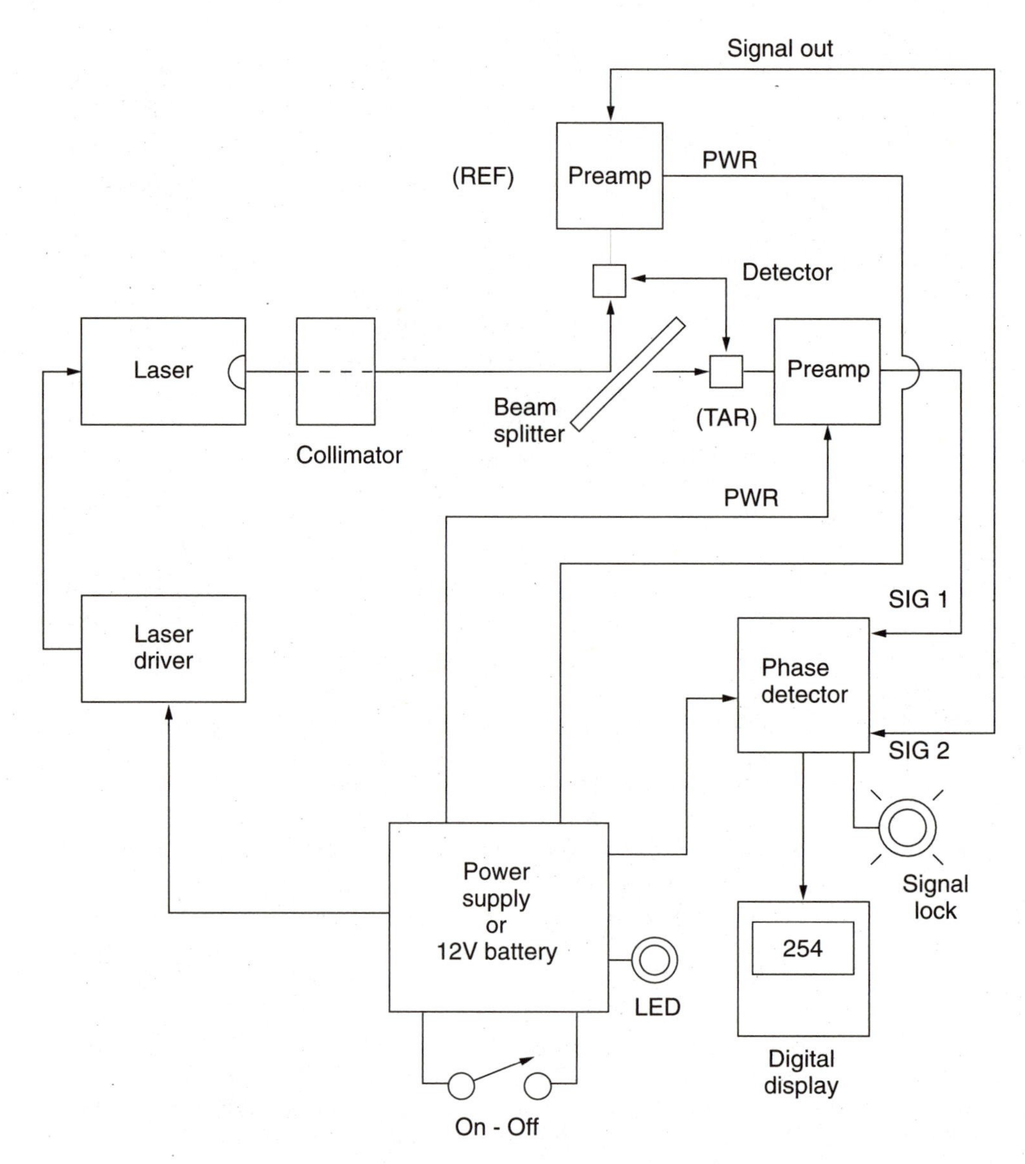

12-10 Laser rangefinder.

The laser diode driver used was a 83-kHz oscillator kit, available from Meredith Industries. The laser diode driver was mounted near the laser diode to minimize the radiated EMI from being picked up by the high-gain detector modules. The IC in the laser diode driver controls the drive current to the laser diode in order to stabilize the power output over any temperature variations. You can easily adjust both the threshold and peak current on the driver PC board. All the assemblies mentioned so far are readily available for you to construct your own laser range finder.

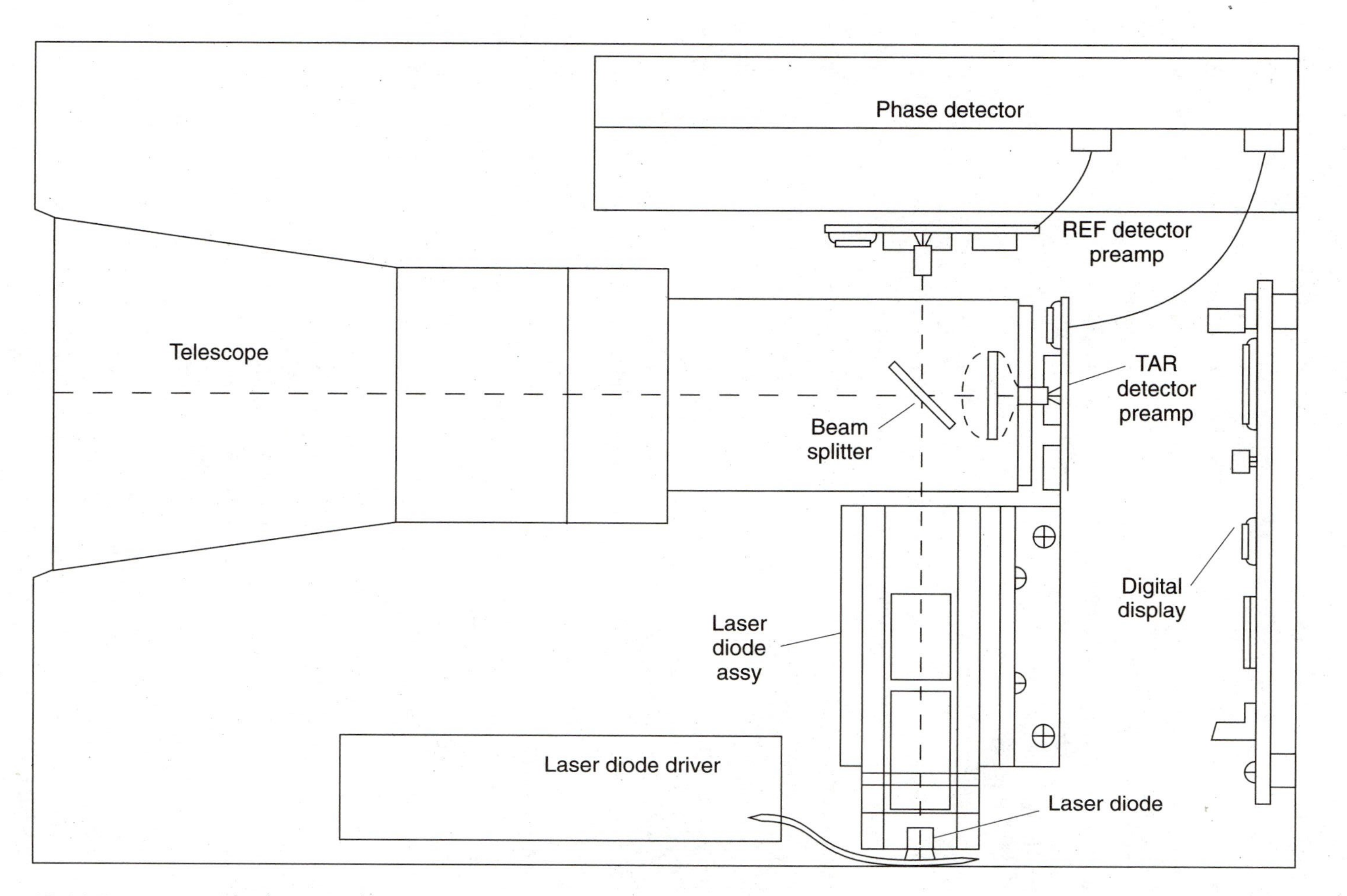

12-11 Laser range finder parts layout.

The next building blocks for the laser range finder are the two 20-MHz detector/preamplifiers: reference (REF) and target (TAR). The detector/preamplifier modules consist of a wide-band silicon photodiode with a 5-kHz to 20-MHz bandwidth with less than a 50-nanosecond rise time. The detector preamplifier used after the photodiode is a low-noise device with a dynamic range of 54 dB, which produces a 700-millivolt peak-to-peak single-ended output or a 1.4-volt peak-to-peak differential output. The 50- ohm output from the preamplifier can drive a short length of coaxial cable connected to the phase detector assembly. The two detector modules can be powered from a 9- to 12-volt dc power supply or battery. The 20-MHz detector/preamplifier modules can be purchased completely assembled from MWK Industries (see the Appendix). As mentioned earlier, two detector/preamplifiers are necessary for the laser range finder, REF and TAR.

In operation, the laser beam is split with a beam-splitter. One laser beam is sent to the REF detector and the second beam is sent to the TAR detector assembly. The laser beam is sent to the REF detector to establish a reference timing signal. The laser beam exits the telescope assembly and goes out to the distant target and returns back to the telescope, and the incoming beam is sent to the TAR detector/preamplifier module. You can then use the difference in time between the REF beam and the TAR beam to compute the timing difference, which corresponds to the distance the laser beam has traveled.

The output from both detector/preamplifiers is fed directly into the phase comparator building block. The phase detector compares the timing differences between the REF and TAR detector/preamplifiers, and produces a dc voltage output that is proportional to the range of the target. The phase comparator is accurate to .1 meter over a 0°C to 40°C temperature range, provided the target return beam voltage is 5 millivolts or greater. A threshold adjustment is provided by a potentiometer on the phase comparator PC board. The phase comparator turns off an LED whenever the return signal voltage is less than 5 millivolts. If the TAR detector's signal voltage is too high, the capture LED will flash. The phase comparator has an internal calibration circuit that is active whenever the return signal is less than 5 millivolts. You can adjust the auto-zero circuit and gain as needed to provide a reliable output. The phase comparator module is powered from 12 volts dc, and is available from Laser Sensor Technology (see the Appendix).

The telescope used for the laser range finder is a 50-mm-diameter scope with a 200-mm focal length. Properly aligning the laser diode with the beam-splitter assembly is crucial for good long-range performance. The laser diode must be centered in the telescope and be coaxial with the detector's field of view. The telescope assembly consists of the telescope, a beam-splitter, and the TAR and REF detector photodiodes. You can construct your own telescope assembly, as shown, or you can purchase a completely assembled telescope assembly from Laser Sensor Technology.

The last building block of the laser range finder is the digital display unit, which can be any 4½-digit dc LCD panel meter that can operate from 12 volts dc.

The telescope and laser diode assemblies can be combined and assembled in a metal enclosure, such as a Hammond 1401A series chassis box, which measures 6 × 10 × 5 inches.

Once all the range-finder components are assembled in the enclosure, refer to Figure 12-12, which illustrates the front-panel layout of the indicators, controls, and connectors. A second gun-site telescope is placed on the range-finder enclosure to aid in aligning the laser beam on the distant target. A 12-volt gel-cell battery provides power to the entire laser range finder via a wall-cube power supply with a series diode to recharge the battery.

This laser range finder will operate to about 500 meters using the 5-milliwatt laser diode. An optional 30-milliwatt laser diode would allow you to operate the range finder out to 1000 meters (these distances are one way; multiply them by 2 for round-trip paths). For long-distance applications, a retroreflector is required on the target object. A simple reflecting surface will work at short distances to about 5 meters.

Applications that range off natural objects without reflectors require higher-power lasers than are used in this project. Ranging a 2-km distance would require a 100-watt to 1-kilowatt peak pulsed power output helium-neon laser, which is beyond the means of most experimenters. Be very careful when operating any type of laser range finder using mirrors or reflectors in order to prevent eye damage. The range finder discussed in this chapter uses an expanded low-power laser beam that passes through a telescope to collimate the beam for low-beam divergence. This improves eye safety, but you still need to be cautious, especially when aligning the telescope and laser assemblies yourself.

You can build the laser range finder from the described subassemblies to keep cost down; if money is not an object, you can buy a completely assembled laser range finder from Laser Sensor Technology (see the Appendix).

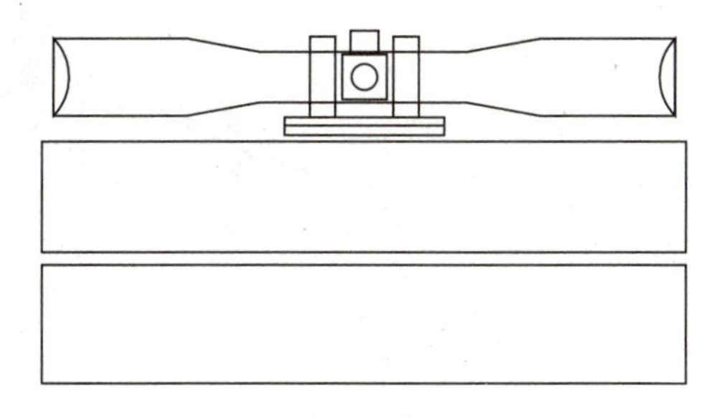

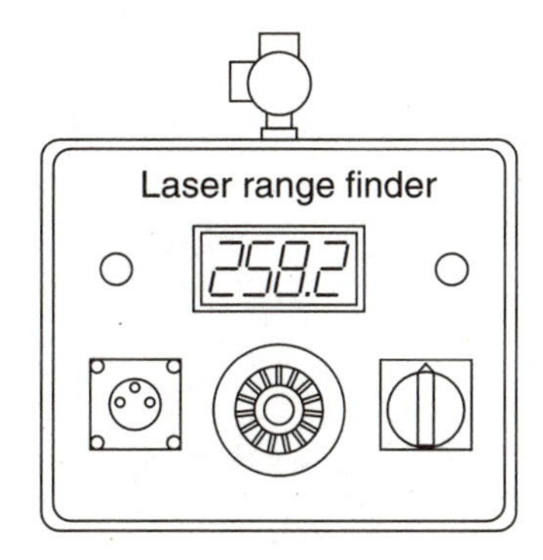

12-12 Rangefinder front panel.

Appendix: Sources and vendors

Acculex Corporation
440 Myles Standish Blvd.
Taunton, MA 02780
LCD digital voltmeters

All Electronics
14928 Oxnard St.
Van Nuys, CA 91411
800-826-5432
818-997-1806
Electronic parts

Allegro Electronic Systems
3 Mine Mountain Rd.
Cornwall Bridge, CT 06754
HeNe lasers, laser communications systems, and analog-to-digital components

American Science & Surplus
3605 Howard St.
Skokie, IL 60076
708-982-0870
Science items, surplus optics, etc.

Aromat Corporation
5 Mount Royal Ave.
Marlboro, MA 01752
508-481-1995
Optical relays

Andover Corporation
4 Commercial Dr.
Salem, NH 03079
UV filters

AT&T Microelectronics
555 Union Blvd.
Allentown, PA 18103
800-372-2447
Optical relays

Burr-Brown Semiconductor
6730 S. Tuscon Blvd.
Tuscon, AZ 85706
602-746-1111
Filters, op-amps, and comparators

Centronics, Inc.
2088 Anchor Court
Newbury Park, CA 91320
805-449-5902
Silicon photodiodes and UV diodes

Cherry Semiconductor
2000 South Country Trail
East Greenwich, RI 02818
401-885-3600
Optical transceivers and smoke-detector ICs

Clarostat Sensors & Controls
1500 International Parkway
Richardson, TX 75081
800-448-2900
Resistive sensors

Crystal Semiconductor
P.O. Box 17847
Austin, TX 78744
800-888-5016
Op-amps, filters, and A/D converters

Digikey Corporation
701 Brooks Ave.
Thief River Falls, MN 56701
Good selection of electronic parts

Dolan Jenner
P.O. Box 1020
Woburn, MA 01801
Bifurcated light pipes

Dyna Art Designs
3535 Stillmeadow Lane
Lancaster, CA 93536
PC-board desktop fabrication and toner transfer paper

Edmund Scientific
101 East Gloucester Pike
Barrington, NJ 00807
609-573-6250
Educational and scientific equipment, lasers, and optics

Electronic Goldmine
P.O. Box 5408
Scottsdale, AZ 85261
602-451-7454
Electronic parts, kits, and overruns

Eltec Instruments
350 Fentress Blvd.
Dayton Beach, FL 32020
800-874-7780
Pyroelectric sensors and high-megohm resistors

Fenwall Electronics
63 Fountain St.
Framingham, MA 01701
Temperature sensors

Hamamatsu, Inc.
360 Foothill Rd.
Bridgewater, NJ 08807
Photodiode sensors, image matrix sensors, and position-sensitive detectors

Hamtronics
65 Moul Rd.
Hilton, NY 14468
716-392-9430
Radio kits

Herbach & Rademan, Inc.
P.O. Box 122
17 Canal St
Bristol, PA 19007
215-788-5583
Optics and surplus electronics

Hewlett-Packard
350 to 370 W. Trimble Rd.
San Jose, CA 95131
800-752-9000
Optics

Hosfelt Electronics, Inc.
2700 Sunset Blvd.
Steubenville, OH 43952
800-524-6464
Electronic parts

Images Company
Box 140742, Dept. N
Staten Island, NY 10314
718-698-8305
Stepper motors and holography supplies

Images Concepts BBS
404-466-3932

Industrial Fiber Optics
P.O. Box 3576
Scottsdale, AZ 85271
602-804-1227
Educational laser and fiber-optic kits

Laser Sensor Technology
3873 Easy Circle
Marietta, GA 30066
770-928-2867
Electronic range finders

LNS Technologies
20993 Foothill Blvd., Suite 307
Hayward, CA 94541
800-886-7150
Laser diodes and drivers, and kits

Marlin P. Jones & Associates
P.O. Box 12685
Lake Park, FL 33403
407-848-8236
Electronic parts, surplus, and kits

Meredith Instruments
5035 N. 55th Ave., #5
P.O. Box 1724
Glendale, AZ 85301
Lasers, drivers, and kits

Metrologic, Inc.
P.O. Box 307
Bellmawr, NJ 08099
800-439-3876
Educational kits, lasers, and books

Micro Coatings
1 Liberty Way
Westford, MA 01886
UV interference filters

Midwest Laser Products
P.O. Box 2187
Bridgeview, IL 60455
708-460-9595
Used lasers, new lasers, and laser supplies

Motorola, Inc.
P.O. Box 20912
Phoenix, AZ 85036
Optical handbooks and optoisolators

Mouser Electronics
2401 Highway 287 N
Mansfield TX 76063
800-346-6873
Electronic parts

MWK Industries
1269 W. Pomona
Corona, CA 91720
909-278-0563
Lasers, optics, kits, books, and laser light-show equipment

Parts Express
365 Blair Rd.
Avenel, NJ 07001
Electronic parts

Prairie Digital
846 17th St.
Industrial Park
Prairie du Sac, WI 53578
Inexpensive A/D cards

Ramsey Electronics
793 Canning Parkway
Victor, NY 14564
800-446-2295
Kits

Sensor Magazine
Helmers Publishing
174 Concord Street
P.O. Box 874
Peterborough, NH 03458

Sharp Electronics Corporation
5700 N.W. Pacific Rim Blvd., M/S 20
Camas, WA 98607
206-834-2500
Optical products

Society of Amateur Scientists
4941D Clairmont Square, Suite 129
San Diego, CA 91777
619-239-8807
Interesting new magazine

Teltone Corp.
10801 120th Ave. N.E.
Kirkland, WA 98033
800-426-3926
Touch-tone detectors

Texas Instruments
P.O. Box 809066
Dallas, TX 75380
800-336-5236
Semiconductors, optoisolators, and phototransistors

Time Space Scientific
101 Highland Dr.
Chapel Hill, NC 27514
Inductors

Toko America
1250 Feehanville Dr.
Mount Prospect, IL 60056
312-297-0070
RF coils

Twardy Technologies
P.O. Box 2221
Darien, CT 6820
UV interface filters

Index